Reviews of Environmental Contamination and Toxicology

VOLUME 242

More information about this series at http://www.springer.com/series/398

Reviews of Environmental Contamination and Toxicology

Editor
Pim de Voogt

Founding Editor
Francis A. Gunther

Volume 242

Springer

ISSN 0179-5953 ISSN 2197-6554 (electronic)
Reviews of Environmental Contamination and Toxicology
ISBN 978-3-319-51242-6 ISBN 978-3-319-51243-3 (eBook)
DOI 10.1007/978-3-319-51243-3

Printed on acid-free paper

This Springer imprint is published by Springer Nature
The registered company is Springer International Publishing AG
The registered company address is: Gewerbestrasse 11, 6330 Cham, Switzerland

Foreword

International concern in scientific, industrial, and governmental communities over traces of xenobiotics in foods and in both abiotic and biotic environments has justified the present triumvirate of specialized publications in this field: comprehensive reviews, rapidly published research papers and progress reports, and archival documentations These three international publications are integrated and scheduled to provide the coherency essential for nonduplicative and current progress in a field as dynamic and complex as environmental contamination and toxicology. This series is reserved exclusively for the diversified literature on "toxic" chemicals in our food, our feeds, our homes, recreational and working surroundings, our domestic animals, our wildlife, and ourselves. Tremendous efforts worldwide have been mobilized to evaluate the nature, presence, magnitude, fate, and toxicology of the chemicals loosed upon the Earth. Among the sequelae of this broad new emphasis is an undeniable need for an articulated set of authoritative publications, where one can find the latest important world literature produced by these emerging areas of science together with documentation of pertinent ancillary legislation.

Research directors and legislative or administrative advisers do not have the time to scan the escalating number of technical publications that may contain articles important to current responsibility. Rather, these individuals need the background provided by detailed reviews and the assurance that the latest information is made available to them, all with minimal literature searching. Similarly, the scientist assigned or attracted to a new problem is required to glean all literature pertinent to the task, to publish new developments or important new experimental details quickly, to inform others of findings that might alter their own efforts, and eventually to publish all his/her supporting data and conclusions for archival purposes.

In the fields of environmental contamination and toxicology, the sum of these concerns and responsibilities is decisively addressed by the uniform, encompassing, and timely publication format of the Springer triumvirate:

Reviews of Environmental Contamination and Toxicology [Vol. 1 through 97 (1962–1986) as Residue Reviews] for detailed review articles concerned with any aspects of chemical contaminants, including pesticides, in the total environment with toxicological considerations and consequences.

Bulletin of Environmental Contamination and Toxicology (Vol. 1 in 1966) for rapid publication of short reports of significant advances and discoveries in the fields of air, soil, water, and food contamination and pollution as well as methodology and other disciplines concerned with the introduction, presence, and effects of toxicants in the total environment.

Archives of Environmental Contamination and Toxicology (Vol. 1 in 1973) for important complete articles emphasizing and describing original experimental or theoretical research work pertaining to the scientific aspects of chemical contaminants in the environment.

The individual editors of these three publications comprise the joint Coordinating Board of Editors with referral within the board of manuscripts submitted to one publication but deemed by major emphasis or length more suitable for one of the others.

Coordinating Board of Editors

Preface

The role of *Reviews* is to publish detailed scientific review articles on all aspects of environmental contamination and associated (eco)toxicological consequences. Such articles facilitate the often complex task of accessing and interpreting cogent scientific data within the confines of one or more closely related research fields.

In the 50+ years since *Reviews of Environmental Contamination and Toxicology* (formerly *Residue Reviews)* was first published, the number, scope, and complexity of environmental pollution incidents have grown unabated. During this entire period, the emphasis has been on publishing articles that address the presence and toxicity of environmental contaminants. New research is published each year on a myriad of environmental pollution issues facing people worldwide. This fact, and the routine discovery and reporting of emerging contaminants and new environmental contamination cases, creates an increasingly important function for *Reviews*. The staggering volume of scientific literature demands remedy by which data can be synthesized and made available to readers in an abridged form. *Reviews* addresses this need and provides detailed reviews worldwide to key scientists and science or policy administrators, whether employed by government, universities, nongovernmental organizations, or the private sector.

There is a panoply of environmental issues and concerns on which many scientists have focused their research in past years. The scope of this list is quite broad, encompassing environmental events globally that affect marine and terrestrial ecosystems; biotic and abiotic environments; impacts on plants, humans, and wildlife; and pollutants, both chemical and radioactive; as well as the ravages of environmental disease in virtually all environmental media (soil, water, air). New or enhanced safety and environmental concerns have emerged in the last decade to be added to incidents covered by the media, studied by scientists, and addressed by governmental and private institutions. Among these are events so striking that they are creating a paradigm shift. Two in particular are at the center of ever increasing media as well as scientific attention: bioterrorism and global warming. Unfortunately, these very worrisome issues are now superimposed on the already extensive list of ongoing environmental challenges.

The ultimate role of publishing scientific environmental research is to enhance understanding of the environment in ways that allow the public to be better informed or, in other words, to enable the public to have access to sufficient information. Because the public gets most of its information on science and technology from internet, TV news, and reports, the role for scientists as interpreters and brokers of scientific information to the public will grow rather than diminish. Environmentalism is an important global political force, resulting in the emergence of multinational consortia to control pollution and the evolution of the environmental ethic. Will the new politics of the twenty-first century involve a consortium of technologists and environmentalists, or a progressive confrontation? These matters are of genuine concern to governmental agencies and legislative bodies around the world.

For those who make the decisions about how our planet is managed, there is an ongoing need for continual surveillance and intelligent controls to avoid endangering the environment, public health, and wildlife. Ensuring safety-in-use of the many chemicals involved in our highly industrialized culture is a dynamic challenge, because the old, established materials are continually being displaced by newly developed molecules more acceptable to federal and state regulatory agencies, public health officials, and environmentalists. New legislation that will deal in an appropriate manner with this challenge is currently in the making or has been implemented recently, such as the REACH legislation in Europe. These regulations demand scientifically sound and documented dossiers on new chemicals.

Reviews publishes synoptic articles designed to treat the presence, fate, and, if possible, the safety of xenobiotics in any segment of the environment. These reviews can be either general or specific, but properly lie in the domains of analytical chemistry and its methodology, biochemistry, human and animal medicine, legislation, pharmacology, physiology, (eco)toxicology, and regulation. Certain affairs in food technology concerned specifically with pesticide and other food-additive problems may also be appropriate.

Because manuscripts are published in the order in which they are received in final form, it may seem that some important aspects have been neglected at times. However, these apparent omissions are recognized, and pertinent manuscripts are likely in preparation or planned. The field is so very large and the interests in it are so varied that the editor and the editorial board earnestly solicit authors and suggestions of underrepresented topics to make this international book series yet more useful and worthwhile.

Justification for the preparation of any review for this book series is that it deals with some aspect of the many real problems arising from the presence of anthropogenic chemicals in our surroundings. Thus, manuscripts may encompass case studies from any country. Additionally, chemical contamination in any manner of air, water, soil, or plant or animal life is within these objectives and their scope.

Manuscripts are often contributed by invitation. However, nominations for new topics or topics in areas that are rapidly advancing are welcome. Preliminary communication with the Editor-in-Chief is recommended before volunteered review manuscripts are submitted. *Reviews* is registered in WebofScience™.

Inclusion in the Science Citation Index serves to encourage scientists in academia to contribute to the series. The impact factor in recent years has increased from 2.5 in 2009 to almost 4 in 2013. The Editor-in-Chief and the Editorial Board strive for a further increase of the journal impact factor by actively inviting authors to submit manuscripts.

Amsterdam, The Netherlands
January 2015

Pim de Voogt

Contents

Contributors

Mohd. Sayeed Akhtar Department of Botany, Gandhi Faiz-E-Aam College, Shahjahanpur, Uttar Pradesh, India

Azizullah Azizullah Department of Botany, Kohat University of Sciences and Technology (KUST), Kohat, Khyber Pakhtunkhwa, Pakistan

H. Barchańska Department of Inorganic, Analytical Chemistry and Electrochemistry, Faculty of Chemistry, Silesian University of Technology, Gliwice, Poland

Muhammad K. Daud Department of Biotechnology and Genetic Engineering, Kohat University of Science and Technology, Kohat, Khyber Pakhtunkhwa, Pakistan

Donat-Peter Häder (Prof. emeritus), Möhrendorf, Germany

Khalid Rehman Hakeem Faculty of Forestry, Universiti Putra Malaysia, Serdang, Selangor, Malaysia

Department of Biological Sciences, Faculty of Science, King Abdulaziz University, Jeddah, Saudi Arabia

Faridah Hanum Ibrahim Faculty of Forestry, Universiti Putra Malaysia, Serdang, Selangor, Malaysia

Narayanan Kannan Institute for Graduate Studies, Taylor's University (Lakeside Campus), Subang Jaya, Selangor Darul Ehsan, Malaysia

Han Yeong Kaw Key Laboratory of Nature Resources of Changhai Mountain and Functional Molecular (Yanbian University) Ministry of Education, Jilin, China

Sanghyo Kim Department of Bionanotechnology, Gachon University, Seongnam-Si,, Gyeonggi-Do Republic of Korea

Gil Medical Center, Graduate Gachon Medical Research Institute, Incheon, Republic of Korea

M. Kurek Department of Inorganic, Analytical Chemistry and Electrochemistry, Faculty of Chemistry, Silesian University of Technology, Gliwice, Poland

Waheed Murad Department of Botany, Kohat University of Sciences and Technology (KUST), Kohat, Khyber Pakhtunkhwa, Pakistan

Farhat Nabeela Department of Botany, Kohat University of Sciences and Technology (KUST), Kohat, Khyber Pakhtunkhwa, Pakistan

Munir Ozturk Botany Department and Centre for Environmental Studies, Ege University, Izmir, Turkey

Hazir Rahman Department of Microbiology, Kohat University of Science and Technology, Kohat, Khyber Pakhtunkhwa, Pakistan

Shafiq ur Rehman Department of Botany, Kohat University of Sciences and Technology (KUST), Kohat, Khyber Pakhtunkhwa, Pakistan

Muhammad Sabir Institute of Soil and Environmental Sciences, University of Agriculture, Faisalabad, Pakistan

Shakirullah Khan Shakir Department of Botany, Kohat University of Sciences and Technology (KUST), Kohat, Khyber Pakhtunkhwa, Pakistan

M. Turek Department of Inorganic, Analytical Chemistry and Electrochemistry, Faculty of Chemistry, Silesian University of Technology, Gliwice, Poland

Buddolla Viswanath Department of Bionanotechnology, Gachon University, Seongnam-Si, Republic of Korea

Toxic Metal Pollution in Pakistan and Its Possible Risks to Public Health

Shakirullah Khan Shakir, Azizullah Azizullah, Waheed Murad, Muhammad K. Daud, Farhat Nabeela, Hazir Rahman, Shafiq ur Rehman, and Donat-Peter Häder

Contents

S.K. Shakir • A. Azizullah (✉) • W. Murad • F. Nabeela
S. ur Rehman
Department of Botany, Kohat University of Sciences and Technology (KUST), 26000 Kohat, Khyber Pakhtunkhwa, Pakistan
e-mail: shakir_kust@yahoo.com; azizswabi@hotmail.com; whdicp@yahoo.com; fnfaree@gmail.com; drshafiq@yahoo.com

M.K. Daud
Department of Biotechnology and Genetic Engineering, Kohat University of Science and Technology, Kohat 26000, Pakistan
e-mail: mkdaud@yahoo.com

H. Rahman
Department of Microbiology, Kohat University of Science and Technology, Kohat 26000, Pakistan
e-mail: hazirrahman@hotmail.com

D.-P. Häder
Neue Straße 9, 91096 Möhrendorf, Germany
e-mail: donat@dphaeder.de

P. de Voogt (ed.), *Reviews of Environmental Contamination and Toxicology, Volume 242*, Reviews of Environmental Contamination and Toxicology 242,
DOI 10.1007/398_2016_9

1 Introduction

Environmental pollution has increased many folds in recent years and in some places has reached levels that are toxic to living things. Among pollutant types, toxic heavy metals and metalloids are among the chemicals that pose the highest threat to biological systems (Jjemba 2004). Unlike organic pollutants, which are biodegradable, heavy metals are not degraded into less hazardous end products (Gupta et al. 2003). Low concentrations of some heavy metals are essential for life, but some of them like Hg, As, Pb and Cd are biologically non-essential and very toxic to living organisms. Even the essential metals may become toxic if they are present at a concentration above the permissible level (Puttaiah and Kiran 2008). For example, exposure to Zn and Fe oxides produce gastric disorder and vomiting, irritation of the skin and mucous membranes. Intake of Ni, Cr, Pb, Cd and Cu cause heart problems, leukemia and cancer, while Co and Mg can cause anemia and hypertension (Drasch et al. 2005). Similarly, various studies indicated that overexposure to heavy metals in air can cause cardiovascular disorders (Miller et al. 2007; Schwartz 2001), asthma (Wiwatanadate and Liwsrisakun 2011), bronchitis/emphysema (Pope 2000), and other respiratory diseases (Dominici et al. 2006).

In Pakistan, heavy metals contamination in soils, urban road dusts, surface water, ground water, wastewater and air has been increased with the rapid increase in population, industrialization and urbanization. Health problems from exposure to heavy metal pollution have been reported in different parts of the country (Haq et al. 2012; Khan et al. 2010, Khan et al. 2013b). Many studies reported heavy metals pollution above the permissible limits in food, soil and water in different parts of Pakistan as reviewed by Waseem et al. (2014) and Azizullah et al. (2011). However, most of these primary as well as review studies are confined to describing the levels of toxic metals in one or more environmental compartments and no special consideration has been given to health risk assessment. Moreover, most of the aforementioned studies have focused on a single or just a few exposure pathways; thus, a comprehensive review of toxic metals pollution and their health risks to public in Pakistan is urgently needed. Therefore, this review article is focused on toxic metals contamination in soils, water and air in Pakistan with special emphasis on their risks to public health. Data was collected from available literature and was analyzed for non-carcinogenic health risk assessment. In addition, toxic metals in human samples, major sources of toxic metal pollution and health problems due to heavy metal exposure in the country are discussed in detail.

2 Toxic Metal Pollutants in Pakistan

2.1 *Arsenic (As)*

Arsenic is a major pollutant of the environment and a serious threat to public health in many countries of the world (Islam-Ul-Haque et al. 2007). In Pakistan, arsenic contamination in both surface and ground water exceeds the permissible limits and is a serious problem in many parts of the country (Tables 2 and 3). For example, in Multan (Punjab), Dadu and Ganbat (Sindh) more than 50 % of the samples exceeded the permissible limits of NSDWQ-Pak. Some areas in Sindh had its concentration exceeding 200 μg L^{-1} (PCRWR 2008). In Northern Punjab, over 20 % of the people are exposed to arsenic contamination in their drinking water (Ahmad et al. 2004). Only few data are available regarding this metalloid contamination in ambient air and soil in the country as rare attention has ever been given to this issue by researchers (Tables 1 and 5), though problems exists where it has been investigated. For example, its concentration in solid aerosols exceeded its permissible level (1 μg m^{-3}) in Lahore and Rawalpindi where its concentrations were recorded in the range of 0.25–2.23 and 0.19–3.12 μg m^{-3}, respectively (Qadir 2002).

2.2 *Cadmium (Cd)*

Cadmium is among the most toxic heavy metals causing environmental pollution in many parts of the world. In Pakistan, a reasonable ratio of studies revealed Cd pollution in soil beyond the permissible limits, particularly in areas where wastewater is used for irrigation (Table 1). As compared to Cd contamination in soil, the situation is alarming in water as its concentration in surface and ground water exceeds the permissible limits of NSDWQ and WHO in most of the studied areas in the country (Tables 2 and 3). Analysis of wastewater samples from various regions in the country also revealed that in some cases Cd concentration exceeded or reached the maximum permissible level of the Food and Agriculture Organization (FAO) (1985) (Table 4). The problem is further aggravated as Cd in ambient air in many cities of the country like Faisalabad, Karachi, Lahore, Rawalpindi and Peshawar exists manifold higher than its permissible level (5 μg m^{-3}) of WHO (Table 5).

Table 1 Heavy metal concentrations in soil and sediment of different regions in Pakistan

S. No	Soil sampling location	Cd	Co	Cr	Cu	Fe	Hg	Mn	Ni	Pb	Zn	References
1	Soil samples from Peshawar city	–	2–7	10–19	30–21	1015–10056	–	–	12–81	3–10	20–30	Hamidullah et al. (2000)
2	Soil of sewage–irrigated farm in SITE, Karachi	0.04–4	0.6–1.6	0.89–2.8	68–177	11514–12228	–	309–439	0.73–4	10.7–26.8	470–870	Yousafzai et al. (2001)
3	Fields irrigated with sewage effluents in Faisalabad	–	–	–	17.53	131.87	–	105.00	0.88	4.40	166.48	Butt et al. (2005)
4	Effluents–irrigated fields, Peshawar	0.10–0.14	–	0.13–0.39	–	–	–	–	1.09–8.28	4.28–8.03	–	Midrar-Ul-Haq et al. (2005)
5	Soil from Industrial Estate, Hattar, Khyber Pakhtunkhwa	0.17	2.75	5.96	–	15.02	–	1.21	1.51	4.46	2.20	Manzoor et al. (2006)
6	Charsada	–	50	2.26	0.80	1197	–	28.76	43	69	5.36	Khan et al. (2008a)
7	Pabbi, Nowshera	–	23	2.15	0.70	180	–	23.46	41	53	3.51	
8	Nowshera	–	40	1.80	0.92	202	–	24.34	47	71	6.45	
9	Soil from different location of Peshawar	–	49	2.96	0.72	190	–	26.6	39	71	5.86	
10	Bara, Khyber Agency	–	41	2.35	0.56	166	–	25.39	31	51	4.10	
11	Fields irrigated with tube well water in Rawalpindi	0.01–0.04	–	0.13–0.22	–	–	–	–	0.04–0.22	0.5–0.75	–	Latif et al. (2008)
12	Fields irrigated with sewage in Rawalpindi	0.09–0.14	–	0.43–0.49	–	–	–	–	0.42–0.44	0.15–3	–	
13	Soil of the Swat valley	0.72–3.49	–	–	1.73–6.7	–	–	–	–	–	–	Nafees et al. (2009)
14	Soil of Hudaira, Lahore	0.31	–	–	6.20	34.16	–	14.00	0.55	1.20	18.43	Kashif et al. (2009)
15	Fields irrigated by water of the Hudaira drain, Lahore	0.61	–	–	9.41	51.44	–	27.20	1.07	3.75	24.90	
16	Motorway M-2 side dust Pakistan	0.42–0.52	–	12.60–17.25	–	695–731	–	–	12.42–17.49	49.22–56.36	66.48–99.33	Ahmad and Erum (2010)
17	Soil of industrial area of Islamabad	–	–	40.2–927.2	8.9–357.4	–	–	–	41.4–59.3	2.0–29.0	61.9–172.6	Malik et al. (2010)

18	River, canal and well water–irrigated fields of an urban area in Karachi	1.435	–	21.385	7.635	9521.0	31.005	4.703	30.92	26.02	31.005	Sharif et al. (2010)
19	Industrial sewage–irrigated fields of an urban area in Karachi	2.083	–	40.96	11.61	10134.0	1.248	466.11	38.67	30.87	59.09	
20	Effluents/wastewater-irrigated fields of Adiala, Rawalpindi	0.4–1.7	–	1.4–28.8	2.2–15.8	7526–11949	–	–	87.6–153.5	16.4–28.4	30.6–38.4	Mushtaq and Khan (2010)
21	Effluents/wastewater–irrigated field of Pirwadhai, Rawalpindi	1.7–2.9	–	28.8–124.5	12.9–74.2	13170–24771	–	–	25.6–44.0	17.8–89.9	13.9–200.0	
22	Effluents/wastewater–irrigated fields of Taxilla	1.9–3.5	–	21.6–144.2	20.2–30.2	16856–26889	–	–	19.7–51.6	19.4–27.5	7.7–30.0	
23	Effluents/wastewater–irrigated fields in Wah factory	1.9–2.5	–	65.4–154.1	10.7–38.2	22176–30579	–	–	28.0–56.9	20.3–38.4	12.4–197.0	
24	Soil from geological source of Gilgit	0.3–2.3	–	–	24–147	–	–	–	24–57	29–138	172–1193	Khan et al. (2010)
25	Soil from road side along different regions of national highway, Pakistan	0.84	5.95	–	12.98	3970	90.67	0.017	8.92	36.46	56.72	Khan et al. (2011b)
26	Agricultural soil irrigated with waste water in central Punjab	0.12–4.22	–	246–1980	34–61	–	–	–	32.1–75.2	24–38	–	Iqbal et al. (2011)
27	Samples from four provinces of Pakistan	–	–	–	–	–	0–0.007					Abbas et al. (2012)
28	Soils samples from various areas in Peshawar	0.89	–	77.25	15.5	49759	–	596	54	11	48	Siddiqui and Khattak (2012)
29	Sewage–irrigated fields in Peshawar	0.24	–	0.35	–	–	–	–	0.65	0.65	–	Parveen et al. (2012)
30	Road side soils of Faisalabad	–	–	–	11.25–73.20	–	–	–	13.38–29.70	0.00–27.05	19.2–336.8	

(continued)

Table 1 (continued)

S. No	Soil sampling location	Cd	Co	Cr	Cu	Fe	Hg	Mn	Ni	Pb	Zn	References
31	Mangrove sediments of paradise point in Karachi	0.002–0.11	0.009–0.045	0.005–0.085	0.009–0.04	1.96–8.32	–	–	–	0.01–0.045	0.01–0.1	Khattak et al. (2012)
32	Mangrove sediments of Hawks bay, Karachi	0.004–0.008	0.009–0.025	0.0011–0.031	0.007–0.031	1.23–5.31	–	–	–	0.011–0.08	0.00–0.033	
33	Mangrove sediments of Sand pit, Karachi	0.00–0.006	0.009–0.045	0.01–0.026	0.009–0.041	1.54 4.56	–	–	–	0.008–0.032	0.00–0.065	
34	Contaminated soil from leather industrial area of Kasur	26.3	22.05	2443		136.9	–	9.42	34.2	18.21	14.3	Afzal et al. (2013)
35	Soil irrigated with waste water from Gadoon industrial estate Pakistan	–	0.001–0.271	0.243–0.936	0.202–1.236	–	–	3.766–9.955	0.38–0.544	0.488–1.259	1.596–6.288	Amin et al. (2014)
36	Soil from Lahore irrigated with tube well water and waste water	1.17 3.15	3.65 8.01	14.04 21.01	15.28 28.74	– –	– –	21.24 28.83	14.5 39.01	7.4 15.38	34.13 50.84	Mahmood and Malik (2014)
37	Sediments of river Panjkora, lower Dir	4.57–16.32	–	–	35.1–45.13	–	–	–	4.57–15.79	6.58–11.63	24.37–28.43	Ahmad et al. (2014b)
38	Sediments from coast of Pakistan	0.14	1.1	171	64.2	–	–	–	34	45	68	Ali et al. (2014)
	Permissible limits of US. EPA	3	23	–	–	35	1	80	50	400	200	USEPA (1997)

Data were extracted from papers published in national and international journals and are arranged chronologically. All values are given in mg kg^{-1}. Values given are mean values or the minimum to maximum range with mean values in parenthesis

Table 2 Occurrence of heavy metals in ground water of different regions in Pakistan

S. No	Sampling location	As (μgL^{-1})	Cd	Co	Cr	Cu	Fe	Hg	Mn	Ni	Pb	Zn	References
1	Various areas in Karachi	80.00	0.04	–	0.34	0.09		0.01		0.50	2.0	4.02	Rahman et al. (1997)
2	Tube well water in Hasan Abdal	–	0.001	–	0.04	0.02	0.07	–	0.02	0.03	0.03	0.009	Lone et al. (2003)
3	Shallow wells in the vicinity of Palosi Drain, Peshawar	–	0.0–0.04	–	–	0.26–0.34	0.0–0.30	–	0.071–0.21	0.0–0.68	0.27–0.38	0.047–0.34	Ilyas and Sarwar (2003)
4	Deep wells water in the vicinity of Palosi Drain, Peshawar	–	0.0–0.056	–	–	0.558–0.598	0.46–0.99	–	0.0–0.306	0.0–0.52	0.0–0.49	0.0–0.082	
5	Well water in the district Multan	60–1000	–	–	–	–	0.07–2.7	–	0.06–1.0	–	–	–	Nickson et al. (2005)
6	Well water from the district Muzaffargarh	00–400	–	–	–	–	0.0–1.65	–	0.0–0.69	–	–	–	
7	Tube well water Korangi, Karachi	–	0.041	–	0.03	0.032	2.39	–	0.124	0.656	0.24	0.048	Saif et al. (2005)
8	Well water samples from Charsadda and Risalpur in Khyber Pakhtunkhwa	–	0.01–0.07 (0.04)	–	0.01–0.30 (0.14)	0.004–0.67 (0.21)	0.01–0.43 (0.16)	–	0.08–2.56 (1.22)	0.002–3.66 (0.96)	0.02–0.73 (0.33)	0.002–0.27 (0.13)	Midrar-Ul-Haq et al. (2005)
9	Well water from Korangi, Karachi	–	0.02–0.04 (0.03)	–	0.003–0.07 (0.03)	0.01–0.21 (0.10)	0.51–2.39 (1.22)	–	0.07–0.12 (0.10)	0.01–2.19 (1.20)	0.10–0.24 (0.15)	0.04–0.52 (0.26)	
10	Groundwater in Chakera, Faisalabad	–	–	–	0.01	1.30	3.70	–	1.06	0.13	0.12	1.84	Mahmood and Maqbool (2006)
11	Tube wells water in Gadoon, Khyber Pakhtunkhwa	–	0.007–0.025	–	0.017–0.111	0.87–0.893	0.004–0.037	–	0.059–0.164	0.03–0.066	0.21–1.20	0.007–0.066	Nasrullah et al. (2006)
12	Tap water, Hattar Industrial Estate, Hattar, Khyber Pakhtunkhwa	–	–	–	–	0.030	0.50	–	0.15	0.00	0.001	0.050	Sial et al. (2006)
13	Tube and dug wells water from Hayatabad, Peshawar	–	0.01–0.06 (0.03)	–	0.03–0.16 (0.09)	0.25–0.45 (0.36)	0.01–0.20 (0.08)	–	0.06–0.22 (0.12)	0.38–1.75 (0.88)	0.20–0.97 (0.66)	0.01–0.23 (0.10)	Tariq et al. (2006a, b)

(continued)

Table 2 (continued)

S. No	Sampling location	As (μgL^{-1})	Cd	Co	Cr	Cu	Fe	Hg	Mn	Ni	Pb	Zn	References
14	Water from Industrial Estate, Hattar, Khyber Pakhtunkhwa	–	0.02	0.08	0.22	–	0.11	–	0.04	0.8	0.26	0.18	Manzoor et al. (2006)
15	Kalalanwala, Punjab	32–1900	–	–	–	–	–	–	–	–	–	–	Farooqi et al. (2007)
16	Well water from a residential area in Kasur city	–	0.001–0.027 (0.007)	–	0.05–9.80 (2.12)	–	0.02–11.8 (2.57)	–	0.01–0.17 (0.04)	0.001–0.24 (0.08)	0.003–0.26 (0.11)	0.01–1.08 (0.14)	Tariq et al. (2008)
17	Tube well water in Rawalpindi	–	0.001–0.030	–	0.01–0.43	–	–	–	–	0.01–0.30	0.01–0.40	–	Latif et al. (2008)
18	Drinking water from different districts of Punjab	6–12 (8.5)	–	–	–	–	–	–	–	–	–	–	Toor and Tahir (2009)
19	Groundwater from different areas in Sialkot city	–	–	–	0.0–0.30 (0.03)	0.01–0.17 (0.06)	0.0–0.83 (0.30)	–	0.0–0.09 (0.03)	0.01–0.22 (0.10)	0.11–0.81 (0.49)	0.0–0.81 (0.16)	Ullah et al. (2009)
20	Hand pumps water in the vicinity of Manchar Lake, Sindh	23.3–96.3	–	–	–	–	–	–	–	–	–	–	Arain et al. (2009)
21	Ground water from Talika Daur, Nawabshah, Sindh	–	00–0.017	00–0.05	–	00–0.188	0.075–1.35	–	00–0.517	0.013–0.092	00–0.053	0.013–0.262	Majidano and Khuhawar (2009)
22	Drinking water of Peshawar	–	–	–	0.21	0.11	0.68	–	0.01	–	0.15	0.09	Bacha et al. (2010)
23	Drinking water from Karachi	–	–	0.002	0.011	0.121	0.106	–	0.081	0.037	0.006	–	Karim (2011)
24	Ground water from Gera Ghazi khan	0.37–29.3	–	–	–	–	–	–	–	–	–	–	Malana and Khosa (2011)
25	Drinking water from Sindh Hyderabad, Tharimirwa and Gumbat	8.7 28.4 112	– – –	– – –	– – –	– – –	– – –	– – –	– – –	– – –	– – –	– – –	Baig et al. (2011)
26	Ground water from (Jijaldubair, Bisham and Alpuri) Kohistan	– – –	0.001 0.002 0.00028	0.0006 0.0002 0.0001	0.021 0.001 0.001	0.738 0.027 0.005	– – –	– – –	0.005 0.019 0.012	0.004 0.002 0.004	0.002 0.024 0.002	0.651 1.376 0.826	Muhammad et al. (2011)

27	Drinking water of Mardan	–	00	–	00–0.02	00–0.01	–	–	00	00–0.1	00–0.03	00–0.016	Hussain et al. (2012)
28	Drinking water of Tehsil Tangi, Khyber Pakhtunkhwa	–	0.002–0.0041	–	0.0049–0.0355	1.7–27.7	0.0347–0.4071	–	–	0.00086–0.0094	0.0144–0.1079	1.490–2.373	Khan et al. (2012a, b)
29	Drinking water of Tehsil Charsadda, Khyber Pakhtunkhwa	–	0.0007–0.0046	–	0.002–0.010	0.053–0.039	0.0596–0.3466	–	–	0.0007–0.0048	0.0147–0.0993	0.0607–0.8037	
30	Drinking water of Tehsil Shabqadar, Khyber Pakhtunkhwa	–	0.0006–0.0044	–	0.0021–0.0043	0.0065–0.0707	0.060–0.712	–	–	0.0003–0.0024	0.0577–0.117	0.159–0.575	
31	Ground water of Winder, Baluchistan	–	0.010	0.007	0.052	0.055	0.411	–	0.041	0.149	0.070	0.076	Naseem et al. (2013)
32	Drinking water of Rawalpindi and Islamabad	–	0	0.01	–	0.04	0.02	–	–	–	0.166	0.01	Mehmood et al. (2013)
33	Under ground water from Zhob River basin, Balochistan	–	–	–	0.06–0.35	0.07–0.25	0.5–1.1	–	0.06–0.25	–	0.02–0.06	0.03–0.75	Umar et al. (2013)
34	Drinking water from polluted localities of Dir and Pehawar	4.19–58.92	–	–	–	–	–	0.00–0.0008	–	–	–	–	Ishaq et al. (2013)
35	Ground water at leather industrial area of Kasur	–	0.04	0.15	1.32	–	0.556	–	0.073	0.108	0.143	0.135	Afzal et al. (2013)
36	Tube well water from Lahore	–	0.01	0.029	0.17	6.27	–	–	0.13	0.93	0.21	0.17	Mahmood and Malik (2014)
37	Drinking water of Islampura, Swat	–	–	–	0.18	–	–	–	–	0.06	1.04	1.20	Hussain et al. (2014)
	National standards for drinking water quality, Pakistan	50	0.01	–	0.05	2.0	–	0.001	0.5	0.02	0.05	5	NSDWQ-Pak (2008)
	WHO Standards for drinking water	10	0.003	–	0.05	2.0	–	0.006	0.5	0.07	0.01	3	WHO (2011)

Data have been extracted from papers published in national and international journals and arranged chronologically. All values are given in mg L^{-1} unless otherwise mentioned. Values given are mean values or the range from minimum to maximum with mean values in *brackets*

Table 3 Heavy metal occurrence in surface waters of Pakistan

S. No	Samples locations	As (μg L^{-1})	Cd	Cr	Cu	Fe	Hg	Mn	Ni	Pb	Zn	References
1	Tarbela Water Reservoir	620	0.004	0.003	0.04	0.012	0.014	0.018	0.061	0.107	0.028	Ashraf et al. (1991)
2	Chashma Water Reservoir	750	0.003	0.071	0.04	0.004	0.017	0.0004	0.065	0.058	.0029	
3	Lloyd Water Reservoir	620	0.002	0.002	0.04	0.012	0.14	0.018	0.061	0.107	0.028	
4	Samples at different sites from River Ravi	0.32–1.8	0.0002–0.0014	0.0003–0.0014	0.001–0.006	0.042–0.127	0.0003–0.0009	0.0013–0.0097	0.0012–0.0013	0.0004–0.0017	0.01–0.034	Tariq et al. (1994)
5	Samples from different sites in Gadoon Amazai, Khyber Pakhtunkhwa	–	–	–	–	5.58	–	1.70	1.13	0.68	1.76	Javed and Mahmood (2001)
6	Palosi drain, Peshawar	–	0.0–0.004	–	0.0	0.37–0.75	–	0.017–0.242	0.0–0.18	0.0–0.34	0.0–0.239	Ilyas and Sarwar (2003)
7	Surface water from different areas in Khyber Pakhtunkhwa	–	0.002–0.09 (0.02)	0.01–0.12 (0.04)	0.01–0.77 (0.20)	0.01–1.29 (0.19)	–	0.01–1.11 (0.22)	0.01–1.52 (0.21)	0.02–0.38 (0.16)	0.003–0.08 (0.04)	Midrar-Ul-Haq et al. (2005)
8	Malir River in Karachi	–	0.002–0.07 (0.04)	0.03–0.29 (0.10)	0.01–0.84 (0.31)	0.13–2.91 (0.78)	–	0.05–0.57 (0.33)	0.02–1.06 (0.59)	0.09–0.32 (0.19)	0.06–0.29 (0.16)	
9	River Chenab	7.00	–	–	–	0.18	–	0.28	–	–	–	Nickson et al. (2005)
10	Canal water, Akbarpura area, Nowshera, Khyber Pakhtunkhwa	–	0.09–0.14	0.13–0.17	0.59–0.73	0.90–1.02	–	0.61–0.71	0.33–0.39	0.34–0.43	0.04–0.05	Nazif et al. (2006)
11	Bara River, Akbarpura area, Khyber Pakhtunkhwa	–	0.15–0.20	0.16–0.29	0.90–1.20	1.29–1.75	–	0.77–0.85	0.53–0.72	0.43–0.62	0.02–0.06	
12	Kalar Kahar lake, Chakwal, Punjab	–	0.01–0.05	–	0.01–1.20	0.20–5.46	–	–	0.04–0.25	0.01–0.30	0.44–2.82	Raza et al. (2007)

13	MNVD, Sehwan, Jamshoro, Sindh	–	0.0001–0.018	–	0.003–0.0084	0.0116–0.0185	–	–	0.0002–0.0052	0.004–0.0096	0.0041–0.0072	Mastoi et al. (2008)
14	Manchar Lake Jamshoro, Sindh	–	0.0001–0.002	–	0.0006–0.0196	0.0073–0.0178	–	–	0.0004–0.0096	0.0057–0.014	0.0046–0.0348	Mastoi et al. (2008)
15	Hudaira drain, Lahore	–	0.18	0.07	0.45	7–9	–	0.85	0.93	0.03	1.7	Kashif et al. (2009)
16	Manchar Lake Jamshoro, Sindh	35–157	–	–	–	–	–	–	–	–	–	Arain et al. (2009)
17	River Kabul, Khyber Pakhtunkhwa	–	0.03	0.02	0.08	–	–	0.18	0.65	0.09	0.2	Khan et al. (2011a)
18	Surface water from (Jijaldubair, Bisham and Alpuri) Kohistan	– – –	0.0006 0.0008 0.00013	0.007 0.0009 0.0028	0.115 0.021 0.031	– – –	– – –	0.013 0.18 0.10	0.0046 0.0028 0.0059	0.0036 0.008 0.0034	0.040 0.040 0.021	Muhammad et al. (2011)
19	Water of river Kabul at Peshawar	–	0.03	0.5	0.07	–	–	–	0.2	0.52	–	Ullah et al. (2013)
20	Water samples from river Panjkora, lower Dir		ND	–	0.1	–	–	–	ND	0.1	0.06–0.18	Ahmad et al. (2014b)
	National standards for drinking water quality, Pakistan	50	0.01	0.05	2.0	–	0.001	0.5	0.02	0.05	5	NSDWQ-Pak (2008)
	WHO standards	10	0.003	0.05	2.0	–	0.006	0.50	0.07	0.01	3	WHO (2011)

Data have been extracted from papers published in national and international journals and arranged chronologically. All values are given in mg L^{-1} unless otherwise mentioned. Values given are mean values or the range from minimum to maximum with mean values in *brackets*. *ND* stands for not detected

Table 4 Heavy metal composition of sewage and waste water in different cities of Pakistan

S. No	Samples locations	Cd	Co	Cr	Cu	Fe	Hg	Mn	Ni	Pb	Zn	References
1	Sewage of Gutter Baghicha SITE, Karachi	0.004	–	0.032	0.025	0.312	–	0.076	0.015	0.015	0.048	Yousafzai et al. (2001)
2	Industrial effluents of Faisalabad	–	–	0.05	0.09	0.16	–	0.14	0.03	0.24	0.14	Ensink et al. (2004)
3	Municipal wastewater samples of Haroonabad	0.01	–	0.23	0.35	0.22	–	0.07	0.14	0.04	ND	
4	Sewage water at Hassan Abdal	0.031	–	0.71	0.70	2.67	–	0.53	2.23	0.42	0.42	Lone et al. (2003)
5	Sewage effluents from Faisalabad	0.02–0.04	–	0.40–0.69	–	–		–	0.22–0.35	0.08–0.14	–	Farid (2003)
6	Municipal corporation Faisalabad	0.06	0.007				–	0.39				Murtaza et al. (2003)
7	Wastewater of Hayatabad (Peshawar)	ND	–	1.62	0.15	–	–	–	–	0.69	0.32	Pak-EPA/OECC (2004)
8	Industrial effluents of Hayatabad Industrial Estate, Peshawar	0.04	–	0.06	0.36	0.42	–	0.16	1.25	0.70	0.01	Tariq et al. (2006a, b)
9	Karnailwala, Faisalabad		–	1.02	0.19	1.79	–	0.31	0.48	0.22	0.79	Hussain et al. (2006)
10	Judgewala, Faisalabad		–	0.92	0.21	1.49	–	0.34	0.47	0.19	0.14	
11	Marzipura, Faisalabad	–	–	0.98	0.18	1.84	–	0.31	0.46	0.21	0.09	
12	Uchkera, Faisalabad	–	–	0.84	0.18	2.25	–	0.29	0.47	0.21	0.09	
13	Industrial effluents of Gadoon Amazai Industrial Estate, Khyber Pakhtunkhwa	0.003–0.043	–	0.004–0.28	0.386–1.136	0.008–4.511	–	0.014–0.113	0.009–0.794	0.20–2.84	0.003–0.097	Nasrullah et al. (2006)
14	Effluents from industrial Estate, Hattar, Khyber Pakhtunkhwa	0.4	0.46	0.71	–	2.14	–	0.16	0.26	0.28	0.26	Manzoor et al. (2006)
15	Sewage in Rawalpindi	0.001–0.030	–	0.01–0.84	–	–	–	–	0.01–0.208	0.01–0.79	–	Latif et al. (2008)
16	Untreated industrial effluents of Rawalpindi	0.001–0.040	–	0.07–1.15	–	–	–		0.43–0.49	0.03–0.27		

17	Industrial effluents from Gujranwala, Peshawar and Hattar	–	–	–	0.32–0.53	0.04–0.269	–	–	–	0.09–0.28	2.0–2.01	Rehman et al. (2008)
18	Wastewater samples from Adiala, Rawalpindi	0.02–0.05	–	0.00–0.13	0.14–0.45	0.31–9.12	–	–	0.00–0.54	0.08–0.24	0.01–0.35	Mushtaq and Khan (2010)
19	Wastewater samples from Pirwadhai, Rawalpindi	0.01–0.03	–	0.01–0.25	0.00–0.04	0.04–0.34	–	–	0.13–0.56	0.00–0.08	0.00–0.17	
20	Wastewater samples from Taxilla	0.00–0.04	–	0.03–0.15	0.00–0.10	0.10–1.40	–	–	0.02–0.40	0.00–0.12	0.00–0.02	
21	Wastewater samples from Wah factory	0.00–0.13	–	0.09–0.34	0.01–0.30	0.03–6.54	–	–	0.01–0.40	0.00–0.14	0.00–0.80	
22	Sewage irrigation water of Peshawar	0.75	–	0.03	–	–		–	0.61	0.25	–	Parveen et al. (2012)
23	Industrial waste water of Kot Lakhpat, Lahore	0.00	–	16.80–22.27	0.00	0.00–2.870	–	–	0.00	0.00–0.321	0.00–1.137	Azeem (2010)
24	Industrial waste water in central Punjab	0.001–0.019	–	0.01–0.2	0.1–0.45	–	–	–	0.1–0.45	1–3	–	Iqbal et al. (2011)
25	Sewage water and sludge of various regions in Pakistan	–	–	–	–	–	0–0.009					Abbas et al. 2012
26	Waste water of river Kabul at Peshawar	0.5	–	1.81	0.5	–	–	–	1.5	1.3	–	Ullah et al. (2013)
27	Tannery effluents from leather industries, Kasur	0.16	0.47	128.77	–	5.8	–	0.62	0.98	1.09	1.03	Afzal et al. (2013)
28	Waste water from Lahore	0.19	0.023	0.33	18.15	–	–	0.63	2.88	0.43	0.83	Mahmood and Malik (2014)
	FAO Maximum permissible value in irrigation water.	0.1	0.05	0.1	0.2	5.0	0.3	0.2	0.2	5.0	2.0	FAO (1985)
	Proposed irrigated water quality guideline	0.01	0.05	0.01	0.2	5.0	0.01	0.01	0.2	0.1	2.0	WWF-Pak (2007)

Data have been extracted from papers published in national and international journals and arranged chronologically. All values are given in mg L^{-1} unless otherwise mentioned. Values given are mean values or the range from minimum to maximum with mean values in *brackets*

Table 5 Heavy metal concentrations in air in different regions of Pakistan

S. No	Sampling location	As	Cd	Co	Cr	Cu	Fe	Mn	Ni	Pb	Zn	References
1	Particulate matter of Gutter Baghicha SITE, Karachi	–	0.04–0.33	0.6–1.6	0.89–2.4	8.2–14.3	341–448	5.6–7.7	0.73–4	10.7–26.8	376–1260	Yousafzai et al. (2001)
2	Ambiant air samples of Lahore	0.25–2.23	–	–	–	0.52–6.72	–	–	–	0.9–7.85	0.92–5.82	Qadir (2002)
3	Ambient air samples of Rawalpindi–Islamabad	0.19–3.12	–	–	–	0.53–8.33	–	–	–	0.71–10.93	1.08–4.48	
4	Atmosphere in Islamabad	–	0.006	0.020	0.059	–	1.761	0.055	0.017	0.128	1.021	Shah and Shaheen (2007)
5	Particulate matter in atmosphere in Islamabad	–	0.004	0.033	0.01	0.31	2.46	0.08	0.024	0.14	2.31	Shah and Shaheen (2008)
6	$PM_{2.5}$ and PM_{10} in atmosphere in Lahore	0.018–0.022	0.069–0.077	0.0006–0.0013	0.012–0.03	0.073–0.045	1.7–8.2	0.16–0.3	0.004–0.018	3.6–4.4	10–11	Schneidemesser et al. (2010)
7	Islamabad Margalla Road	–	26.0	–	–	–	–	–	–	5.1	13.5	Awan et al. (2011)
8	Gujranwala–Shaikhupura Road	–	326.5	–		–	–	–		91.9	35.2	
9	Faisalabad Abdullahpur Crossing Junction	–	320.8	–	–	–	–	–	–	88.6	25.5	
10	Islamabad Sector H–12	–	8.2	–	–	–	–	–	–	71.3	–	
11	Air of M.A. Jinnah road, Karachi	–	89	–	–	1.15	16.87	–	0.55	4.20	2.83	Majid et al. (2011)
12	Air of Lahore	–	59	–	–	0.66	8.24	–	0.36	1.10	1.64	
13	Air of Rawalpindi	–	82	–	–	4.98	15.03	–	1.47	3.90	5.03	
14	Air of Peshawar	–	85	–		0.69	8.56	–	0.561	4.10	1.45	

15	Air sample of Peshawar	–	0.085	–	0.556	0.69	8.56	0.199	0.561	4.10	1.45	Alam et al. (2011)
16	Air sample of Rawalpindi	–	82	–	218	4.98	15.03	375	1475	3.90	5.03	
17	Air sample of Lahore	–	59	–	218	0.66	8.24	205	360	1.10	1.64	
18	Air sample of M.A. Jinnah, Karachi	–	89	–	501	1.15	16.87	471	545	4.20	2.83	
19	Air sample of SPARCO, Karachi	–	48	–	156	0.27	3.39	96	230	1.55	0.76	
20	Air sample of Sea view, Karachi	–	51	–	616	0.94	7.34	162	474	0.14	1.64	
21	Air samples from four cities of Pakistan	8.2–339.8	–	–	–	–	–	–	–	1.2–98.9	0.1–94.3	Awan et al. (2013)
	WHO standards	1.00	5.00	–	1.00	–	–	0.15	1.00	0.5	–	WHO (1999)

Data have been extracted from papers published in national and international journals and arranged chronologically. All values are given in μg m^{-3} unless otherwise mentioned. Values given are mean values or the range from minimum to maximum with mean values in *brackets*

2.3 Cobalt (Co)

Cobalt, a commonly occurring heavy metal, is one of the major environmental pollutants. In Pakistan, it exceeds the permissible limits in some cases in soil, water and air (Tables 1–5). Its highest concentrations in soil were reported by Khan et al. (2008a) in various regions of Khyber Pakhtunkhwa which exceeded the permissible limits of USEPA (1997). Co contamination in water does not seem to be a serious problem in Pakistan as evident from the reviewed data (Tables 2 and 3). However, in wastewater Co concentration exceeds the permissible limit of FAO (1985) in some cases in the country (Table 4). Very few studies have given attention to Co in the air and therefore data on Co contamination in the air can rarely be found for the country (Table 5).

2.4 Chromium (Cr)

Chromium is one of the most common heavy metals found in the earth crust and water. According to SEPA (1995) Cr in soil should not exceed 250 mg kg^{-1}. This limit is violated in few cases in Pakistan, particularly in industrial areas or in wastewater irrigated land (Table 1). However, Cr seems to be a serious cause of water pollution in different regions of Pakistan as the permissible level of WHO and NSDWQ-Pak (0.05 mg L^{-1}) is violated in many parts of the country (Tables 2 and 3). In comparison to ground water, Cr concentrations in surface water seem to be low in the country but still above the permissible level in many regions (Table 3). In contrast to the individual studies, a country-wide study by a Governmental organization in 23 major cities revealed that only 1 % of the water samples exceeded the safe limits for Cr (PCRWR 2005). Similarly, municipal and industrial wastewater in different regions of the country showed large variation in Cr concentration and exceeded the permissible limits in some cases (Table 4). Cr is also a serious pollutant of the atmosphere in different regions of Pakistan as its presence in the air of major cities like Karachi, Lahore and Rawalpindi was found many fold higher than the permissible limit (1 μg m^{-3}) of WHO (Table 5).

2.5 Copper (Cu)

Copper is the third most commonly used metal in the world. In Pakistan, the presence of Cu shows large variations in different environmental components (Tables 1–5). In soil, its concentrations show a variation in different parts of the country (Table 1), which exceed the permissible level of 50 mg kg^{-1} is some cases (Khan et al. 2015). In drinking water the permissible concentration of Cu recommended by WHO and NSDWQ-Pak is 2 mg L^{-1}. In water, both ground and surface water, Cu usually occurs below the permissible limits as reviewed by Azizullah et al. (2011). However, in some regions as for example in Tehsil Tangi in Khyber Pakhtunkhwa and Lahore, elevated concentrations of Cu have been observed (Khan et al. 2012a; Mahmood and Malik 2014) (Tables 2 and 3). The concentration of Cu in municipal and industrial effluents in the country was mostly found within the permissible limits (Table 4). Like soil and water, Cu has also been reported in air of Pakistan in variable concentrations (Table 5).

2.6 *Iron (Fe)*

In Pakistan, soil contains high contents of Fe in different regions (Table 1). The highest concentration of Fe was found in the soil of Peshawar followed by the soil of Adiala, Rawalpindi, Taxilla and Wah (Mushtaq and Khan 2010; Siddiqui and Khattak 2012). Over all survey of published literature revealed that most of the studied areas in Pakistan had Fe concentrations above the permissible limits in soils as shown in Table 1. Same is the situation regarding iron in ground water in the country having iron beyond the permissible limit (Table 2). According to a country-wide study by PCRWR, 40 % of surface water and 28 % of ground water samples had Fe beyond the permissible limits (PCRWR 2005). In surface water too, Fe concentrations crossed the allowable level in many regions in the country (Table 3). As compared to ground and surface water, concentrations of Fe in wastewater, with some exceptions, have mostly been found within the permissible limits (Table 4). In the air also contamination of Fe has usually been reported (Table 5). Major cities of the country like Karachi, Lahore, Rawalpindi and Peshawar had higher concentration of Fe in the air (Yousafzai et al. 2001; Majid et al. 2011; Alam et al. 2011). However, there are no standards of WHO, EPA or any other international organization for iron in ambient air for comparison purpose (Caravanos et al. 2011).

2.7 *Mercury (Hg)*

Mercury is a commonly occurring heavy metal having the property of an accumulative persistent toxin (Weiss and Wright 2001). Data regarding Hg concentration in soil are limited in Pakistan and only four peer reviewed reports were found among which three reported Hg in soil higher than the permissible level of USEPA (1 mg kg^{-1}) (Table 1). Its higher concentration in soil was usually observed in road side areas of national highways in the country (Khan et al. 2011b). A very few and old studies exist regarding Hg concentrations in water of Pakistan revealing its concentration beyond the safe limits (Tables 2 and 3). However, recent data on Hg contamination in water can hardly be found in the country. Similarly, very limited studies gave attention to Hg determination in air, wastewater and sewage in the country (Table 4).

2.8 Manganese (Mn)

Manganese naturally occurs both in soil, water and air. It contributes 0.1 % of Earth's crust and is ubiquitous in the environment (Affum et al. 2011). In Pakistan, Mn is a major polluting agent of soil as it was usually observed exceeding the permissible limit in various regions (Table 1). Results of the summarized studies indicate that Mn in ground water of Pakistan is mostly within the permissible limits with a very few exceptions (Tables 2 and 3). Similarly, in surface water and wastewater also Mn exceeded the permissible limits in very few cases (Tables 3 and 4). In comparison to water, most of the reports showed that Mn in air exceeds the permissible limit of WHO (0.15 $\mu g\ m^{-3}$) in different regions of the country (Table 5) as for example in Karachi, Rawalpindi and Lahore (Alam et al. 2011).

2.9 Nickel (Ni)

In Pakistan, Ni was mostly found within the acceptable range in soil (Table 1) with few exceptions as in the soil of Adiala, Rawalpindi and industrial area of Islamabad (Mushtaq and Khan 2010; Malik et al. 2010). However, Ni can be regarded as a serious pollutant of water in the country (Tables 2 and 3) as in many cases it crossed the permissible limits of WHO (0.07 $mg\ L^{-1}$) and NSDWQ (0.02 $mg\ L^{-1}$) for drinking water. The overall situation of Ni contamination in ground as well as in surface water makes it a problem of serious concern in Pakistan and needs special attention (Tables 2 and 3). The situation is further aggravated by high level contamination of Ni in wastewater in the country (Table 4). Similar to its contamination in soil and water, Ni concentrations in air also show large variation in different parts of Pakistan (Table 5). However, some studies revealed high concentration of Ni in air at various sites in different cities of the country (Alam et al. 2011; Faiz et al. 2009).

2.10 Lead (Pb)

Lead is one of the ubiquitously distributed most abundant toxic elements in soil. Pb concentrations in soils of various regions in Pakistan are usually found below the maximum permissible limit of USEPA (Table 1). However, both ground and surface water had Pb contamination beyond the permissible limits of WHO and NSDWQ-Pak in many cases (Tables 2 and 3). A country wide study by PCRWR (2007) indicated that surface water is comparatively more polluted with Pb than ground water in Pakistan (Tables 2 and 3). In contrast to ground and surface water, Pb concentrations in wastewater (Table 4) have mostly been found below the permissible limits of FAO (1985). Like water, its elevated concentrations have been reported in air in some of the studied regions in the country (Table 5). For example, its higher concentrations were observed even in the air of the capital Islamabad (Faiz et al. 2009; Awan et al. 2011).

2.11 Zinc (Zn)

According to USEPA (1997) and SEPA (1995), Zn in soil should not exceed 200 or 300 mg kg^{-1}, respectively. This criterion is usually fulfilled in the soils of Pakistan (Table 1) with few exceptions as for example in Gilgit, in industrial sewage irrigated soil in Karachi and road side soil of Faisalabad where Zn was observed above the permissible level (Yousafzai et al. 2001; Khan et al. 2010; Parveen et al. 2012). Large variation in the concentration of Zn occurs in both ground and surface waters in different regions of Pakistan (Tables 2 and 3), but its concentration is mostly well below the permissible limits of WHO (3 mg L^{-1}) and NSDWQ (5 mg L^{-1}), with the exceptions of some old studies in various areas in Karachi (Rahman et al. 1997). Similarly, wastewaters from different areas in Pakistan have Zn below the threshold value (Table 4). In comparison to soil and water, Zn was found in higher concentrations in air of Pakistan as revealed by different studies (Table 5). The highest concentration of Zn was found in particulate matter of Gutter Baghicha SITE, Karachi and Islamabad Expressway (Yousafzai et al. 2001; Faiz et al. 2009).

3 Health Risk Assessment of Toxic Metals

Risk assessment is a function of hazard and exposure which is defined as "the process of estimating the probability of occurrence of an event and the probable magnitude of adverse health effect on human exposures to environmental hazards over a specified time period" (Paustenbach 2002; Wongsasuluk et al. 2013). According to Lee et al. (2005), risk assessment consists of hazard identification, exposure assessment, dose response (toxicity) and risk characterization. The health risk assessment of each potentially toxic metal is usually based on the quantification of the risk level and is expressed in terms of a carcinogenic or a non-carcinogenic health risk. To assess the human health risk of heavy metals, it is necessary to calculate the level of human exposure to that metal by tracing the route of exposure of pollutant to human body. There exist many routes of exposure to toxic metals that depend upon the contaminated media, i.e. soil, water and air. Humans are sometimes indirectly exposed to toxic metals by using food (e.g. crops and vegetables) having metal contamination (Khan et al. 2008c, 2013a).

The health risks due to toxic metals are primarily associated with the exposure dose and time. Different organizations such as NSDWQ, USEPA and WHO have established the maximum permissible levels of toxic metals through different media. However, it is documented that toxic metals even if within the permissible limits may pose adverse effects to public health, particularly when more than one toxic metal are present. Therefore, we calculated the hazard quotient (HQ) and Hazards Index (HI) of individual metals as well as collective HI for toxic metals reviewed in this article. To assess the human health risks due to toxic metals, different indices like chronic daily intake (CDI), non-metalloids hazard quotient (HQ) and hazard index (HI) were calculated in the present using the data for toxic metals and metalloids shown in Tables 1–5. CDI represents daily exposure of population to hazardous substances in $mg\,kg^{-1}$ day. The minimum allowable exposure concentration (also known as reference dose RfD) is the exposure limit ($mg\,kg^{-1}$ day) of humans to a toxic metal. RfD is an estimated per day exposure of metal to the human body that has no hazardous effect during its life time. People exposed to toxic metals above the reference doses have risks of health problems (USEPA 2011). HQ is the ratio of CDI to RfD and reflect the risk posed by a metal. A value of HQ above 1.0 for a metal is considered unsafe from the health point of view. HI reflects the potential chronic risk from multiple metals in a medium, i.e. collective HQs of all the analyzed toxic metals in a given medium. Health risk assessment can be used to illustrate the health effects of human exposures to hazardous metals and can play an important role in decision-making and remedial actions. It can be helpful in monitoring and to educate the public in understanding the environmental quality guidelines to avoid hazardous environmental risks. Uncertainties are inevitable in calculating the health risk assessment because of the incomplete and imprecise information, errors in measurement and other limitations in the available data. For estimation of risk assessment of the toxic metals at a specific site, deterministic techniques can be used. Similarly, for the minimization of uncertainty and errors, probabilistic approaches can be used to evaluate the risk levels. For this purpose Monte Carlo simulation was used to calculate the output value and the probability of its occurrence which can provide a risk assessment by considering the uncertainty and bounding the true value of the risk (Bruce et al. 2007; Li et al. 2013a, b).

In the present study, different risk indices were calculated as follow.

Chronic daily intake (CDI): The CDI ($mg\,kg^{-1}$-day) was calculated by Eqs. (1)–(4) for air, water and soil (Li et al. 2013a, b; Liu et al. 2013; Khan et al. 2014).

$$\text{Soil Ingestion } CDI_{\text{Iingest-soil}} = C \times IR \times EF \times ED/BW \times AT \tag{1}$$

$$\text{Soil dermal } CDI_{\text{Dermal-soil}} = C \times IR \times SA \times EF \times ED/BW \times AT \tag{2}$$

$$\text{Water Oral intake } CDI_{\text{oral-water}} = C \times IR \times EF \times ED/BW \times AT \tag{3}$$

$$\text{Air Inhalation } CDI_{\text{Inh-Air}} = C \times IR \times ET \times EF \times ED/BW \times AT \tag{4}$$

where CDI — chronic daily intake; C—concentration of metal in soil ($mg\,kg^{-1}$); in water ($mg\,L^{-1}$) and in air ($mg\,m^{-3}$); EF—exposure frequency 365 days (USEPA 2011); ED—exposure duration 70 year (USEPA 2011); ET—Daily exposure time

24 h/day (UDOE 2011); AT—averaging time for non-carcinogens 25,550 days (USEPA 2011); BW—average adult body weights were assumed to be 73 kg (Khan et al. 2013a; USEPA 2011); PEF—particle emission factor $1.36 \times 10^9\ m^3\ kg^{-1}$ (USEPA 2002); CF—units conversion factor 10^{-6} kg mg^{-1} (USEPA 2002); IR—ingestion rate of water 2 L/day (Wang et al. 2005), soils 100 mg/day and air $1.3\ m^3\ h^{-1}$ (Li et al. 2013a, b; Liu et al 2013); SA—exposure skin area 5700 cm^2 (USEPA 2011); AF—adherence factor 0.07 mg cm^{-2} (USEPA 2011); ABS—dermal absorption fraction: 0.03 (As) and 0.001 (other metals) (USEPA 2011).

Hazard quotient (HQ): The non-carcinogen health risk was assessed by calculating hazard quotient (HQ) using Eq. (5) (Huang et al. 2008; Khan et al. 2014).

$$HQ = CDI/RfD \tag{5}$$

where CDI is chronic daily intake and RfD represents the oral reference dose of As, Cd, Co, Cr, Cu, Fe, Hg, Mn, Ni, Pb and Zn, as 3.0E-04, 5.0E-04, 3.0E-02, 3.0E-03, 4.0E-02, 8.0E-01, 3.0E-04, 1.4E-01, 2.0E-02, 3.5E 02 and 3.0E 01 mg kg^{-1}-day, respectively (USEPA 2011; Shah et al. 2012; Khan et al. 2014).

Health risk of multiple metals: The total potential chronic risk for multiple metals has been represented as hazard index (HI) calculated according to Eq. (6) (Huang et al. 2008; Bermudez et al. 2011).

$$HI = HQ_{Cd} + HQ_{Cr} + \ldots + HQ_n \tag{6}$$

where HI is the aggregate hazard index and HQ_{Cd}, HQ_{Cr} and H_{Qn} are the calculated HQs of Cd, Cr … *n* metal, respectively. A value of HI above 1 is assumed to be unsafe for human health and can cause non-carcinogenic health problems (USEPA 2011).

3.1 Health Risks due to Heavy Metals in Pakistan

The chronic daily intake (CDI), non-carcinogenic hazard quotient (HQ) and hazard index (HI) calculated for various metals and metalloids in Pakistan are summarized in Tables 7–11 and Fig. 1.

3.1.1 Arsenic

Arsenic exposure affects more than 150 million people around the world because of its elevated concentration in drinking water sources (Brammer and Ravenscroft 2009). The present analysis reveals that more than 70 % of people in the studied regions of Pakistan are exposed to higher doses of As through drinking water and in some case through air but very few available data on As in the air make it difficult to conclude its risks via air (Tables 9–11). Although As might be a problem in soil it

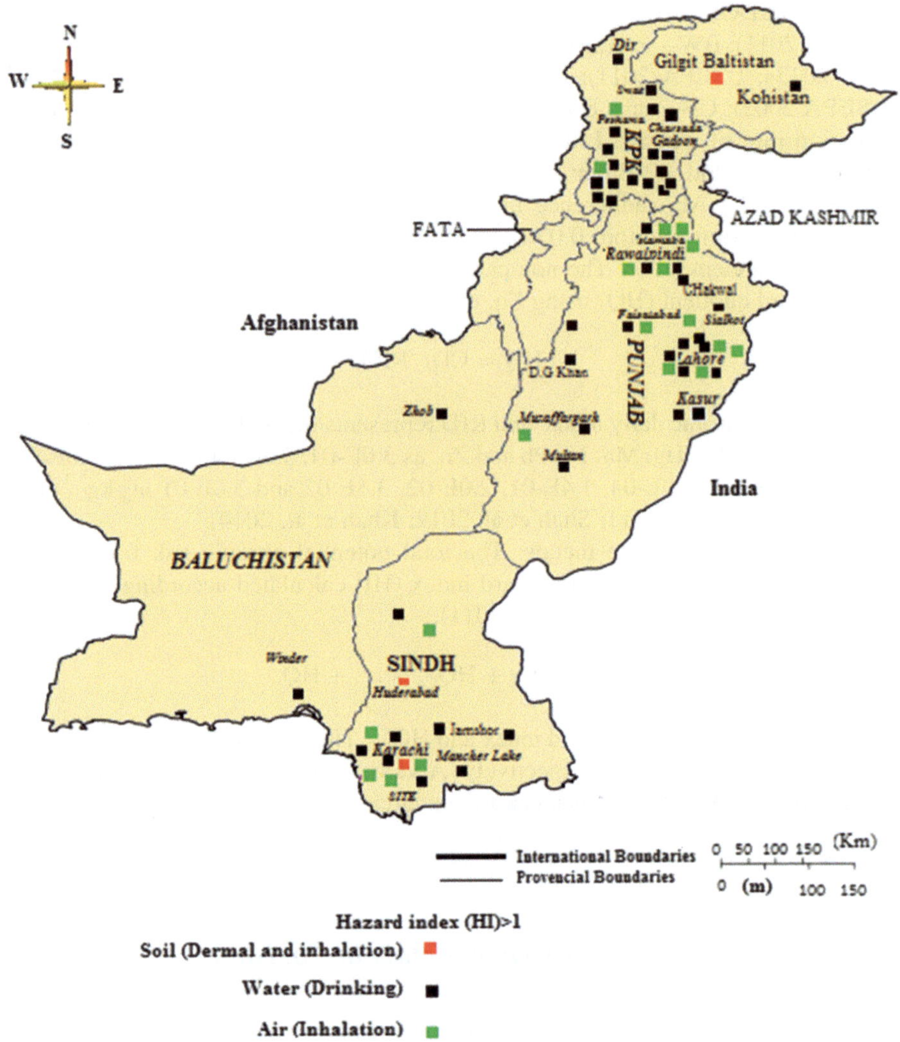

Fig. 1 Map of Pakistan showing areas with Hazard index > 1 due to heavy metal contamination in soil, water and air. Only areas with a possible risk are shown here

was not analyzed in most of the reviewed studies, hence was not included in the present risk analysis. The risk analysis of available data for As contamination in water, in most of the cases, revealed higher HQ values which exceeded the threshold value of 1. The highest value of HQ (173.516) was found for ground water of Kalalanwala in east Punjab followed by well water of Multan (96). Similarly, the analyzed risk value in Chashma Water Reservoir was 68.49 while it was 56.621 in Tarbela and Lloyd Water Reservoir. Furthermore, the risk assessment of well water in district Multan indicated an HQ value of 36.530 (Tables 9 and 10 and Fig. 1).

In most of the cases (about 80 %), the HQs for As were found above the threshold value (Tables 9 and 10). In air, the risk analysis, based on the available four studies in different regions, indicated that three out of four regions (66.66 %) have crossed the threshold limit for HQ (Table 11). For example, the data collected by Awan et al. (2013) when analyzed the risk showed that the air has higher concentration of As with an HQ of 6.05 followed by 5.56 and 3.97 in the ambient air samples of Rawalpindi–Islamabad and Lahore, respectively (Table 11 and Fig. 1).

Exposure to arsenic through drinking water or the food chain can accumulate in the human body and causes various problems (Smith et al. 1992). It is evident from elevated levels of arsenic in human samples from populations of different regions in Pakistan (Table 6). Exposure to high concentrations of arsenic creates health problems like "pins and needles" sensation in hands and feet, gastrointestinal irritation, decrease in white and red blood cells production, disrupts the heart rhythms and damages blood vessels (Abernathy et al. 2003). Long-term exposure to high concentrations causes cancer, cardiovascular disease, melanosis, leuko-melanosis, hyperkeratosis, black foot disease, and neuropathy (Caussy 2005; Gary 2012). For example, in Basti Rasul Pur of Rahim Yar Khan District in Pakistan, four persons per household were affected with diseases, particularly with water-linked diseases, and about 42 % patients were found to suffer from arsenicosis (Haq et al. 2012). An inverse relation of arsenicosis was found with income resources of the people as well as with awareness amongst the people in the area. The authors roughly estimated that about 60 % earning of the poor class in this area is being spent on hospitalization (Haq et al. 2012). Recently, a research team found significant correlation between both arsenic skin lesions and arsenic groundwater contamination of wells with proximity to the river Indus (Dawn 2013). In Khairpur district, 13.5 % of the people (based on a survey of 3000 individuals) in the area had arsenic skin lesions (Dawn 2013). Elevated concentrations of arsenic in water have also been the cause of health problems in many other parts of the country (Arain et al. 2009; Kazi et al. 2009). Abbas and Cheema (2014) also found a positive correlation between As concentration in drinking water and in hair samples of female population in Sheikhupura, Pakistan. A recent study by Abbas and Cheema (2015) revealed that high concentration of As in water causes various health problems in Sheikhupura, Pakistan. Among the residents of this area, 17, 14, 5, 11, 12, 27, 8 and 22 % of respondents had respiratory diseases, liver disorder, reduced weight, reduced bulk of limbs, skin infection, blood pressure, diabetes, and muscular pain, respectively. Wadhwa et al. (2015) found higher levels of metals and metalloids like As, Cd and Ni in the scalp hair samples of cancer patients of Pakistani females.

3.1.2 Cadmium (Cd)

Cd was identified as being responsible for wide-spread intoxication (viz., itai-itai disease) of residents in Japan in the 1950s and after that time Cd pollution has been an issue of great concern all over the world. In Pakistan, the concentration of Cd

Table 6 Heavy metal concentrations in human samples (blood, hairs and nails) collected from different regions of Pakistan

S. No	Sampling location	As	Cd	Co	Cr	Cu	Fe	Mn	Ni	Pb	Zn	References
1	Samples collected from population of Karachi	–	0.05–0.054		0.03–0.04	0.86–0.99	400.4–482.1	–	–	0.19–0.22	6.20–6.61	Mumtaz et al. (1999)
2	Samples of hairs from residence of Lahore	0.31–1.12	0.08	–	–	–	–	–	–	3.53	–	Anwar (2005)
3	Samples of Toenails from residence of Lahore	0.70–4.47	0.05	–	–	–	–	–	–	2.11	–	
4	Blood samples from child living around automobile and battery repair work shops	–	–	–	–	–	–	–	–	0.010–0.012	–	Ahmad et al. (2009)
5	Hairs samples from two villages of Hyderabad	0.28–6.40										Baig et al. (2010)
6	Scalp hairs of Hyderabad inhabitants	0.72–4.94	–	–	–	–	–	–	–	–	–	Shah et al. (2011)
7	Blood samples collected from lower Dir	–	–	–	00–0.0015	00–0.0042	0.124–0.542	0.0005–0.0088	00–0.0007	00–0.0014	0.0001–0.018	Jan et al (2011)
8	Blood samples collected from Peshawar	–	–	–	00–0.018	0.0001–0.034	0.203–0.596	00–0.013	00–0.0007	00–0.0007	0.0001–0.087	
9	Blood samples of welders, flood victims, fishermen and dyers from Karachi	–	–	–	–	–	0.23–1.18	0.22–1.8	–	–	–	Ali et al. (2012)
10	Nail samples of Jewelers from Multan	–	45.31–143.51	–	–	38.62–130.79	–	–	–	4.2–20.9	–	Shan and Ikram (2012)
11	Hairs samples of Jewelers from Multan	–	25.43–104.72	–	–	23.25–99.34	–	–	–	10.5–27.9	–	
12	Nail samples of automobile workers from Multan	–	7.83–38.60	–	–	43.62–157.67	–	–	–	19.35–72.45	–	

13	Hairs samples of automobile workers from Multan	–	4.62–25.87	–	–	38.25–121.56	–	–	–	24.54–85.79	–	
14	Nail samples of industrial workers from Multan	–	0.21–0.49	–	–	43.25–93.80	–	–	–	29.31–79.5	–	
15	Hairs samples of industrial workers from Multan	–	0.27–0.51	–	–	60.25–106.34	–	–	–	38.14–103.79	–	
16	Hairs samples from individuals related to different occupations	–	89.20–143.57	–	–	80.76–110.79	–	–	–	–	–	Ayyub et al. (2013)
17	Nail samples from individuals related to different occupations	–	77.54–104.71	–	–	77.65–99.36	–	–	–	–	–	
18	Blood samples from children and adults in district Shangla	–	0.006–0.15	0.005–0.45	0.356–7.82	3.07–16.92	154.6–420.20	–	0.006–0.88	0.92–7.8	15.61–47.49	Ahmad et al. (2014a)
19	Human milk samples from district Shangla	–	ND–0.139	ND–3.8	8.57–15.72	ND–0.086	46.62–76.28	–	0.006–0.142	ND–6.03	ND–4.82	
20	Samples of human seminal plasma of Pakistani population	–	0.0001–0.0081	0.00008–0.0002	0.01–0.02	0.03–0.23	2.7–3.49	0.005–0.04	0.0019–0.01	0.011–0.071	111.06–150.13	Zafar et al. (2014)

Data have been extracted from papers published in national and international journals and arranged chronologically. All values are given in mg L^{-1} in blood and μg g^{-1} in nails and hairs samples unless otherwise mentioned. Values given are mean values or the range from minimum to maximum with mean values in *brackets*. *ND* stands for not detected varies in different regions and different compartments of the environment but generally pose a problem. Risk evaluation of the reported data revealed that Cd poses a higher risk in the studied areas through water and air while the situation remains safe in the case of exposure through soil (Tables 7–11 and Fig. 1). In water, 58 % of the studied regions were found with HQ values for Cd higher than 1 (Tables 9 and 10). Similarly, in air 71 % of the reported studies analyzed for risk assessment showed HQ values above the threshold of 1 (Table 11). A higher risk (HQ, 10.958) was found due to the surface water of Bara River, Akbarpura area, Khyber Pakhtunkhwa followed by Hudaira drain, Lahore (HQ 9.863). However, the water of these two sources are not directly consumed by humans. HQ values of 7.671 and 2.74 for Cd were obtained for the water of Akbarpura canal in Nowshera area of Khyber Pakhtunkhwa and Kalar Kahar lake, Chakwal, Punjab, respectively (Table 10 and Fig. 1). Similarly, risk analysis for ground water revealed that deep wells water in the vicinity of Palosi Drain, Peshawar, poses a risk with HQ value of 3.068 followed by Tube well water of Korangi, Karachi with an HQ value of 2.247 (Table 9). The highest health risk of Cd through air was found in Gujranwala–Shaikhupura Road and Faisalabad Abdullahpur crossing junction where HQ reached 349 and 343, respectively. Similarly, in big cities of the country like Karachi (M.A. Jinnah road) and Peshawar are at risk of Cd through air where the HQ of Cd through air reached 95.1 and 90.8, respectively (Table 11). The overall analysis for Cd risk analysis reveal that the situation is very critical as in about 72 % of the collected data HQ values are beyond the threshold value of 1 (Table 11 and Fig. 1).

The higher exposure to Cd through water and air is also evident from the reported concentrations of Cd in human samples to an unsatisfactory level in the country (Table 6). Overexposure to Cd can cause various health problems like kidney damage (Barbier et al. 2005), reproductive problems (Frery et al. 1993; Johnson et al. 2003; Piasek and Laskey 1999), cancer (Waalkes et al. 1988), bone damage (Kazantzis 1979), vomiting and diarrhea (Nordberg 2004). It may also cause other health problems like nausea, muscle cramps, salvation, sensory disturbances, liver injury, convulsions shocks, and renal failure (Nordberg 2004). Long-time exposure to Cd also results in lung damage and fragile bones (WHO 2007). Exposure to Cd can cause cellular damage of intestine and liver (Aziz et al. 2014). Ahmad et al. (2014c) reported that inhalation of heavy metals like Cu, Pd and Cd) can cause cataract disease of eyes. According to Hussain et al. (2014), long term exposure to Cd contamination can results in various disorders like renal dis-function, lung cancer and respiratory problems. A combination of Cd and Al has been regarded to be the cause of kidney disorder in Southern parts of Pakistan (Panhwar et al. 2015).

3.1.3 Cobalt (Co)

At low concentrations, Co is an essential micronutrient as it is required for vitamin B_{12} synthesis and has a role in enzymatic functions, but its overexposure can cause serious health problems (Gál et al. 2008). The risk assessment due to Co

Table 7 Exposure to heavy metals and health risk assessment via ingestion of contaminated soils in different regions of Pakistan

Soil samples	Cd	Co	Cr	Cu	Fe	Hg	Mn	Ni	Pb	Zn	$HI_{Ingestion}$
1	– –	9.59E-06 3.20E-04	2.60E-05 8.68E-03	2.88E-05 7.19E-04	1.38E-02 1.72E-02	– –	– –	1.11E-04 5.55E-03	1.37E-05 3.91E-03	4.11E-05 1.37E-04	3.65E-02
2	5.48E-06 1.10E-02	2.19E-06 7.31E-05	3.84E-06 1.28E-03	– –	– –	– –	6.01E-04 4.30E-03	– –	– –	– –	1.66E-02
3	– –	– –	– –	2.40E-05 6.00E-04	1.81E-04 2.26E-04	– –	1.44E-04 1.03E-03	1.21E-06 6.03E-05	6.03E-06 1.72E-03	2.28E-04 7.60E-04	4.40E-03
4	1.92E-07 3.84E-04	– –	5.34E-07 1.78E-04	– –	– –	– –	– –	– –	– –	– –	5.62E-04
5	2.33E-07 4.66E-04	3.77E-06 1.26E-04	8.16E-06 2.72E-03	– –	2.06E-05 2.57E-05	– –	1.66E-06 1.18E-05	2.07E-06 1.03E-04	6.11E-06 1.75E-03	3.01E-06 1.00E-05	5.21E-03
6	2.33E-07 4.66E-04	6.85E-05 2.28E-03	3.10E-06 1.03E-03	1.10E-06 2.74E-05	1.64E-03 2.05E-03	– –	3.94E-05 2.81E-04	5.89E-05 2.95E-03	9.45E-05 2.70E-02	7.34E-06 2.45E-05	3.61E-02
7	– –	3.15E-05 1.05E-03	2.95E-06 9.82E-04	9.59E-07 2.40E-05	2.47E-04 3.08E-04	– –	3.21E-05 2.30E-04	5.62E-05 2.81E-03	7.26E-05 2.07E-02	4.81E-06 1.60E-05	2.62E-02
8	– –	5.48E-05 1.83E-03	2.47E-06 8.22E-04	1.26E-06 3.15E-05	2.77E-04 3.46E-04	– –	3.33E-05 2.38E-04	6.44E-05 3.22E-03	9.73E-05 2.78E-02	8.84E-06 2.95E-05	3.43E-02
9	– –	6.71E-05 2.24E-03	4.05E-06 1.35E-03	9.86E-07 2.47E-05	2.60E-04 3.25E-04	– –	3.64E-05 2.60E-04	5.34E-05 2.67E-03	9.73E-05 2.78E-02	8.03E-06 2.68E-05	3.47E-02
10	– –	5.62E-05 1.87E-03	3.22E-06 1.07E-03	7.67E-07 1.92E-05	2.27E-04 2.84E-04	– –	3.48E-05 2.48E-04	4.25E-05 2.12E-03	6.99E-05 2.00E-02	5.62E-06 1.87E-05	2.56E-02
11	5.48E-08 1.10E-04	– –	3.01E-07 1.00E-04	– –	– –	– –	– –	– –	– –	– –	2.10E-04
12	1.92E-07 3.84E-04	– –	6.71E-07 2.24E-04	– –	– –	– –	– –	– –	– –	– –	
13	4.78E-06 9.56E-03	– –	– –	– –	– –	– –	– –	– –	– –	– –	9.56E-03

(continued)

Table 7 (continued)

Soil samples	Cd	Co	Cr	Cu	Fe	Hg	Mn	Ni	Pb	Zn	$HI_{Ingestion}$
14	4.25E-07 8.49E-04	– –	– –	8.49E-06 2.12E-04	4.68E-05 5.85E-05	– –	1.92E-05 1.37E-04	7.53E-07 3.77E-05	1.64E-06 4.70E-04	2.52E-05 8.42E-05	1.85E-03
15	8.36E-07 1.67E-03	– –	– –	1.29E-05 3.22E-04	7.05E-05 8.81E-05	– –	3.73E-05 2.66E-04	1.47E-06 7.33E-05	5.14E-06 1.47E-03	3.41E-05 1.14E-04	4.00E-03
16	7.12E-07 1.42E-03	– –	2.36E-05 7.88E-03		– –	– –	– –	– –	– –	– –	9.30E-03
17	1.97E-06 3.93E-03	– –	1.27E-03 4.23E-01	4.90E-04 1.22E-02	– –	– –	– –	– –	– –	– –	4.40E-01
18	4.11E-06 8.22E-03	– –	2.93E-05 9.76E-03	1.05E-05 2.61E-04	1.30E-02 1.63E-02	4.25E-05 1.42E-01	6.44E-06 4.60E-05	4.24E-05 2.12E-03	3.56E-05 1.02E-02	4.25E-05 1.42E-04	1.89E-01
19	2.85E-06 5.71E-03	– –	5.61E-05 1.87E-02	1.59E-05 3.98E-04	1.39E-02 1.74E-02	1.71E-06 5.70E-03	6.39E-04 4.56E-03	5.30E-05 2.65E-03	4.23E-05 1.21E-02	8.09E-05 2.70E-04	6.74E-02
20	2.33E-06 4.66E-03	– –	3.95E-05 1.32E-02	2.16E-05 5.41E-04	– –	– –	– –	– –	– –	– –	1.83E-02
21	3.97E-06 7.95E-03	– –	1.71E-04 5.68E-02	1.02E-04 2.54E-03	– –	– –	– –	– –	– –	– –	6.73E-02
22	4.79E-06 9.59E-03	– –	1.98E-04 6.58E-02	4.14E-05 1.03E-03	– –	– –	– –	– –	– –	– –	7.65E-02
23	3.42E-06 6.85E-03	– –	2.11E-04 7.04E-02	5.23E-05 1.31E-03	– –	– –	– –	– –	– –	– –	7.85E-02
24	3.15E-06 6.30E-03	– –	– –	2.01E-04 5.03E-03	– –	– –	– –	– –	– –	– –	1.13E-02
25	1.15E-06 2.30E-03	8.15E-06 2.72E-04	– –	1.78E-05 4.45E-04	5.44E-03 6.80E-03	1.24E-04 4.14E-01	2.33E-08 1.66E-07	1.22E-05 6.11E-04	4.99E-05 1.43E-02	7.77E-05 2.59E-04	4.39E-01
26	5.78E-06 1.16E-02		2.71E-03 9.04E-01	– –	– –	– –	– –	– –	– –	– –	9.16E-01
27	3.29E-07 6.58E-04	– –		– –	– –	– –	– –	– –	– –	– –	6.58E-04

28	1.22E-06 2.44E-03	– –	1.06E-04 3.53E-02	2.12E-05 5.31E-04	6.82E-02 8.52E-02		8.16E-04 5.83E-03	7.40E-05 3.70E-03	1.51E-05 4.31E-03	6.58E-05 2.19E-04	1.38E-01
29	3.29E-07 6.58E-04	– –	4.79E-07 1.60E-04	– –	– –	– –	– –	8.90E-07 4.45E-05	8.90E-07 2.54E-04		1.12E-03
30	– –	– –	– –	– –	– –	– –	– –	4.07E-05 2.03E-03	– –	– –	2.03E-03
31	1.51E-07 3.01E-04	6.16E-08 2.05E-06	1.16E-07 3.88E-05	– –	– –	– –	– –	– –	– –	– –	3.42E-04
32	1.10E-08 2.19E-05	3.42E-08 1.14E-06	4.25E-08	4.25E-08 1.06E-06	– –	– –	– –	– –	– –	– –	3.83E-05
33	8.22E-09 1.64E-05	– –	3.56E-08 1.19E-05		6.25E-06 7.81E-06	– –	– –	– –	– –	– –	3.61E-05
34	3.60E-05 7.21E-02	3.02E-05 1.01E-03	3.35E-03 1.12E+00	– –	1.88E-04 2.34E-04	– –	1.29E-05 9.22E-05	4.68E-05 2.34E-03	2.49E-05 7.13E-03	1.96E-05	1.20E+00
35	– –	3.71E-07 1.24E-05	1.28E-06 4.27E-04	– –	– –	– –	– –	– –	– –	– –	4.40E-04
36	4.32E-06 7.21E-02	1.10E-05 3.66E-04	2.88E-05 9.59E-03	3.94E-05 9.84E-04	– –	– –	3.95E-05 2.82E-04	5.34E-05 2.67E-03	2.11E-05 6.02E-03	6.96E-05	2.88E-02
37	2.24E-05	– –	– –	– –	– –	– –	– –	– –	– –	– –	4.47E-02
38	1.92E-07 3.84E-04	1.51E-06 5.02E-05	2.34E-04 7.81E-02	8.79E-05 2.20E-03	– –	– –	– –	4.66E-05 2.33E-03	6.16E-05 1.76E-02	9.32E-05	1.01E-01

In each cell, the first value represents CDI (Chronic daily intake mg kg^{-1}-day) and the second value is Hazard Quotient (HQ) of individual metal. Hazard index (HI) of multiple metals in a given area is shown in the *last column* at the *right side* of the table. The serial number in the *first column* shows heavy metal locations in soil with reference to Table 1. Mean values have been used for the determination of CDI, HQ and HI while in case where data is available in minimum to maximum range, risk was calculated for the maximum value

Table 8 Exposure to heavy metals and health risk assessment via dermal contact and ingestion of contaminated soil in different regions of Pakistan

Soil samples Locations	Cd	Co	Cr	Cu	Fe	Hg	Mn	Ni	Pb	Zn	HI Dermal	HI_{dermal} + $HI_{Ingestion}$ (HI_{Total})
1	– –	3.83E-05 1.28E-03	1.04E-04 3.46E-02	1.15E-04 2.87E-03	5.50E-02 6.87E-02	– –	– –	4.43E-04 2.21E-02	5.47E-05 1.56E-02	1.64E-04 5.47E-04	1.46E-01	1.82E-01
2	2.19E-05 4.37E-02	8.75E-06 2.92E-04	1.53E-05 5.10E-03		6.68E-02 8.35E-02	– –	2.40E-03 1.71E-02	– –	– –	– –	– –	1.66E-02
3	– –	– –		9.58E-05 2.40E-03	7.21E-04 9.01E-04	– –	5.74E-04 4.10E-03	4.81E-06 2.40E-04	2.40E-05 6.87E-03	9.10E-04 3.03E-03	1.75E-02	2.19E-02
4	7.65E-07 1.53E-03	– –	2.13E-06 7.11E-04	– –	– –	– –	– –	– –	– –	– –	2.24E-03	2.80E-03
5	9.29E-07 1.86E-03	1.50E-05 5.01E-04	3.26E-05 1.09E-02	– –	8.21E-05 1.03E-04	– –	6.61E-06 4.72E-05	8.25E-06 4.13E-04	2.44E-05 6.96E-03	1.20E-05 4.01E-05	2.08E-02	2.60E-02
6	9.29E-07 1.86E-03	2.73E-04 9.11E-03	1.24E-05 4.12E-03	4.37E-06 1.09E-04	6.54E-03 8.18E-03	– –	1.57E-04 1.12E-03	2.35E-04 1.18E-02	3.77E-04 1.08E-01	2.93E-05 9.77E-05	1.44E-01	1.80E-01
7	– –	1.26E-04 4.19E-03	1.18E-05 3.92E-03	3.83E-06 9.57E-05	9.84E-04 1.23E-03	– –	1.28E-04 9.16E-04	2.24E-04 1.12E-02	2.90E-04 8.28E-02	1.92E-05 6.39E-05	1.04E-01	1.31E-01
8	– –	2.19E-04 7.29E-03	9.84E-06 3.28E-03	5.03E-06 1.26E-04	1.10E-03 1.38E-03	– –	1.33E-04 9.50E-04	2.57E-04 1.28E-02	3.88E-04 1.11E-01	3.53E-05 1.18E-04	1.37E-01	1.71E-01
9	– –	2.68E-04 8.93E-03	1.62E-05 5.39E-03	3.94E-06 9.84E-05	1.04E-03 1.30E-03	– –	1.45E-04 1.04E-03	2.13E-04 1.07E-02	3.88E-04 1.11E-01	3.20E-05 1.07E-04	1.38E-01	1.73E-01
10	– –	2.24E-04 7.47E-03	1.28E-05 4.28E-03	3.06E-06 7.65E-05	9.07E-04 1.13E-03	– –	1.39E-04 9.91E-04	1.69E-04 8.47E-03	2.79E-04 7.96E-02	2.24E-05 7.47E-05	1.02E-01	1.28E-01
11	2.19E-07 4.37E-04	– –	1.20E-06 4.01E-04	– –	– –	– –	– –	– –	– –	– –	8.38E-04	1.05E-03
12	7.65E-07 1.53E-03	– –	2.68E-06 8.93E-04	– –	– –	– –	– –	– –	– –	– –	2.42E-03	2.42E-03
13	1.91E-05 3.82E-02	– –	– –	– –	– –	– –	– –	– –	– –	– –	3.82E-02	4.77E-02

14	1.69E-06 3.39E-03	– –	– –	3.39E-05 8.47E-04	1.87E-04 2.33E-04	– –	7.65E-05 5.47E-04	3.01E-06 1.50E-04	6.56E-06 1.87E-03	1.01E-04 3.36E-04	7.38E-03	9.22E-03
15	3.33E-06 6.67E-03	– –	– –	5.14E-05 1.29E-03	2.81E-04 3.51E-04	– –	1.49E-04 1.06E-03	5.85E-06 2.92E-04	2.05E-05 5.86E-03	1.36E-04 4.54E-04	1.60E-02	2.00E-02
16	2.84E-06 5.68E-03	– –	9.43E-05 3.14E-02	– –	– –	– –	– –	– –	– –	– –	3.71E-02	4.64E-02
17	7.84E-06 1.57E-02	– –	5.07E-03 1.69E+00	1.95E-03 4.88E-02	– –	– –	– –	– –	– –	– –	1.75E+00	2.19E+00
18	1.64E-05 3.28E-02	– –	1.17E-04 3.90E-02	4.17E-05 1.04E-03	5.20E-02 6.50E-02	1.69E-04 5.65E-01	2.57E-05 1.84E-04	1.69E-04 8.45E-03	1.42E-04 4.06E-02	1.69E-04 5.65E-04	7.53E-01	9.41E-01
19	1.14E-05 2.28E-02	– –	2.24E-04 7.46E-02	6.35E-05 1.59E-03	5.54E-02 6.92E-02	6.82E-06 2.27E-02	2.55E-03 1.82E-02	2.11E-04 1.06E-02	1.69E-04 4.82E-02	3.23E-04 1.08E-03	2.69E-01	3.36E-01
20	9.29E-06 1.86E-02	– –	1.57E-04 5.25E-02	8.64E-05 2.16E-03	– –	– –	– –	– –	– –	– –	7.32E-02	9.16E-02
21	1.59E-05 3.17E-02	– –	6.80E-04 2.27E-01	4.06E-04 1.01E-02	– –	– –	– –	– –	– –	– –	2.69E-01	3.36E-01
22	1.91E-05 3.83E-02	– –	7.88E-04 2.63E-01	1.65E-04 4.13E-03	– –	– –	– –	– –	– –	– –	3.05E-01	3.82E-01
23	1.37E-05 2.73E-02	– –	8.42E-04 2.81E-01	2.09E-04 5.22E-03	– –	– –	– –	– –	– –	– –	3.13E-01	3.92E-01
24	1.26E-05 2.51E-02	– –	– –	8.03E-04 2.01E-02	– –	– –	– –	– –	– –	– –	4.52E-02	5.66E-02
25	4.59E-06 9.18E-03	3.25E-05 1.08E-03	– –	7.09E-05 1.77E-03	2.17E-02 2.71E-02	4.96E-04 1.65E+00	9.29E-08 6.64E-07	4.88E-05 2.44E-03	1.99E-04 5.69E-02	3.10E-04 1.03E-03	1.75E+00	2.19E+00
26	2.31E-05 4.61E-02	– –	1.08E-02 3.61E+00	– –	– –	– –	– –	– –	– –	– –	3.65E+00	4.57E+00
27	1.31E-06 2.62E-03	– –	– –	– –	– –	– –	– –	– –	– –	– –	2.62E-03	3.28E-03
28	4.86E-06 9.73E-03	– –	4.22E-04 1.41E-01	8.47E-05 2.12E-03	2.72E-01 3.40E-01		3.26E-03 2.33E-02	2.95E-04 1.48E-02	6.01E-05 1.72E-02	2.62E-04 8.75E-04	5.49E-01	6.86E-01

(continued)

Table 8 (continued)

Soil samples Locations	Cd	Co	Cr	Cu	Fe	Hg	Mn	Ni	Pb	Zn	HI Dermal	HI_{dermal} + $HI_{Ingestion}$ (HI_{Total})
29	1.31E-06 2.62E-03	– –	1.91E-06 6.38E-04	– –	– –	– –	– –	3.55E-06 1.78E-04	3.55E-06 1.02E-03	– –	4.45E-03	5.57E-03
30	– –	– –	– –	– –	– –	– –	– –	1.62E-04 8.12E-03	– –	– –	8.12E-03	1.02E-02
31	6.01E-07 1.20E-03	2.46E-07 8.20E-06	4.65E-07 1.55E-04	– –	– –	– –	– –	– –	– –	– –	1.37E-03	1.71E-03
32	4.37E-08 8.75E-05	1.37E-07 4.55E-06	1.69E-07 5.65E-05	1.69E-07 4.24E-06	– –	– –	– –	– –	– –	– –	1.53E-04	1.91E-04
33	3.28E-08 6.56E-05	– –	1.42E-07 4.74E-05	– –	2.49E-05 3.12E-05	– –	– –	– –	– –	– –	1.44E-04	1.80E-04
34	1.44E-04 2.87E-01	1.21E-04 4.02E-03	1.34E-02 4.45E+00	– –	7.48E-04 9.35E-04	– –	5.15E-05 3.68E-04	1.87E-04 9.35E-03	9.95E-05 2.84E-02	7.82E-05 2.61E-04	4.78E+00	9.98E+00
35	– –	1.48E-06 4.94E-05	5.12E-06 1.71E-03	– –	– –	– –					1.75E-03	2.19E-03
36	1.72E-05 3.44E-02	4.38E-05 1.46E-03	1.15E-04 3.83E-02	1.57E-04 3.93E-03	– –	– –	1.58E-04 1.13E-03	2.13E-04 1.07E-02	8.41E-05 2.40E-02	2.78E-04 9.26E-04	1.15E-01	1.44E-01
37	8.92E-05 1.78E-01	– –	– –	– –	– –	– –	– –	– –	– –	– –	1.78E-01	2.23E-01
38	7.65E-07 1.53E-03	6.01E-06 2.00E-04	9.35E-04	3.51E-04 8.77E-03	– –	– –	– –	1.86E-04 9.29E-03	2.46E-04 7.03E-02	3.72E-04 1.24E-03	4.03E-01	5.04E-01

In each cell, the first value represents CDI (Chronic daily intake mg kg^{-1}-day) and the second value is Hazard Quotient (HQ) of individual metal. Hazard index (HI) of multiple metals in a given area is shown in the *last column* at the *right side* of the table. The serial number in the first column shows heavy metal locations in soil with reference to Table 1. Mean values have been used for the determination of CDI, HQ and HI while in case where data is available in minimum to maximum range, risk was calculated for the maximum value

Table 9 Exposure to heavy metals via drinking ground water and their health risk assessment in different regions of Pakistan

Ground water sample locations	As	Cd	Co	Cr	Cu	Fe	Hg	Mn	Ni	Pb	Zn	Hazard Index
1	0.002	0.001	–	0.009	0.002	–	0.000	–	0.014	0.055	0.110	30.285
	7.306	2.192	–	3.105	0.062	–	0.913	–	0.685	15.656	0.367	
2	–	0.0003	–	0.001	0.001	0.002	–	0.001	0.001	0.001	0.000	0.717
	–	0.055	–	0.365	0.014	0.002	–	0.004	0.041	0.235	0.001	
3	–	0.001	–	–	0.009	0.008	–	0.006	0.019	0.010	0.009	6.413
	–	2.192	–	–	0.233	0.010	–	0.041	0.932	2.975	0.031	
4	–	0.002	–	–	0.016	0.027	–	0.008	0.014	0.013	0.002	8.127
	–	3.068	–	–	0.410	0.034	–	0.060	0.712	3.836	0.007	
5	0.027	–	–	–	–	0.074	–	0.027	–	–	–	91.612
	91.324	–	–	–	–	0.092	–	0.196	–	–	–	
6	0.011	–	–	–	–	0.045	–	0.019	–	–	–	36.721
	36.530	–	–	–	–	0.057	–	0.135	–	–	–	
7	–	0.001	–	0.001	0.001	0.065	–	0.003	0.018	0.007	0.001	5.430
	–	2.247	–	0.274	0.022	0.082	–	0.024	0.899	1.879	0.004	
8	–	0.001	–	0.004	0.006	0.004	–	0.033	0.026	0.009	0.004	7.768
	–	2.192	–	1.279	0.144	0.005	–	0.239	1.315	2.583	0.012	
9	–	0.001	–	0.001	0.003	0.033	–	0.003	0.033	0.004	0.007	4.981
	–	1.644	–	0.365	0.068	0.042	–	0.020	1.644	1.174	0.024	
10	–	–	–	0.000	0.036	0.101	–	0.029	0.004	0.003	0.050	2.601
	–	–	–	0.091	0.890	0.127	–	0.207	0.178	0.939	0.168	
11	–	–	–	0.003	0.024	0.001	–	0.004	0.002	0.033	0.002	11.148
	–	–	–	1.014	0.612	0.001	–	0.032	0.090	9.393	0.006	
12	–	–	–		0.001	0.014	–	0.004	–	0.000	0.001	0.079
	–	–	–		0.021	0.017	–	0.029	–	0.008	0.005	
13		0.001	–	0.002	0.010	0.002	–	0.003	0.024	0.018	0.003	9.120
		1.644	–	0.822	0.247	0.003	–	0.023	1.205	5.166	0.009	

(continued)

Table 9 (continued)

Ground water sample locations	As	Cd	Co	Cr	Cu	Fe	Hg	Mn	Ni	Pb	Zn	Hazard Index
14	– –	0.001 1.096	0.002 0.073	0.006 2.009	– –	0.003 0.004	– –	0.001 0.008	0.022 1.096	0.007 2.035	0.005 0.016	6.337
15	0.052 173.516	– –	– –		– –		– –	– –	– –	– –	– –	173.516
16	– –	0.001 1.479	– –	0.058 19.361	– –	0.070 0.088	– –	0.001 0.008	0.002 0.110	0.003 0.861	0.004 0.013	21.919
17	– –	0.001 1.644	– –	0.012 3.927	– –	– –	– –	– –	0.008 0.411			5.982
18	0.0002 0.776		– –	– –	– –	– –	– –	– –		– –	– –	0.776
19		– –	– –	0.001 0.274	0.002 0.041	0.008 0.010	– –	0.001 0.006	0.003 0.137	0.013 3.836	0.004 0.015	4.318
20	0.003 8.795	– –	– –	– –	– –		– –	– –		– –	– –	8.795
21	– –	– –	– –	– –	0.005 0.129	0.037 0.046	– –	0.014	0.003 0.126	0.004 1.198	0.007 0.024	1.624
22	– –	– –	– –	0.006 1.918	0.003 0.075	0.019 0.023	– –	0.000		0.004 1.174	0.002 0.008	3.201
23	– –	– –	0.000 0.002	0.000 0.100	0.003 0.083	0.003 0.004	– –	0.002	0.001 0.051	0.000 0.047		0.302
24	0.001 2.676	– –	– –	– –	– –	– –	– –	– –	– –	– –	– –	2.676
25	0.003 10.228	– –	– –	– –	– –	– –	– –	– –	– –	– –	– –	10.228
26	– –	0.00001 0.110	0.000 0.001	0.001 0.192	0.020 0.505	– –	– –	0.001 0.004	0.000 0.005	0.001 0.188	0.038 0.126	1.130

27	– –	– –	– –	0.001 0.183	0.000 0.007	– –	– –	– –	0.003 0.137	0.001 0.235	0.000 0.001	0.563
28	– –	– –	– –	0.001 0.324	0.759 18.973	0.011 0.014	– –	– –	0.000 0.013	0.003 0.845	0.065 0.217	20.385
29	– –	– –	– –	0.000 0.091	0.001 0.027	0.009 0.012	– –	– –	0.000 0.007	0.003 0.777	0.022 0.073	0.987
30	– –	– –	– –	0.000 0.039	0.002 0.048	0.020 0.024	– –	– –	0.000 0.003	0.003 0.916	0.016 0.053	1.084
31	– –	0.0004 0.548	– –	0.001 0.475	0.002 0.038	0.011 0.014	– –	0.001 0.008	0.004 0.204	0.002 0.548	0.002 0.007	1.842
32	– –	– –	0.000 0.006	– –	0.001 0.027	0.001 0.001	– –	– –	– –	0.005 1.299	0.000 0.001	1.335
33	– –	– –	– –	0.010 3.196	0.007 0.171	0.030 0.038	– –	0.007 0.049	– –	0.002 0.470	0.021 0.068	3.992
34	0.002 5.381	– –	– –	– –	– –	– –	0.000 0.073		– –	– –		5.454
35	– –	0.001 2.192	0.004 0.137	0.036 12.055	– –	0.015 0.019	– –	0.002 0.014	0.003 0.148	0.004 1.119	0.004 0.012	15.697
36	– –	0.0004 0.548	0.001 0.026	0.005 1.553	0.172 4.295	– –	– –	0.004 0.025	0.025 1.274	0.006 1.644	0.005 0.016	9.380
37	– –	– –	– –	0.005 1.644	– –	– –	– –	– –	0.002 0.082	0.028 8.141	0.033 0.110	9.977

In each cell, the first value represents CDI (Chronic daily intake mg kg^{-1}-day) and the second value is Hazard Quotient (HQ) of individual metal. Hazard index (HI) of multiple metals in a given area is shown in the *last column* at the *right side* of the table. The serial number in the first column shows heavy metal locations in soil with reference to Table 2. Mean values have been used for the determination of CDI, HQ and HI while in case where data is available in minimum to maximum range, risk was calculated for the maximum value

Table 10 Exposure to heavy metal by drinking surface water and their health risk assessment in different regions of Pakistan

Surface water samples locations	As	Cd	Cr	Cu	Fe	Hg	Mn	Ni	Pb	Zn	Hazard Index
1	0.017 56.621	0.0001 0.219	8.22E-05 0.027	0.001 0.027	0.0003 0.0004	4.0E-04 1.278	0.0005 0.003	0.002 0.080	0.002 0.837	0.0007 0.002	59.101
2	0.020 68.49	8.2E-05 0.164	0.002 0.648	0.001 0.027	0.0001 1.3E-04	0.0004 1.55	1.1E-05 7.8E-05	0.0018 0.089	0.0016 0.454	7.95E-05 0.0005	71.429
3	0.017 56.621	5.47E-05 0.109	5.48E-05 0.018	1.09E-03 0.027	0.0003 0.0004	0.0038 12.785	0.000493 0.003523	0.002 0.083	0.003 0.837	0.0008 0.0025	70.489
4	4.9E-05 0.164	3.83E-05 0.077	3.84E-05 0.018	0.0007 0.004	0.0035 0.004	2.47E-05 0.082	0.0003 0.0019	3.56E-05 0.002	4.6E-05 0.013	0.0009 0.003	0.364
5	– –	– –	– –	– –	0.152 0.191	– –	0.046 0.332	0.030 1.547	0.019 5.323	0.048 0.160	7.555
6	– –	0.0001 0.219	– –	– –	0.02 0.025	– –	0.007 0.047	0.005 0.246	0.009 2.661	0.006 0.021	3.222
7	– –	0.0005 1.095	0.001 0.365	0.0055 0.136	0.005 0.006	– –	0.006 0.043	0.006 0.288	0.004 1.252	0.0022 0.007	3.195
8	– –	0.0011 2.191	0.003 0.913	0.008 0.212	0.021 0.026	– –	0.009 0.064	0.016 0.808	0.005 1.487	0.0011 0.003	5.707
9	0.0002 0.639	– –	– –	– –	0.0049 0.0061	– –	0.0076 0.055	– –	– –	0.004 0.014	0.71484
10	– –	0.0038 7.671	0.005 1.552	0.02 0.5	0.028 0.035	– –	0.019 0.139	0.01 0.534	0.012 3.366	– –	13.79781
11	– –	0.005 10.958	0.008 2.648	0.033 0.822	0.048 0.06	– –	0.023 0.166	0.019 0.986	0.017 4.853	0.001 0.0045	20.5

12	– –	0.0013 2.74	– –	0.033 0.822	0.149 0.187	– –	– –	0.007 0.343	0.008 2.348	0.002 0.005	6.445
13	– –	0.0005 0.987	– –	0.0002 0.006	0.0005 0.0006	– –	– –	0.0001 0.007	0.0003 0.075	0.077 0.257	1.332
14	– –	5.48E-05 0.109		0.0005 0.013	0.0005 0.0006	– –	– –	0.0003 0.013	0.0004 0.109	0.0002 0.0006	0.247
15	– –	0.005 9.863	0.002 0.639	0.012 0.308	0.246 0.308	– –	0.023 0.166	0.025 1.274	0.0008 0.234	0.001 0.003	12.797
16	0.004 14.337	– –	– –	– –	– –	– –	– –	– –	– –	0.046 0.155	14.493
17	– –	0.0008 1.643	0.0005 0.182	0.002 0.054	– –	– –	0.005 0.035	0.017 0.890	0.002 0.704	0.0055 0.018	3.52968
18	– –	2.19E-05 0.044	7.67E-05 0.025	0.003 0.079	– –	– –	0.005 0.035	0.0002 0.008	0.0002 0.063	0.0011 0.0036	0.257
19	– –	0.0008 1.643	0.013 4.567	0.0019 0.047	– –	– –	– –	0.005 0.274	0.014 4.070	– –	10.602
20	– –	– –	– –	0.003 0.068	– –	– –	– –	– –	0.003 0.783	0.005 0.0164	0.86771

In each cell, the first value represents CDI (Chronic daily intake mg kg^{-1}-day) and the second value is Hazard Quotient (HQ) of individual metal. Hazard index (HI) of multiple metals in a given area is shown in the *last column* at the *right side* of the table. The serial number in the first column shows heavy metal locations in soil with reference to Table 3. Mean values have been used for the determination of CDI, HQ and HI while in case where data is available in minimum to maximum range, risk was calculated for the maximum value

Table 11 Exposure to heavy metal via inhalation of contaminated air and their health risk assessment in different regions of Pakistan

Air sample locations	As	Cd	Co	Cr	Cu	Fe	Mn	Ni	Pb	Zn	Hazard index (HI)
1		1.41E-04 2.82E-01	6.84E-04 2.28E-01	1.03E-03 3.42E-01	6.11E-03 1.53E-01	1.91E-01 2.39E-01	3.29E-03 2.35E-02	1.71E-03 1.71E-01	1.15E-02 3.27E+00	5.39E-01 1.79E+00	1.75E+01
2	1.19E-03 3.97E+00	– –	– –	– –	3.59E-03 8.98E-02	– –	– –	– –	4.19E-03 1.20E+00	3.11E-03 1.03E-02	5.27E+00
3	1.67E-03 5.56E+00	– –	– –	– –	4.45E-03 1.11E-01	– –	– –	– –	5.84E-03 1.67E+00	2.39E-03 7.97E-03	7.34E+00
4	– –	3.21E-06 6.41E-03	1.07E-05 3.56E-03	3.15E-05 1.05E-02		9.41E-04 1.18E-03	2.94E-05 2.10E-04	9.08E-06 9.08E-04	6.84E-05 1.95E-02	5.45E-04 1.81E-03	1.42E-01
5	– –	2.14E-06 4.27E-03	1.76E-05 5.88E-03	5.34E-06 1.78E-03	1.66E-04 4.14E-03	1.31E-03 1.64E-03	4.27E-05 3.05E-04	1.28E-05 1.28E-03	7.48E-05 2.14E-02	1.23E-03 4.11E-03	1.87E-01
6	1.18E-05 3.92E-02	4.11E-05 8.23E-02	6.95E-07 2.32E-04	1.60E-05 5.34E-03	2.40E-05 6.01E-04	4.38E-03 5.48E-03	1.60E-04 1.14E-03	9.62E-06 9.62E-04	2.35E-03 6.72E-01	5.88E-03 1.95E-02	1.36E+00
7	– –	1.39E-02 2.78E+01	– –	– –	– –	– –	– –	– –	2.72E-03 7.78E-01	7.21E-03 2.40E-02	2.86E+01
8	– –	1.74E-01 3.49E+02	– –	– –	– –	– –	– –	– –	4.91E-02 1.40E+01	1.88E-02 6.26E-02	3.63E+02
9	– –	1.71E-01 3.43E+02	– –	– –	– –	– –	– –	– –	4.73E-02 1.35E+01	1.36E-02 4.54E-02	3.56E+02
10	– –	4.38E-03 8.76E+00	– –	– –	– –	– –	– –	– –	3.81E-02 1.09E+01	– –	1.96E+01
11		4.75E-02 9.51E+01	– –	– –	6.14E-04 1.54E-02	9.01E-03 1.13E-02	– –	2.94E-04 2.94E-02	2.24E-03 6.41E-01	1.51E-03 5.03E-03	9.58E+01

12	– –	3.15E-02 6.30E+01	– –	– –	3.53E-04 8.82E-03	4.40E-03 5.50E-03	– –	1.92E-04 1.92E-02	5.88E-04 1.68E-01	8.76E-04 2.92E-03	6.32E+01
13	– –	4.38E-02 8.76E+01	– –	– –	2.66E-03 6.65E-02	8.03E-03 1.00E-02	– –	7.85E-04 7.85E-02	2.08E-03 5.95E-01	2.69E-03 8.95E-03	8.84E+01
14	– –	4.54E-02 9.08E+01	– –	– –	3.69E-04 9.22E-03	4.57E-03 5.72E-03	– –	3.00E-04 3.00E-02	2.19E-03 6.26E-01	7.75E-04 2.58E-03	9.15E+01
15	– –	4.54E-05 9.08E-02	– –	2.97E-04 9.90E-02	3.69E-04 9.22E-03	4.57E-03 5.72E-03	1.06E-04 7.59E-04	3.00E-04 3.00E-02	2.19E-03 6.26E-01	7.75E-04 2.58E-03	1.22E+00
16	– –	4.38E-02 8.76E+01	– –	1.16E-01 3.88E+01	2.66E-03 6.65E-02	8.03E-03 1.00E-02	2.00E-01 1.43E+00	7.88E-01 7.88E+01	2.08E-03 5.95E-01	2.69E-03 8.95E-03	8.75E+02
17	– –	3.15E-02 6.30E+01	– –	1.16E-01 3.88E+01	3.53E-04 8.82E-03	4.40E-03 5.50E-03	1.10E-01 7.82E-01	1.92E-01 1.92E+01	5.88E-04 1.68E-01	8.76E-04 2.92E-03	4.87E+02
18	– –	4.75E-02 9.51E+01	– –	2.68E-01 8.92E+01	6.14E-04 1.54E-02	9.01E-03 1.13E-02	2.52E-01 1.80E+00	2.91E-01 2.91E+01	2.24E-03 6.41E-01	1.51E-03 5.03E-03	1.05E+03
19	– –	2.56E-02 5.13E+01	– –	8.33E-02 2.78E+01	1.44E-04 3.61E-03	1.81E-03 2.26E-03	5.13E-02 3.66E-01	1.23E-01 1.23E+01	8.28E-04 2.37E-01	4.06E-04 1.35E-03	2.63E+02
20	– –	2.72E-02 5.45E+01	– –	3.29E-01 1.10E+02	5.02E-04 1.26E-02	3.92E-03 4.90E-03	8.65E-02 6.18E-01	2.53E-01 2.53E+01	7.48E-05 2.14E-02	8.76E-04 2.92E-03	4.79E+02
21	1.85E-01 6.05E+02	– –	– –	– –	– –	– –	– –	– –	5.28E-2 1.50E+1	5.03E-02 1.68E-01	6.20E+02

In each cell, the first value represents CDI (Chronic daily intake mg/kg-day) and the second value is Hazard Quotient (HQ) of individual metal. Hazard index (HI) of multiple metals in a given area is shown in the *last column* at the *right side* of the table. The serial number in the first column shows heavy metal locations in soil with reference to Table 5. Mean values have been used for the determination of CDI, HQ and HI while in case where data is available in minimum to maximum range, risk was calculated for the maximum value contaminations in various regions of Pakistan indicates that it poses no health risk in any medium and the values were found below the threshold value. Although in comparison to other heavy metals, toxicity of Co is low to living organisms, exposure to high concentrations can cause various health effects. Health problems like asthma, pneumonia, and wheezing have been found in workers inhaling air with a high concentration of Co (ATSDR 1992). According to a report of the International Agency for Research on Cancer, cobalt is a possible carcinogen to humans at high concentration (IARC 2006). However, the Pakistani population seems to be on no risk from Co as revealed by the present risk analysis (Tables 9–11).

3.1.4 Chromium (Cr)

The HQ values obtained for Cr (Tables 9–11) indicate that it is posing a risk in 40 and 30 % of the studied areas through ground and surface water, respectively. It was not found to pose any risk through soil, but was found to pose a risk through air as HQ values for about 50 % of the available data on air were calculated above the risk limit. In ground water, the highest HQs of Cr (12.05–19.36) were calculated for well water in industrial and residential areas of Kasur city. For surface water of river Kabul at Peshawar, the Cr HQ was calculated as 4.56 while for surface water of Bara River, Akbarpura area, HQ for Cr was 2.486 (Table 10). The situations seems very alarming in the case of air as Cr is posing a high risk to public through air. For example, in the biggest city of the country, Karachi, the HQ calculated for Cr in the air was 110 at Sea view point, 89.2 at M.A. Jinnah road and 27.8 at SPARCO (Table 11). Similarly, in other big cities of the country like Rawalpindi and Lahore, Cr is posing a risk through air where its HQ reaches 38.8 (Table 11 and Fig. 1). A high exposure to Cr has resulted in high level of Cr in human samples as revealed by some studies (Table 6).

Cr plays an important role in the metabolism of carbohydrates in the body (Cefalu and Hu 2004), but some compounds of Cr can cause various health problems like skin diseases, cancers, irritation and diseases related to digestive, excretory, respiratory and reproductive systems (Anonymous 2008a, b). Chromium is also associated with allergic dermatitis in humans (Scragg 2006). A study conducted on Cr effects on health of workers in tannery industries at Sialkot in Pakistan revealed that 13 % workers had skin rashes, 12 % had chronic bronchitis, 8 % had gastritis and 3 % were with conjunctivitis. Similarly, 54 % workers had blood Cr levels above the upper limit set by the Agency for Toxic Substance and Drug Registry (Khan et al. 2013). The workers had hematological, hepatic and renal function impairments. The workers had significantly higher Cr levels in their biological fluids and suffered from adverse health effects due to enhanced oxidative stress and inflammatory changes (Khan et al. 2013a).

3.1.5 Copper (Cu)

The hazardous risk assessment for Cu (Tables 7–11) reveals that it does not pose any risk to humans through any media except in only 4 % cases through water. The only data of Mahmood and Malik (2014) on the tube well water from Lahore region when analyzed for health risk gave an HQ of 4.29 (Table 9 and Fig. 1). Cu was reported in human samples in various concentrations in the Pakistani population (Table 6). Cu is an essential element in moderate amounts but exposure to high concentrations can cause various problems. Short-term exposure to Cu may cause health problems like feeling of illness, common cold with sensation of chill and headache as well as irritation of eyes. Repeated exposure to Cu can cause irritation of skin and discoloration of both skin and hair (Eck and Wilson 1989). Likewise, exposure to high level of Cu can cause eye problems such as cataract etc. (Ahmad et al. 2014c).

3.1.6 Mercury (Hg)

Although the available data for Hg in Pakistan are comparatively few, risk assessment of the few available data showed no risk from Hg except from surface water (Tables 7–11). For example, the risk analysis of the Lloyd water reservoir for Hg gave a HQ of 12.78. Similarly, assessment of health risk for the waters of Chashma and Tarbella reservoirs showed that the resulting HQs (1.55 and 1.278, respectively) were above the alarming level of 1. However, it is worth mentioning that very few data are available on this toxic pollutant in the country and it is very difficult to conclude on the risk posed by this element.

Hg is a very serious pollutant which interferes with various processes in the cells of the nervous system and disrupts the neurotransmitter production as well as decreases the production of hormones like thyroid and testosterone (Fatoki and Awofolu 2003). Long-time exposure to even low levels of Hg may cause disorders like pneumonitis and bronchial irritation (Gary 2012). In northern Pakistan, including Gilgit and Chitral, overexposure to Hg has resulted in health problems like tiredness and headache, kidney diseases, cognitive impairment, skin rashes, sensory problems, chest pain, tremor, cough and sputum, night blindness, hyporeflexia and neurasthenia along with slow growth of children (Khan et al. 2012b). Similarly, Hg was found at higher concentrations in the biological samples of night blindness children in the studied region of Pakistan hence it can have a role in eyes related disorders (Afridi et al. 2015). Therefore, it is strongly recommended that researcher should focus on pollution level of Hg in different environmental compartments to identify if there is any risk from this pollutant.

3.1.7 Manganese (Mn)

The risk analysis for Mn indicates that it does not pose any major risk in Pakistan except through air in some cases (Tables 7–11). Analysis of the available data reveal that in 20 % of the studied areas the HQ for Mn in air exceeded the threshold limit of 1. For example, Mn HQ for air of Rawalpindi and Karachi (M.A. Jinnah) were 1.8 and 1.43, respectively. Few studies have been conducted in the country on Mn concentration in human samples (Table 6) but reported high concentration in some cases as for example in blood samples from Lower Dir and Peshawar (Jan et al. 2011). Mn is an essential trace element but overexposure can cause health problems like nervous system disturbance and permanent neurological disorder (Crossgrove and Zheng 2004; Barbeau 1984; Inoue et al. 1996). Overexposure to Mn can interferes with the absorption of dietary iron. Therefore, long-term exposure to high concentration of Mn may result in iron-deficiency anemia (Chua and Morgan 1996). Mn, As and Pb collectively can cause oxidative stress, influence neurotransmitter and affect the hematopoietic system in people exposed to them (Andrade et al. 2015).

3.1.8 Nickel (Ni)

Risk analysis of Ni hazards through different media (Tables 7–11) reveals that it does not pose any risk in Pakistan upon exposure through soil. However, it is posing health risks in 20 and 36 % of the studied regions through water and air, respectively (Tables 9–11). The highest HQ of Ni through ground water was obtained for well water in Korangi, Karachi (1.644) followed by different sites in Gadoon Amazai in Khyber Pakhtunkhwa (1.543) (Tables 9 and 10 and Fig. 1). Ni seems to pose a higher risk through air as the HQ values exceed the threshold limits many folds. For example, in Karachi its HQ for the air of M.A. Jinnah road was 29.1 while at Sea View point it was calculated as 25.3. Similarly, in the twin city of capital Islamabad, i.e. Rawalpindi its HQ through air reached as high as 78.8 while in the provincial capital Lahore, its HQ through air crossed 19. Ni contamination is also reflected by Ni occurrence in blood samples collected from different parts of the country (Table 6). Exposure to high concentration of Ni can cause various health complexities such as lung fibrosis, cardiovascular diseases, kidney problems, nickel allergy and cancer of the respiratory tract (Oller et al. 1997; McGregor et al. 2000). According to Wadhwa et al. (2015), a positive correlation between Ni level in scalp hairs and cancer in females of Pakistan was found.

3.1.9 Lead (Pb)

The obtained HQ values for Pb in air, soil and water (Tables 7–11) reveal that no risk is expected from this toxic metal through soil in Pakistan, but is posing risk in 33, 50 and 38 % of the studied locations through air, ground water and surface water, respectively (Tables 7–11). Evaluation of risk assessment revealed that Pb posed the highest risk (15.65 HQ) through ground water in various areas of Karachi followed by well water in Gadoon Amazai and Islampura, Swat (HQs of 9.393 and 8.141, respectively) in Khyber Pakhtunkhwa (Table 9). The highest HQs for Pb through surface water were obtained for different sites in Gadoon Amazai (5.32) and for Bara River, Akbarpura (4.853) in Khyber Pakhtunkhwa (Table 10). Pb is also posing a potential health risk through inhalation of contaminated air as the HQs of Pb in many cities was found many fold above the threshold value. For example, an HQ of 14 for Pb was calculated for air near Gujranwala–Shaikhupura. Similarly, HQs of 13.5 and 10.9 were also obtained for Pb concentration in air in Faisalabad Abdullahpur Crossing Junction and Islamabad Sector H-12, respectively (Table 11).

Due to environmental contamination, elevated levels of Pb in human samples like blood, hairs and nails are found in the Pakistani population (Table 6). Children living around automobile and battery repair workshops were found having high levels of Pb (100.19–120.03 $\mu g\ L^{-1}$) in their blood (Ahmad et al. 2009). Other studies also found Pb in human samples in different cities including Karachi, Lahore, Peshawar and Dir (Anwar 2005; Jan et al. 2011; Mumtaz et al. 1999). Pb is a hazardous cumulative body poison and can result in chronic health problems such as brain and nerve disorders, kidney damage, digestive problems, cardiovascular problems and hypertension (SDWF 2006; Riess and Halm 2007; Gidlow 2004). Pb accumulates in the skeleton and results in adverse health effects including subencephalopathic, neurological and behavioral effects (Lehloesa and Muyima 2000). In Pakistan, Pb has been the cause of adverse effects on hematopoietic, renal, and hepatic functions in people of Gujrawanla city, particularly in children. Children of lead-related occupational workers had significantly increased frequency (31 %) of lead poisoning (Khan et al. 2010). Overexposure to lead in occupational workers in a lead-related environment in Rawalpindi city adversely affected hematological, renal, and hepatic functions in workers (Khan et al. 2008b). A community-based study hypothesized that high levels of Hg in blood can be the cause of hypertension in the Pakistani population (Rahman et al. 2006).

3.1.10 Iron (Fe)

Risk assessment of Fe (Tables 7–11 and Fig. 1) does not reveal any risk in any part of the country through any media. In human blood Fe concentration in the populations of Karachi, Lower Dir and Peshawar were found to the maximum level (Jan et al. 2011; Mumtaz et al. 1999). Fe is an essential element needed for the normal growth but exposure to higher concentrations can cause heart diseases (Milman et al. 2001; Rasmussen et al. 2001), cancer (Beckman et al. 1999; Parkkila et al. 2001), liver and neurodegenerative disorders (Berg et al. 2001; Sayre et al. 2000) and diabetes (Ellervik et al. 2001; Huang 2003; Parkkila et al. 2001). However, the Pakistani population is safe from Fe toxicity as revealed by the present risk assessment based on the available data.

3.1.11 Zinc (Zn)

Zinc, in optimum concentration, is an essential element for living organisms but exposure to excessive concentrations can be detrimental to human health (Solomons and Ruz 1998; Singh and Kumar 2006). Zn can enter into the body through inhalation, skin, or by ingestion. Inhalation of Zn can cause metal fume fever with symptoms of fever, nausea, fatigue, and cough (Rohrs 1957). Oral uptake of Zn at high concentrations can cause abdominal pain, nausea, and vomiting (Porea et al. 2000). However, health risk assessment of Zn in the present study revealed that it does not pose any serious health problem in any region of Pakistan through any medium except in a single case through air pollution in particulate matter of Gutter Baghicha SITE, Karachi where a HQ of 1.79 was obtained (Tables 7–11). In human samples of Pakistani population also Zn was not usually observed in higher concentration.

3.1.12 Collective Hazard Index (HI)

Analysis of the collective hazard index (HI) of all the toxic metals indicates that the public in most of areas in Pakistan is exposed to toxic metals at a risky level through water or air (Tables 7–11 and Fig. 1). In case of ground and surface water, 86 and 80 % of the analyzed data gave HI above the risk level (Tables 9 and 10). Similarly, in 91 % of the studies on air HI values were found higher than the risk level. Contrary to air and water, most of the HI values for toxic metals in soil were found within the safe limits and only 8 % regions were found at risk possessing higher HI values beyond the permissible limits (Tables 7 and 8 and Fig. 1).

As a conclusion, people in Pakistan are mostly exposed to risky levels of heavy metals through water and air while exposure through soil is comparatively little (Tables 7–11 and Fig. 1). Among the investigated heavy metals, As, Cd and Pb in water and air pose the most potent risk to public health in the country. Central Khyber Pakhtukhwa, Northern and Central Punjab and Southern Sindh are the more risky areas with respect to heavy metal pollution in water and air. However, it does not mean that the rest of country is safe in term of metal pollution, but actually there is a lack of data in most parts of the country.

4 Sources of Heavy Metal Pollution in Pakistan

4.1 Natural Sources

Toxic metals reach the environment via both natural and anthropogenic activities. The most important natural source of toxic metals in soil and ground water is geologic parent materials. In soil, the level of metals depends on the type of underlying rocks and the environmental conditions that foments weathering.

Geologic parent materials generally have a high content of Cr, Mn, Sn, Cd, Hg, Pb, Co, Ni, Cu, and Zn which lead to elevated levels of heavy metals in the environment (Kafayatullah et al. 2001; Shah et al. 2010). Aerosols produced in marine ecosystems and forest fires are also important sources of metal pollution in the environment. Emission from natural vegetation is another source of metals release in the soil, water and atmosphere through leaching processes from leaves and stems (Nagajyoti et al. 2010).

In addition to natural sources, toxic metals are discharged into the environment through urban wastes, industrial effluents, and vehicular emissions (Chaoyang et al. 2009; Chang et al. 2009; Xia et al. 2011). In Pakistan, a diverse range of sources such as industrial effluents and emissions, agricultural practices, domestic effluents, atmospheric deposition, vehicular emissions, mining, smelting and different natural phenomena are contributing to toxic metal pollution.

4.2 Industrial Sources

Industries release toxic metals in different ways which contaminate the environment. In Pakistan, various industries like petrochemicals, textiles, pharmaceuticals, ceramics, food industries, steels, oil mills, leather tannings, fertilizer factories, and sugar industries are the main sources of environmental pollution (Sial et al. 2006; WWF 2007). Similarly, effluents and sewage sludge of textiles, plastics processing, paper processing, microelectronics and wood preservation are major contributors of metal pollution (Nagajyoti et al. 2010). The power stations supplying energy through coal burning, petroleum combustion and nuclear reactions add toxic metals to the environment in different ways. High tension lines of power supply also contribute toxic metals such as Se, B, Cd, Cu, Zn, Cs and Ni to the environment (Verkleji 1993). According to a survey of the United Nations Industrial Development Organization (UNIDO) in Pakistan, industries for textile, metal works, chemicals, petrochemicals, refining, food processing, pulp and paper are the major contributors of heavy metals pollution in the country (UNIDO 2000). These industries produce wastewater in large amounts with various metals and metalloids like As, Fe, Pb, Hg, Cr, Cd, Cu, Ni, Zn, Co and Mn which contribute to water and soil pollution (Sial et al. 2006; Ullah et al. 2009; Hussain et al. 2006; Murtaza et al. 2008).

According to Sial et al. (2006), in Pakistan out of 6634 registered industries, 1228 are serious sources of pollution in the country. A country report by Murtaza and Zia (2012) revealed that out of 388 cities, only 8 have facilities for wastewater treatment including three in Islamabad, two trickling filters in Karachi, one in Faisalabad and some screening and grit removal systems in Lahore, but they are hardly functional. The situation is worst in Karachi, because the two biggest industrial estates of Pakistan, i.e. Sindh Industrial Trading Estate (SITE) and Korangi Industrial and Trading Estate (KITE) situated there have no effluent treatment plant. Karachi accommodates 70 % of Pakistan's industry, and thus produces approximately 70 % of wastewater (0.242×10^9 m^3 $years^{-1}$) which is discharged into the Arabian Sea without any prior treatment (Murtaza and Zia 2012). According to a report of FAO (2002), the total discharge of sewage from 14 major cities of Pakistan is about 1.83×10^7 m^3 h^{-1}. In Khyber Pakhtunkhwa, about 0.701×10^9 m^3 of industrial effluents containing high levels of pollutants are discharged into the River Kabul annually (SOE 2005). The wastewater of the main drains in Peshawar is carrying higher concentrations of Pb and Cr beyond the permissible limits (Pak-EPA/OECC 2004). The problem of industrial water pollution has remained uncontrolled because there have been little or no incentives for industry to treat their effluents (WWF 2007).

Industrial sectors in Pakistan are concentrated in or around big cities which cause heavy pollution of the local environment. For example, leather tanneries are concentrated in the regions of Kasur, Sialkot and Multan, textile in Faisalabad and Lahore, cutlery in Wazirabad, thermal power plants and fertilizer industry in Multan, light engineering and electrical goods in Gujraat and Gujranwala, sports and surgical objects in Sialkot, steel in Karachi and Lahore. In addition, brick kilns, major sources of air contamination with heavy metals, are spread all over the country. According to Awan et al. (2011), industrial emission is the major source of toxic metals like Cd, Pb and Zn in ambient air of cities like Gujranwala, Faisalabad and Bahawalnagar. These metals in air ultimately return to the soil as precipitations by rain and contaminate water and soil.

4.3 Domestic and Municipal Effluents

Domestic and municipal effluents are major sources of various environmental pollutants including toxic metals. According to Beede and Bloom (1995), 1.3 billion metric tons of domestic and municipal solid waste was generated in 1990 in the world. But the municipal systems in most of the developing countries lack the financial resources and skills needed to cope with this crisis. Several countries have realized that the way they manage their solid wastes does not satisfy the objectives of sustainable development (Abu-Qdais 2006).

In Pakistan, domestic and municipal wastes are either directly disposed to a sewer system, a drainage system or water body, a nearby field or an internal septic tank and pollute the environment in different ways (Murtaza and Zia 2012). There are no proper treatment systems installed in most of the cities in the country hence, municipal wastewater is not subjected to any treatment except in Islamabad and Karachi where only a small proportion of wastewater is treated before being discharged. If all the installed treatment plants of the country were working at their full capacity, they would have still been treating only 8 % of urban municipal and domestic wastewater. These treatment plants treat wastewater through sedimentation ponds to a primary level only but most of them do not work properly, therefore the real figure of treatment has been estimated around 1 % (Murtaza and Zia 2012).

According to a report of Pak-SCEA (2006), in Pakistan treatment facilities exist in major cities, but in most cases these have been installed without the completion of associated sewerage networks; therefore, the treatment plants are often either under-loaded or abandoned. Currently solid waste treatment in Pakistan has not been carried out in a sufficient and proper manner in collection, transportation and disposal or dumping. Therefore, the environmental and sanitary conditions are becoming more serious year by year (Mahar et al. 2007). In the Swat valley the contamination of soil with Cd has been attributed to the disposal of Cd-batteries (Nafees et al. 2009). According to Awan et al. (2011), open burning of municipal wastes is another serious source of air contamination with heavy metals in big cities of Pakistan. Heavy metals in the air return to the soil with rain water and can be a possible source of water and soil contamination. Wastewater irrigation, a common practice in many parts of the country, contaminates the soil, water as well as crops which pose serious health risks (Umar et al. 2013).

4.4 Agricultural Sources

Various inorganic and organic fertilizers, pesticides, liming, irrigation waters and sewage sludges are the most important sources of heavy metal pollution in agricultural soil. The application of animal manure usually results in an increased concentration of heavy metals like Co, Cu, Mn and Zn while the application of sewage results in an increased accumulation of Zn, Cr, Pb, Ni, Cd and Cu in soil (Verkleji 1993). The extensive use of pesticides and chemical fertilizers in agriculture has been a major source of environmental pollution in the world including Pakistan (Khan et al. 2013c). In addition to soil pollution, agrochemicals cause water pollution as these are mixed with the irrigation water in the field and leach through the soil and ultimately reach water bodies. Similarly, these agrochemicals cause air pollution as they reach the atmosphere either during the application process or later on by volatilization. Some fertilizers contain heavy metals as by-products, and their consistent use results in high accumulation of these metals in soil and water (Li and Wu 2008).

As an agricultural country, Pakistan extensively uses pesticides and fertilizers for better agricultural products. According to Hussain et al. (2002), Pakistan is the second largest consumer of pesticides among the South Asian countries. Similarly, in the last few decades fertilizer consumption has increased manifold in the country. Estimated production has risen from one million tons in 1981 to three million tons in 2003 (Khaskheli 2013). Haq et al. (2012) observed that water contamination with heavy metals such as As in Rahim Yar Khan (Southern Punjab) was because of the extensive use of pesticides and fertilizers in the area. Nafees et al. (2009) verified fungicides as a source of heavy metal pollution in soil of the Swat valley.

4.5 Vehicular Emissions

Vehicular emissions are considered as a major source of environmental pollution, particularly atmospheric pollution (Li et al. 2001). Due to the unavailability of technical and financial resources in Pakistan, it is very difficult to measure the contribution of individual shares in air pollution of different cities; however, the road transport sector causes more urban air pollution than any other single human activity (Qadir 2002). It has been noted in the last two decades that air pollution from vehicles exceeded the maximum limits set by various organizations including WHO, USEPA, and is likely to be a major cause of respiratory diseases in Pakistan (Qadir 2002).

The concentration of heavy metals in the urban environment can mainly be attributed to heavy traffic flow and consequent emission of smoke onto the road side soil. It has been observed that metals like Zn, Cu, Ni and Pb had high concentrations in soil and air near road sides (Parveen et al. 2012). Heavy metals are also emitted to the environment from tires of vehicles during transportation, from diesel engines and from aerosol emissions. Lubricants, used for good performance of vehicles, emit different heavy metals like Cd, Cr, Hg, Ni, Pb and Zn into the environment (Nagajyoti et al. 2010). During the burning process of leaded gasoline, a significant amount of Pb is released into the environment. Ilyas (2007) revealed that Pakistan consumes most of the petroleum products for transportation, and the consumption has doubled in the last decade. Air pollution may not remain confined to a specific area but the contaminated particulate matter in the atmosphere may be wind-blown over great distances and ultimately returns to the lithosphere as precipitation by rain or snowfall and cause pollution in areas far away from the site of origin. In Pakistan, vehicle fumes constitute a major source of environmental Pb pollution and are probably the major contributing factor to the high blood Pb level in people who are regularly exposed to high traffic areas (Agha et al. 2005).

5 Conclusions and Recommendations

Based on the results of the exposed pollution and health risk assessments, it is apparent that the soils, water and air surrounding the industrial, mining and road side areas are seriously polluted by heavy metals emitted from vehicles, industries and mining activities.

The risk assessment in terms of HQ and HI revealed that air, soil and water in Pakistan are polluted with toxic metals to a level that pose risks in some areas of the country. Exposure through oral and dermal contact via soil posed limited health hazard in terms of HQ but a higher risk is predicted through air and water. Assessment of health risk in terms of HI shows that people in most of the studied areas are at risk of the combined effect of toxic metals, particularly through air and water. Air, especially from urban and road side areas in the country, is contaminated with toxic metals. As, Cd and Pb in water and air pose higher risks to public health.

Central Khyber Pakhtukhwa, Northern and Central Punjab and Southern Sindh are the more risky areas with respect to heavy metal pollution in water and air. The frequent and high level occurrence of some toxic metals in the environment is alarming in some areas of Pakistan and can cause different types of health problems. However, comparatively little data are available on metal-related health problems in the country due to the lack of proper surveys, poor facilities and no proper diagnosis and record maintenance at hospitals. There are many gaps as no well-planned and regular studies have ever been conducted on the subject matter in the country, particularly on health problems due to heavy metal pollution. It is necessary to intensify research programs for analyzing heavy metal contamination in various compartments of the environment so as to restore and maintain the ecological sanity and protect the public from the risks posed by toxic metals. In addition to natural sources, anthropogenic activates like industrialization, unplanned agricultural practices, improper disposal of municipal and domestic wastes and vehicular emissions are the main sources of pollution in the country. The present review provides quantitative evidence of health risks to the exposed people in the country and demand the critical need for strengthened rules and regulations about industrial emission, mining activities, vehicular emission etc. to protect people from the hazardous effects of toxic metals in Pakistan. The following are our recommendations and suggestions for minimizing soil, water and air pollution in Pakistan:

- Strict environmental laws should be implemented with no compromise on environmental quality.
- Governmental and non-governmental organizations of Pakistan should act to protect the environment from the deteriorating effects of toxic metals.
- The government should initiate and carry out routine monitoring programs to define the distribution and severity of major toxic-metal pollutants in Pakistan.
- Most of the studies conducted in Pakistan were found to be restricted to central Khyber Pakhtukhwa, northern and central Punjab and southern Sindh. There is a urgent need of studies and risk assessment in the rest of areas in the country.
- Local government should renovate old and corroded pipelines that exist in water supply networks.
- The government should initiate a strict monitoring system that requires industry owners to adopt appropriate metal waste treatment measures.
- There should be properly operating systems for the collection and treatment of industrial and municipal wastes.
- A traffic management cell should be installed for the improvement of the existing transport system. Regulations and protocols for continuous checking of vehicles should be implemented.
- The use of environmental friendly fuels should be promoted.
- To reduce the mass transportation, alternative and efficient means, such as metro services should be introduced in the metropolitan cities. The government should renovate the railway system to minimize road transport.
- Active campaigns of public awareness should be launched by governmental organizations through electronic and print media and by conducting awareness seminars to educate people about the importance of protecting the environment.

6 Summary

Environmental pollution is a major threat to public health in Pakistan. Different environmental compartments (viz., air, soil and water) are contaminated with harmful substances that include toxic metals, pesticides and other organic and inorganic pollutants. Some toxic metals exist in the terrestrial or aquatic environment at levels that violate levels established by health agencies. Anthropogenic activities like indiscriminate disposal of industrial and municipal effluents, vehicular and industrial emissions and improper applications of chemicals in agriculture are factors contributing to toxic metal pollution in the country. Beside anthropogenic activities, natural sources are also contributing to the heavy metal pollution in the country. Toxic metal pollution, with or without other pollutant types, may pose risks to human health in Pakistan, although lack of proper survey data and diagnosis, along with poor maintenance of records at hospitals are obstacles in properly correlating health problems with environmental pollution in the country. In this review we present a detailed description of the literature on toxic metal contamination in air, water, soil and human samples in Pakistan. We also made an attempt to determine the possible non-carcinogenic health risks to public in Pakistan due to toxic metal pollution. Moreover, we describe the major sources of toxic metals and their contribution to environmental pollution in Pakistan. This article may assist authorities and the environmental community in recognizing what the main sources of heavy metals pollution in Pakistan are, and in implementing proper monitoring and assessment programs to prevent further environmental deterioration and to reduce risks to public health.

References

Abbas Z, Chaudary MN, Raza A, Mehmood A (2012) Toxicity of mercury in different samples (water and soils) and its exposure in Pakistan. Sci Int (Lahore) 24:421–429

Abbas M, Cheema KJ (2014) Correlation studies of Arsenic level in drinking water and hair samples of females in District Sheikhupura, Pakistan. Int J Curr Microbiol App Sci 3:1077–1085

Abbas M, Cheema KJ (2015) Arsenic levels in drinking water and associated health risk in District Sheikhupura, Pakistan. J Animal Plant Sci 25:719–724

Abernathy CO, Thomas DJ, Calderon RL (2003) Health effects and risk assessment of arsenic. J Nutri 133:1536–1538

Abu-Qdais HA (2006) Techno-economic assessment of municipal solid waste management in Jordan. Waste Manage 27:1666–1672

ATSDR (1992) Toxicological profile for cobalt. Agency for Toxic Substances and Disease Registry Public Health Service. U.S. Department of Health and Human Services, Atlanta, GA

Affum HA, Oduro-Afriyie K, Nartey VK, Addo MA, Nyarko BJB, Mumuni II, Adu PS, Appiah GK, Coleman A (2011) Lichens as a monitor for atmospheric manganese pollution. Res J Environ Earth Sci 3:534–540

Afridi HI, Kazi TG, Talpur FN, Arain SS, Arain SA, Brahman KD, Panhwar AH, Khan N, Arain MS, Ali J (2015) Assessment of selenium and mercury in biological samples of normal and

night blindness children of age groups (3–7) and (8–12) years. Environ Monit Assess 187:82. doi:10.1007/s10661-014-4201-z
Afzal M, Shabir G, Iqbal S, Mustafa T, Khan QM, Khalid ZM (2013) Assessment of heavy metal contamination in soil and groundwater at leather industrial area of Kasur, Pakistan. Clean–Soil, Air, Water 41:1–7
Agha F, Sadaruddin A, Khatoon N (2005) Effect of environmental lead pollution on blood lead levels in traffic police constables in Islamabad, Pakistan. J Pak Med Assoc 55:10
Ahmad T, Kahlown MA, Tahir A, Rashid H (2004) Arsenic an emerging issue, experience from Pakistan. In: Proceeding of the 30th WEDC International Conference, Vientiane, Lao PDR p. 459–466
Ahmad T, Mumtaz A, Ahmad D, Rashid N (2009) Lead exposure in children living around the automobile and battery repair workshops. Biomedica 25:128–132
Ahmad SS, Erum S (2010) Integrated assessment of heavy metals pollution along motorway M-2. Soil Environ 29:110–116
Ahmad N, Rahim M, Haris MRHM (2014a) Toxicological impact assessment of heavy metals in human blood and milk samples collected in district Shangla, Pakistan. Sci Int 1:223–226
Ahmad K, Azizullah A, Shama S, Khattak MNK (2014b) Determination of heavy metal contents in water, sediments, and fish tissues of Shizothorax plagiostomus in river Panjkora at Lower Dir, Khyber Pakhtunkhwa, Pakistan. Environ Monit Assess 186:7357–7366
Ahmad A, Ahmad I, Khan MA, Munshi AB, Siddiqui I, Anzar O, Anzar A (2014c) On-set of cataract and accumulation of Copper, Lead and Cadmium in smokers of Karachi, Pakistan. J Environ Toxicol 4:252–258
Alam K, Blaschke T, Madl P, Mukhtar A, Hussain M (2011) Aerosol size distribution and mass concentration measurements in various cities of Pakistan. J Environ Monit 13:1944–1952
Ali SFA, Mehdi SU, Asif M, Hassan SNU, Mirza EH (2012) Detection of iron and manganese concentrations in human biological fluid with flame atomic absorption spectroscopy (FAAS). International conference on bioscience, biochemistry and pharmaceutical sciences (ICBBPS'2012) Penang, Malaysia
Ali U, Malik RN, Syed JH, Mehmood CT, García LS, Khalid A, Chaudhry MJI (2014) Mass burden and estimated flux of heavy metals in Pakistan coast: sedimentary pollution and eco-toxicological concerns. Environ Sci Pollut Res. doi:10.1007/s11356-014-3612-2
Amin N, Ibrar D, Alam S (2014) Heavy metals accumulation in soil irrigated with industrial effluents of Gadoon Industrial Estate, Pakistan and its comparison with fresh water irrigated soil. J Agric Chem Environ 3:80–87
Andrade VM, Mateus ML, Batoréu MC, Aschner M, Marreilha dos Santos AP (2015) Lead, Arsenic, and Manganese metal mixture exposures: focus on biomarkers of effect. Biol Trace Elem Res 166:13–23
Anonymous (2008a) Economic survey of Pakistan. Finance Division, Government of Pakistan, Islamabad
Anonymous (2008b) Assessment of surface water for drinking quality. Directorate of Land Reclamation Punjab, Irrigation and Power Department, Canal Bank Mughalpura, Lahore, Pakistan
Anwar M (2005) Arsenic, cadmium and lead levels in hair and toenail samples in Pakistan. Environ Sci 12:71–86
Arain MB, Kazi TG, Baig JA, Jamali MK, Afridi HI, Shah AQ (2009) Determination of arsenic levels in lake water, sediment, and food stuff from selected area of Sindh, Pakistan: estimation of daily dietary intake. Food Chem Toxicol 47:242–248
Ashraf M, Tariq J, Jaffar M (1991) Contents of trace metals in fish, sediment and water from three freshwater reservoirs on the Indus River, Pakistan. Fish Res 12:355–364
Awan MA, Ahmed SH, Aslam MR, Qazi IA (2011) Determination of total suspended particulate matter and heavy metals in ambient air of four cities of Pakistan. Iranica J Ener Environ 2:128–132
Awan MA, Ahmed SH, Aslam MR, Qazi IA, Baig MA (2013) Determination of heavy metals in ambient air particulate matter using laser-induced breakdown spectroscopy. Arab J Sci Eng 38:1655–1661

Ayyub A, Ahmad MS, Ikram N, Shan U (2013) Copper and cadmium levels in hairs and nails; An indicator of heavy metal pollution. Professional Med J 20:296–300
Azeem HA (2010) Analysis of industrial waste water from Kot Lakhpat area (Lahore, Pakistan) by atomic absorption spectrometer. Biologia (Pakistan) 55:35–41
Aziz R, Rafiq MT, Yang J, Liu D, Lu L, He Z, Daud MK, Li T, Yang X (2014) Impact assessment of cadmium toxicity and Its bioavailability in human cell lines (Caco-2 and HL-7702). Bio Med Res Inter. Article ID 839538, p 8
Azizullah A, Khattak MNK, Richter P, Häder DP (2011) Water pollution in Pakistan and its impact on public health—a review. Environ Int 37:479–497
Bacha AA, Durrani MI, Paracha PI (2010) Chemical characteristics of drinking water of Peshawar. Pakistan J Nutr 9:1017–1027
Barbeau A (1984) Manganese and extrapyramidal disorders (a critical review and tribute to Dr. George C. Cotzias). Neuro Toxicol 5:13–35
Barbier O, Jacquillet G, Tauc M, Cougnon M, Poujeol P (2005) Effect of heavy metals on, and handling by, the kidney. Nephron Physiol 99:105–110
Baig JA, Kazi TG, Shah AQ, Afridi HI, Kandhro GA, Khan SA (2010) Green analytical procedure for sensitive and selective determination of arsenic in scalp hair samples of arsenic exposed adults of both genders. Pak J Anal Environ Chem 11:23–29
Baig JA, Kazi TG, Shah AQ, Afridi HI, Khan S, Kolachi NF, Kandhro GA, Wadhwa SK, Shah F (2011) Evaluation of toxic risk assessment of arsenic in male subjects through drinking water in southern Sindh Pakistan. Biol Trace Elem Res 143:772–786
Beckman LE, Van Landeghem GF, Sikstrom C, Wahlin A, Markevarn B, Hallmans G (1999) Interaction between haemochromatosis and transferrin receptor genes in different neoplastic disorders. Carcinogenesis 20:1231–1233
Beede DN, Bloom DE (1995) The economics of municipal solid waste. The World Bank research observer 10:113–150
Berg D, Gerlach M, Youdim MB, Double KL, Zecca L, Riederer P, Becker G (2001) Brain iron pathways and their relevance to Parkinson's disease. J Neurochem 79:225–236
Bermudez GM, Jasan R, Pla R, Pignata ML (2011) Heavy metal and trace element concentrations in wheat grains: assessment of potential non carcinogenic health hazard through their consumption. J Hazard Mater 193:264–271
Brammer H, Ravenscroft P (2009) Arsenic in groundwater: a threat to sustainable agriculture in South and South East Asia: review article. Environ Int 35:647–654
Bruce ED, Abusalih AA, Mcdonald TJ, Autenrieth RL (2007) Comparing deterministic and probabilistic risk assessments for sites contaminated by polycyclic aromatic hydrocarbons (PAHs). J Environ Sci Health 42:697–706
Butt MS, Sharif K, Bajwa BE, Aziz A (2005) Hazardous effects of sewage water on the environment; focus on heavy metals and chemical composition of soil and vegetables. Manag Environ Quality 16:338–346
Caravanos J, Clark E, Fuller R, Lambertson C (2011) Assessing worker and environmental chemical exposure risks at an e-waste recycling and disposal site in Accra, Ghana. J Health Pollut 1:16–25
Caussy DM (2005) A field guide for detection, management and surveillance of arsenicosis cases. Regional office for South-East Asia. World Health Organization, New Delhi
Cefalu WT, Hu FB (2004) Role of chromium in human health and in diabetes. Diab Care 27:2741–2751
Chang SH, Wang K, Chang HF, Ni CW, Wu J, Wong R, Lee HS (2009) Comparison of source identification of metals in road-dust and soil. Soil Sediment Contam 18:669–683
Chaoyang W, Cheng W, Linsheng Y (2009) Characterizing spatial distribution and sources of heavy metals in the soils from mining-smelting activities in Shuikoushan, Hunan Province, China. J Environ Sci 21:1230–1236
Chua A, Morgan E (1996) Effects of iron deficiency and iron overload on manganese uptake and deposition in the brain and other organs of the rat. Biol Trace Elem Res 55:39–54

Crossgrove J, Zheng W (2004) Manganese toxicity upon overexposure. NMR Biomed 17:544–553
Dawn daily (2013) Arsenic, lead poisoning causing 'silent epidemics' in Sindh. Newspaper staff reporter. 20 Feb 2013.
Dominici M, Le Blanc K, Mueller I, Slaper-Cortenbach I, Marini F (2006) Minimal criteria for defining multipotent mesenchymal stromal cells. The inter-national society for cellular therapy position statement. Cytotherapy 8:315–317
Drasch G, Schopfer J, Schrauzer GN (2005) Selenium/cadmium ratios in human prostates: indicators of prostate cancer risk of smokers and nonsmokers, and relevance to the cancer protective effects of selenium. Biol Trace Elem Res 103:103–107
Eck P, Wilson L (1989) Toxic metals in human health and disease. Eck Institute of Applied Nutrition and Bioenergetics, Ltd, Phoenix
Ellervik C, Mandrup-Poulsen T, Nordestgaard BG, Larsen LE, Appleyard M, Frandsen M (2001) Prevalence of hereditary haemochromatosis in late-onset type 1 diabetes mellitus: a retrospective study. Lancet 358:1405–1409
Ensink JHJ, Simmons RW, Hoek WV (2004) Wastewater use in Pakistan: the cases of Haroonabad and Faisalabad. In: Scott CA, Faruqi NI, Raschid-Sally L (eds) Wastewater use in irrigated agriculture: confronting the livelihood and environmental realities. CAB International, Wallingford, pp 91–99
Faiz Y, Tufail M, Javed MT, Chaudhry MMA, Siddique N (2009) Road dust pollution of Cd, Cu, Ni, Pb and Zn along Islamabad Expressway. Pak Microchem J 92:186–192
Farid S (2003) Toxic elements concentration in vegetables irrigated with untreated city effluents. Sci Technol Dev 22:58–60
Food and Agriculture Organization (FAO) (1985) Water quality for agriculture. Irrigation and drainage paper 29 revised.
Food and Agriculture Organization of the United Nations (FAO) (2002). Investment in Land and Water. In proceedings of the regional consultation Bangkok, Thailand 2002; 3–5
Farooqi A, Masuda H, Firdous N (2007) Toxic fluoride and arsenic contaminated groundwater in the Lahore and Kasur districts, Punjab, Pakistan and possible contaminant sources. Environ Pollut 145:839–849
Fatoki OS, Awofolu R (2003) Levels of Cd, Hg and Zn in some surface waters from the Eastern Cape Province, South Africa. Water SA 29:375–380
Frery N, Nessmann C, Girard F, Lafond J, Moreau T, Blot P (1993) Environmental exposure to cadmium and human birth weight. Toxicology 79:109–118
Gál J, Hursthouse A, Tatner P, Stewart F, Welton R (2008) Cobalt and secondary poisoning in the terrestrial food chain: data review and research gaps to support risk assessment. Environ Int 34:821–838
Gary F (2012) Common heavy metals: sources and specific effects. http://www.becomehealthynow.com/article/productshair/122/3/
Gidlow DA (2004) Lead toxicity. Occup Med 54:76–81
Gulfraz M, Mussaddeq Y, Khannum R, Ahmad T (2003) Physicochemical parameters and trace metal concentrations in effluents from various industries in vicinity of Lahore. Sci Technol Dev 22:9–11
Gupta VK, Gupta M, Sharma S (2003) Process development for the removal of lead and chromium from aqueous solution using red mud and aluminum industry waste. Water Res 35:1125–1134
Hamidullah S, Saifullah, Shah MT (2000) Heavy metal pollution in soil of Peshawar city NWFP, Pakistan. Geol Bull Univ Peshawar 33:1–9
Haq I, Hanif W, Hasnain G, Durez S (2012) Socio-economic impacts on human life in arsenic affected area of Basti Rasul Pur, Rahim Yar Khan, Pakistan. Sustain Agric Res 1:100
Huang X (2003) Iron overload and its association with cancer risk in humans: evidence for iron as a carcinogenic metal. Mutat Res 533:153–171
Huang M, Zhou S, Sun B, Zhao Q (2008) Heavy metals in wheat grain: assessment of potential health risk for inhabitants in Kunshan, China. Sci Total Environ 405:54–61

Hussain S, Masud T, Ahad K (2002) Determination of pesticides residues in selected varieties of mango. Pakistan J Nutr 1:41–42
Hussain SI, Ghafoor A, Ahmad S, Murtaza G, Sabir M (2006) Irrigation of crops with raw sewage: hazard and assessment in effluent, soil and vegetables. Pak J Agric Sci 43:97–101
Hussain J, Shah J, Khan LWA, Souza RC, Silva JDSD, Junior EPQ, Nascimento IA (2012) Determination of trace elements in the drinking water of Mardan district KPK, Pakistan. Am Eurasian J Agric Environ Sci 12:1091–1094
Hussain R, Ali L, Hussain I, Khattak SA (2014) Source identification and assessment of physicochemical parameters and heavy metals in drinking water of Islampur area, Swat, Pakistan. J Himalayan Earth Sci 47:99–106
Ibrahim M, Ahmad N, Anwar SA (1998) International seminar. In: Degraded soils "process, management and analysis." Univ Agri Faisalabad, Pakistan
Ilyas A, Sarwar T (2003) Study of trace elements in drinking water in the vicinity of Palosi drain, Peshawar. Pak J Biol Sci 6:86–91
Ilyas SZ (2007) A review of transport and urban air pollution in Pakistan. J Appl Sci Environ Manage 11:113–121
International Agency for Research on Cancer (IARC) (2006) Metallic cobalt particles. In: Cobalt in hard-metals and cobalt sulphate, gallium arsenide, indium phosphide and vanadium pentoxide. International Agency for Research on Cancer IARC Monographs on the Evaluation of Carcinogenic Risks to Humans, Lyon, 86
Inoue N, Makita Y, Chang LW (1996) Neurological aspects in human exposures to manganese. In: Chang LW (ed) Toxicology of metals. CRC Press, Boca Raton, FL, pp 415–421
Ishaq M, Jan FA, Khan MA, Ihsanullah I, Ahmad I, Shakirullah M, Roohullah (2013) Effect of mercury and arsenic from industrial effluents on the drinking water and comparison of the water quality of polluted and non-polluted areas: a case study of Peshawar and Lower Dir. Environ Monit Assess 185:1483–1494
Islam-Ul-Haq M, Deedar N, Wajid H (2007) Groundwater arsenic contamination—a multidirectional emerging threat to water scarce areas of Pakistan. 6th International IAHS Groundwater Quality Conference (2–7 December 2007), Fremantle, Western Australia
Iqbal MA, Chaudhary MN, Zaib S, Imran M, Ali K, Iqbal A (2011) Accumulation of Heavy Metals (Ni, Cu, Cd, Cr, Pb) in agricultural soils and spring seasonal plants, irrigated by industrial waste water. J Environ Tech Manag 2.
Jan FA, Ishaq M, Khan S, Shakirullah M, Asim SM, Ahmad I (2011) Bioaccumulation of metals in human blood in industrially contaminated area. J Environ Sci 23(12):2069–2077
Javed M, Mahmood G (2001) Metal toxicity of water in a stretch of river Ravi from Shahdera to Baloki headworks. Pak J Agric Sci 38:37–42
Jjemba PK (2004) Interaction of metals and metalloids with microorganisms in the environment. In: Jjemba PK (ed) Environmental microbiology-principles and applications. Science Publishers, Enfield, pp 257–270
Johnson MD, Kenney N, Stoica A, Hilakivi-Clarke L, Singh B, Chepko (2003) Cadmium mimics the in vivo effects of estrogen in the uterus and mammary gland. Nat Med 9:1081–1084
Kafayatullah Q, Shah MT, Irfan M (2001) Biogeochemical and environmental study of the chromite-rich ultramafic terrain of Malakand area, Pakistan. Environ Geol 40:1482–1486
Karim Z (2011) Risk assessment of dissolved trace metals in drinking water of Karachi, Pakistan. Bullet Environ Cont Toxicol 86:676–678
Kashif SR, Akram M, Yaseen M, Ali S (2009) Studies on heavy metals status and their uptake by vegetables in adjoining areas of Hudiara drain in Lahore. Soil Environ 28:7–12
Kazantzis G (1979) Renal tubular dysfunction and abnormalities of calcium metabolism in cadmium workers. Environ Health Perspect 28:155–159
Kazi TG, Arain MB, Baig JA, Jamali MK, Afridi HI, Jalbani N, Sarfraza RA, Shaha AQ, Niaza A (2009) The correlation of arsenic levels in drinking water with the biological samples of skin disorders. Sci Total Environ 407:1019–1026

Khan MA, Wajid A, Noor S, Khattak FK, Akhtar S, Rehman IU (2008a) Effect of soil contamination on some heavy metals contents of Cannabis sativa. J Chem Soc Pak 30:805
Khan DA, Qayyum S, Saleem S, Khan FA (2008b) Lead-induced oxidative stress adversely affects health of the occupational workers. Toxicol Ind Health 24:611–618
Khan S, Cao Q, Zheng YM, Huang YZ, Zhu YG (2008c) Health risks of heavy metals in contaminated soils and food crops irrigated with wastewater in Beijing China. Environ Pollut 152:686–692
Khan DA, Qayyum S, Saleem S, Ansari WM, Khan FA (2010) Lead exposure and its adverse health effects among occupational worker's children. Toxicol Indus Health 26:497–504
Khan T, Muhammad S, Khan B, Khan H (2011a) Investigating the levels of selected heavy metals in surface water of Shah Alam River (A tributary of River Kabul, Khyber Pakhtunkhwa). J Himalayan Earth Sci 44:71–79
Khan MN, Wasim AA, Sarwar A, Rasheed ZF (2011b) Assessment of heavy metal toxicants in the roadside soil along the N-5, National Highway, Pakistan. Environ Monit Assess 182:587–595
Khan S, Shahnaz M, Jehan N, Rehman S, Shah MT, Din I (2012a) Drinking water quality and human health risk in Charsadda district, Pakistan. J Clean Prod 10:10–16
Khan S, Shah MT, Din IU, Rehman S (2012b) Mercury exposure of workers and health problems related with small-scale gold panning and extraction. J Chem Soc Pak 34:870–876
Khan DA, Mushtaq S, Khan FA, Khan MQA (2013a) Toxic effects of chromium on tannery workers at Sialkot (Pakistan). Toxicol Indus Health 29:209–215
Khan K, Lu Y, Khan H, Zakir S, Khan IS, Khan AA, Wei L, Wang T (2013b) Health risks associated with heavy metals in the drinking water of Swat, northern Pakistan. J Environ Sci (China) 25:1–10
Khan SA, Ali A, Baig MN (2013c) The linkage between agricultural practices and environmental degradation. J Environ Treat Tech 1:19–22
Khan K, Khan H, Lu Y, Ihsanullah I, Nawab J, Khan S, Shah NS, Shamshad I, Maryam A (2014) Evaluation of toxicological risk of foodstuffs contaminated with heavy metals in Swat, Pakistan. Ecotoxicol Environ Safety 108:224–232
Khan ZI, Ahmad K, Ashraf M, Parveen R, Mustafa I, Khan A et al (2015) Bioaccumulation of heavy metals and metalloids in luffa (Luffa cylindrica L.) irrigated with domestic wastewater in Jhang, Pakistan: a prospect for human nutrition. Pak J Bot 47(1):217–224
Khattak MI, Khattak MI, Mohibullah M (2012) Study of heavy metal pollution in mangrove sediments reference to marine environment along the coastal areas of Pakistan. Pak J Bot 44:373–378
Khaskheli MA (2013) Fertilizers and environmental pollution (A report). http://www.pakissan.com/english/issues/fertilizers.and.environmental.pollution.shtml
Latif MI, Lone MI, Khan KS (2008) Heavy metals contamination of different water sources, soils and vegetables in Rawalpindi area. Soil Environ 27:29–35
Lee JS, Chon HT, Kim KW (2005) Human risk assessment of As, Cd, Cu and Zn in the abandoned metal mine site. Environ Geochem Health 27:185–191
Lehloesa LJ, Muyima NYO (2000) Evaluation of the impact of household treatment procedures on the quality of groundwater supplies in the rural community of Victoria district, Eastern Cape: technical note. Water SA 26:285–290
Li PH, Kong SF, Geng CM, Han B, Lu B, Sun RF, Zhao RJ, Bai ZP (2013a) Assessing the hazardous risks of vehicle inspection workers' exposure to particulate heavy metals in their work places. Aerosol Air Qual Res 13:255–265
Li X, Poon CS, Liu PS (2001) Heavy metal contamination of urban soils and street dusts in Hong Kong. Appl Geochem 16:1361–1368
Li DP, Wu ZJ (2008) Impact of chemical fertilizers application on soil ecological environment. J Appl Ecol 19:1158–1165
Li P, Wu HJ, Chen B (2013b) A resource-oriented solid waste management system for a mixed rural-urban area through Monte Carlo simulation-based fuzzy programming. Math Probl Eng. doi:10.1155/2013/780354

Liu X, Song Q, Tang Y, Li W, Xu J, Wu J, Wang F, Brookes PC (2013) Human health risk assessment of heavy metals in soil-vegetable system: A multi-medium analysis. Sci Total Environ, 530–540

Lone MI, Saleem S, Mahmood T, Saifullah K, Hussain G (2003) Heavy metal contents of vegetables irrigated by sewage/tube-well water. Int J Agri Biol 5:533–535

Mahar A, Malik RN, Qadir A, Ahmed T, Khan Z, Khan MA (2007) Review and Analysis of Current Solid Waste Management Situation in Urban Areas of Pakistan. Proceedings of the International Conference on Sustainable Solid Waste Management, 5–7 September Chennai, India, pp 34–41

Mahmood S, Maqbool A (2006) Impacts of wastewater irrigation on water quality and on the health of local community in Faisalabad. Pak J Water Res 10:19–22

Mahmood A, Malik RN (2014) Human health risk assessment of heavy metals via consumption of contaminated vegetables collected from different irrigation sources in Lahore, Pakistan. Arab J Chem 7:91–99

Majid H, Alam K, Madl P (2011) Aerosol size distribution, mass concentration measurement and lung deposition calculations for metropolitan cities of Pakistan. Research thesis, University of Salzburg, Austria

Majidano SA, Khuhawar MY (2009) Distribution of heavy metals in the ground water of Taluka Daur, District Nawabshah, Sindh, Pakistan, and its impacts on human health. J Chem Soc Pak 31:408–414

Malana MA, Khosa MA (2011) Groundwater pollution with special focus on arsenic, Dera Ghazi Khan-Pakistan. J Saudi Chem Soc 15:39–47

Malik RN, Husain SZ, Nazir I (2010) Heavy metal contamination and accumulation in soil and wild plant species from industrial area of Islamabad, Pakistan. Pak J Bot 42:291–301

Manzoor S, Shah MH, Shaheen N, Khalique A, Jaffar M (2006) Multivariate analysis of trace metals in textile effluents in relation to soil and groundwater. J Haz Mat 137:31–37

Mastoi GM, Shah SGS, Khuhawar MY (2008) Assessment of water quality of Manchar Lake in Sindh (Pakistan). Environ Monit Assess 141:287–296

Mumtaz M, Siddique A, Mukhtar N, Mehboob T (1999) Status of trace elements level in blood samples of different age population of Karachi (Pakistan). Tr J Med Sci 29:697–699

Mehmood S, Ahmad A, Ahmed A, Khalid N, Javed T (2013) Drinking water quality in capital city of Pakistan. Sci Report 2:637

McGregor DB, Baan RA, Partensky C, Rice JM, Wilbourn JD (2000) Evaluation of the carcinogenic risks to humans associated with surgical implants and other foreign bodies—a report of an IARC Monographs Programme Meeting. Eur J Cancer 36:307–313

Midrar-Ul-Haq, Khattak RA, Puno HK, Saif MS, Memon KS (2005) Surface and ground water contamination in NWFP and Sindh provinces with respect to trace elements. Int J Agric Biol 7:214–217

Miller KA, Siscovick DS, Sheppard L, Shepherd K, Sullivan JH (2007) Long term exposure to air pollution and incidence of cardiovascular events in women. N Engl J Med 356:447–458

Milman N, Pedersen P, Steig T, Byg KE, Graudal N, Fenger K (2001) Clinically overt hereditary hemochromatosis in Denmark 1948–1985: epidemiology, factors of significance for long-term survival, and causes of death in 179 patients. Ann Hematol 80:737–744

Muhammad S, Shah MT, Khan S (2011) Health risk assessment of heavy metals and their source apportionment in drinking water of Kohistan region, northern Pakistan. Microchem J 98:334–343

Murtaza G, Ghafoor A, Qadir M, Rashid MK (2003) Accumulation and bioavailability of Cd, Co and Mn in soils and vegetables irrigated with city effluents. Pak J Agric Sci 40:18–24

Murtaza G, Ghafoor A, Qadir MJ (2008) Accumulation and implications of cadmium, cobalt and manganese in soils and vegetables irrigated with city effluent. J Sci Food Agric 88:100–107

Murtaza G, Zia MH (2012) Wastewater production, treatment and use in Pakistan. http://www.ais.unwater.org/ais/pluginfile.php/232/mod_page/content/127/pakistan_murtaza_finalcountryreport2012.pdf

Mushtaq N, Khan KS (2010) Heavy metals contamination of soils in response to wastewater irrigation in Rawalpindi region. Pak J Agri Sci 47:215–224

Nafees M, Khan H, Jan MR, Rashid N, Khan F (2009) Soil contamination in Swat Valley caused by Cadmium and Copper. Sarhad J Agric 25:37–43

Nagajyoti PC, Lee KD, Sreekanth TVM (2010) Heavy metals, occurrence and toxicity for plants: a review. Environ Chem Lett 8:199–216

Naseem S, Hamza S, Bashir E, Pirzada T, Talpur MMA (2013) Trace element geochemistry of groundwater of winder Balochistan Pakistan and its appraisal for irrigation water quality. Brit J Appl Sci Tech 3:182–198

Nasrullah, Naz R, Bibi H, Iqbal M, Durrani MI (2006) Pollution load in industrial effluent and ground water of Gadoon Amazai Induatrial Estate (GAIE) Swabi, NWFP. J Agric Biol Sci 1:18–24

Nazif W, Perveen S, Shah SA (2006) Evaluation of irrigation water for heavy metals of Akbarpura area. J Agric Biol Sci 1:51–54

Nickson RT, McArthur JM, Shrestha B, Kyaw-Myint TO, Lowry D (2005) Arsenic and other drinking water quality issues, Muzaffargarh District, Pakistan. Appl Geochem 20:55–68

Nordberg GF (2004) Cadmium and health in the 21st century—historical remarks and trends for the future. Biometals 17:485–489

NSDWQ-Pak (2008) National Standards for Drinking Water Quality, Health Services Academy, Ministry of Health, Government of Pakistan

Oller AR, Costa M, Oberdörster G (1997) Carcinogenicity assessment of selected nickel compounds. Toxicol Appl Pharmacol 143:152–166

Pak-SCEA (2006) Pakistan; Strategic Country Environmental Assessment Report: rising to the challenges

Pak-EPA/OECC (2004) Urban environmental problems in Pakistan (A case study for urban environment in Hayatabad, Peshawar). http://www.Environment.Gov.Pk

Panhwar AH, Kazi TG, Afridi HI, Arain SA, Arain MS, Brahaman KD, Naeemullah, Arain SS (2016) Correlation of cadmium and aluminum in blood samples of kidney disorder patients with drinking water and tobacco smoking: related health risk. Environ Geochem Health 38:1–10. doi:10.1007/s10653-015-9715-y

Parkkila S, Niemelä O, Savolainen ER, Koistinen P (2001) HFE mutations do not account for transfusional iron overload in patients with acute myeloid leukemia. Transfusion 41:828–831

Paustenbach DJ (2002) Human and ecological risk assessment: theory and practice. Wiley, New York

Parveen N, Ghaffar A, Shirazi SA, Bhalli MN (2012) Spatial distribution of heavy metal contamination in road side soils of Faisalabad Pakistan. Pak J Sci 64:309

PCRWR (2005) National Water Quality Monitoring Program. Water quality report 2003–2004. Pakistan Council for Research in Water Resources, Islamabad, Pakistan, http://www.pcrwr.gov.pk/wq_phase3_report/TOC.htm

PCRWR (2007) Water quality report, water quality report 2003–2004 (n.d.). http://www.pcrwr.gov.pk/wq_phase2_report/wq_phase2_chap5.htm

PCRWR (2008) Arsenic contamination in ground water of Central Sindh Phase I. Pakistan Council for Research in Water Resources. PCRWR, Islamabad, Pakistan, pp 19–38, http://www.pcrwr.gov.pk/Arsenic_CS/ACS_TOC.htm

Piasek M, Laskey JW (1999) Effects of in vitro cadmium exposure on ovarian steroidogenesis in rats. J Appl Toxicol 19:211–217

Pope CA (2000) Epidemiology of fine particulate air pollution and human health: biologic mechanisms and who is at risk? Environ Health Perspect 108:713–723

Porea TJ, Belmont JW, Mahoney DH (2000) Zinc-induced anemia and neutropenia in an adolescent. J Pediatr 136:688–690

Puttaiah ET, Kiran BR (2008) Heavy metal transport in a sewage fed Lake of Karnataka, India. Proceedings of Taal. 2007. The 12th World Lake Conference; Jaipur, India 2008. pp 347:347–354

Qadir NF (2002) Air quality in urban areas in Pakistan vs transport planning: issues and management tools. Document 16B, presentation
Rahman A, Lee HK, Khan MA (1997) Domestic water contamination in rapidly growing mega cities of Asia: case of Karachi, Pakistan. Environ Monit Assess 44:339–360
Rahman S, Khalid N, Zaidi ZH, Ahmad S, Iqbal MZ (2006) Non-occupational lead exposure and hypertension in Pakistani adults. J Zhejiang Univ Sci B 7:732–737
Rasmussen ML, Folsom AR, Catellier DJ, Tsai MY, Garg U, Eckfeldt JH (2001) A prospective study of coronary heart disease and the hemochromatosis gene (HFE) C282Y mutation: the atherosclerosis risk in communities (ARIC) study. Atherosclerosis 154:739–746
Raza N, Niazi SB, Sajid M, Iqbal F, Ali M (2007) Studies on relationship between season and inorganic elements of Kallar Kahar Lake (Chakwal), Pakistan. J Res (Sci) 18:61–68
Rehman W, Zeb A, Noor N, Nawaz M (2008) Heavy metal pollution assessment in various industries of Pakistan. Environ Geol 55:353–358
Riess ML, Halm JK (2007) Lead poisoning in an adult: lead mobilization by pregnancy. J Gen Intern Med 22:1212–1215
Rohrs LC (1957) Metal-fume fever from inhaling zinc oxide. AMA Arch Ind Health 16:42–47
Saif MS, Midrar-Ul-Haq, Memon KS (2005) Heavy metals contamination through industrial effluent to irrigation water and soil in Korangi area of Karachi (Pakistan). Int J Agri Biol 7:646–648
Saleemi MA (1993) Environmental assessment and management of irrigation and drainage scheme for sustainable agriculture growth. EPA, Lahore, p 64
Sayre LM, Perry G, Atwood CS, Smith MA (2000) The role of metals in neurodegenerative diseases. Cell Mol Biol (Noisy-le-Grand) 46:731–741
Schneidemesser EV, Stone EA, Quraishi TA, Shafer MM, Schauer JJ (2010) Toxic metals in the atmosphere in Lahore, Pakistan. Sci Total Environ 408:1640–1648
Schwartz J (2001) Air pollution and blood markers of cardiovascular risk. Environ Health Perspect 109:405–409
Scragg A (2006) Environmental biotechnology, 2nd edn. Oxford University Press, Oxford, UK
SDWF (2006) Drinking water quality and health. Safe drinking water foundation. http://www.safewater.org/members/teachers/water%20and%20health.com. Accessed August 2006
SEPA (1995) Environmental quality standard for soils. State Environmental Protection Administration, China. GB15618-1995
Shah AQ, Kazi TG, Baig JA, Afridi HI (2011) Correlation between arsenic concentration in fish and human scalp hair of people living in arsenic-contaminated and noncontaminated areas of Pakistan. Bio Trace Element Res 144:197–204
Shah MH, Shaheen N (2007) Statistical analysis of atmospheric trace metals and particulate fractions in Islamabad, Pakistan. J Hazard Mater 147:759–767
Shah MH, Shaheen M (2008) Annual and seasonal variations of trace metals in atmospheric suspended particulate matter in Islamabad, Pakistan. Water Air Soil Pollut 190:13–25
Shah MT, Begum S, Khan S (2010) Pedo and biogeochemical studies of mafic and ultramafic rocks in the Mingora and Kabal areas, Swat, Pakistan. Environ Earth Sci 60:1091–1102
Shah MT, Ara J, Muhammad S, Khan S, Tariq S (2012) Health risk assessment via surface water and sub-surface water consumption in the mafic and ultramafic terrain, Mohmand agency, northern Pakistan. J Geochem Explor 118:60–67
Shan UA, Ikram N (2012) Heavy metals in human scalp hair and nail samples from Pakistan: influence of working and smoking habits. Int J Chem Biochem Sci 1:54–58
Sharif M, Nergis Y, Farooq MA (2010) Soil contamination of toxic elements irrigated with mixed industrial effluent and its environmental impact on urban area of Karachi, Pakistan. Am Eurasian J Agric Environ Sci 9:584–591
Sharmaa RK, Agrawala M, Marshall F (2007) Heavy metal contamination of soil and vegetables in suburban areas of Varanasi, India. Ecotox Environ Safe 66:258–266
Sial RA, Chaudhary MF, Abbas ST, Latif MI, Khan AG (2006) Quality of effluents from Hattar Industrial Estate. J Zhejiang Uni Sci B 7:974–980

Siddiqui S, Khattak RA (2012) Heavy metals distribution in the soils of Peshawar valley Northern Pakistan. Sci Res Ess 7:544–552

Singh S, Kumar M (2006) Heavy metal load of soil, water and vegetables in Peri-Urban Delhi. Environ Monit Assess 120(1–3):79–91

Smith AH, Hopenhayn-Rich C, Bates MN, Goeden HM, Hertz-Picciotto I, Duggan HM, Wood R, Kosnett MJ, Smith MT (1992) Cancer risks from arsenic in drinking water. Environ Health Perspect 97:259–267

Solomons NW, Ruz M (1998) Trace element requirements in humans: an update. J Trace Elem Exp Med 11:177–195

State of the Environment, report (2005) Draft report. Ministry Environ Govt Pakistan

Stein JA, Tschudy DP, Coroan PC, Coffins A (1999) Metal pollution. Biol Chem, 245–253

Tariq J, Jaffar M, Ashraf M (1994) Trace metal concentration, distribution and correlation in water, sediment and fish from the Ravi River, Pakistan. Fish Res 19:131–139

Tariq MI, Afza S, Hussain I (2006a) Degradation and persistence of cotton pesticides in sandy loam soils from Punjab, Pakistan. Environ Res 100:184–196

Tariq SR, Shah MH, Shaheen N, Jaffar M, Khalique A (2008) Statistical source identification of metals in groundwater exposed to industrial contamination. Environ Monit Assess 138:159–165

Tariq M, Ali M, Shah Z (2006b) Characteristics of industrial effluents and their possible impacts on quality of underground water. Soil Environ 25:64–69

Toor IA, Tahir SNA (2009) Study of arsenic concentration levels in Pakistani drinking water. Polish J Environ Stud 18:907–912

UDOE (2011) Best practices laboratories for the 21st century: modeling exhaust dispersion for specifying. Acceptable exhaust/intake designs. http://www1.eere.energy.gov/femp/pdfs/bp_modeling_508.pdf

Ullah R, Malik RN, Qadir A (2009) Assessment of groundwater contamination in an industrial city, Sialkot, Pakistan. Afr J Environ Sci Technol 3:429–446

Ullah Z, Khan H, Waseem A, Mahmood Q, Farooq U (2013) Water quality assessment of the River Kabul at Peshawar, Pakistan: industrial and urban wastewater impacts. J Water Chem Tech 35:170–176

Umar M, Waseem A, Sabir MA, Kassi AM, Khan AS (2013) The impact of geology of recharge areas on groundwater quality: a case study of Zhob River Basin, Pakistan. Clean–Soil, Air, Water 2:119–127

United Nations Industrial Development Organization (UNIDO) (2000) Industrial policy and the environment in Pakistan. UNIDO, Vienna, Austria

USEPA (US Environmental Protection Agency) (1997) Exposure factors handbook. Volume II–food ingestion factors. Office of Research and Development, Washington, DC, EPA/600//P-95/002Fa. http://www.epa.gov/superfund/health/conmedia/soil/index.htm. http://www.epa.gov/region9/superfund/prg/

USEPA (2002) Supplemental guidance for developing soil screening levels for superfund sites. Office of Solid Waste and Emergency Response, Washington, DC [OSWER 9355.4-24]

USEPA IRIS (US Environmental Protection Agency's Integrated Risk Information System) (2011) http://www.epa.gov/iris/. Accessed 15 Jan 2011

Verkleji JAS (1993) The effects of heavy metals stress on higher plants and their use as bio monitors. In: Markert B (ed) Plant as bioindicators: indicators of heavy metals in the terrestrial environment. VCH, New York, pp 415–424

Waalkes MP, Rehm S, Riggs CW, Bare RM, Devor DE, Poirier LA (1988) Cadmium carcinogenesis in male Wistar [Crl:(WI) BR] rats: dose-response analysis of tumor induction in the prostate and testes and at the injection site. Cancer Res 48:4656–4663

Wadhwa SK, Kazi TG, Afridi HI, Talpur FN, Naeemullah (2015) Interaction between carcinogenic and anti-carcinogenic trace elements in the scalp hair samples of different types of Pakistani female cancer patients. Clinica Chimica Acta 439:178–184

Wang XL, Sato T, Xing BS, Tao S (2005) Health risks of heavy metals to the general public in Tianjin, China via consumption of vegetables and fish. Sci Total Environ 350:28–37

Waseem A, Arshad J, Iqbal F, Sajjad A, Mehmood Z, Murtaza G (2014) Pollution status of Pakistan: a retrospective review on heavy metal contamination of water, soil, and vegetables. Bio Med Res Int 2014, Article ID 813206, http://dx.doi.org/10.1155/2014/813206

Wei B, Yang L (2010) A review of heavy metal contaminations in urban soils, urban road dusts and agricultural soils from China. Microchem J 94(2):99–107

Weiss L, Wright S (2001) Mercury on the road to zero. Recommended strategies to eliminate mercury releases from human activities in Oregon by 2020. Oregon Environmental Council & The Mercury Solution Team, Oregon

WHO (1999) Air quality guidelines for Europe, 2nd edn. World Health Organization Regional Office for Europe. World Health Organization, Copenhagen

WHO (2007) Health risks of heavy metals from long range trans boundary air pollution, DK-2100 Copenhagen 0. World Health Organization, Denmark

WHO (2011) Guidelines for drinking-water quality. 4th ed. http://www.who.int/water_sanitation_health/publications/2011/dwq_chapters/en/

Wiwatanadate P, Liwsrisakun C (2011) Acute effects of air pollution on peak expiratory flow rates and symptoms among asthmatic patients in Chiang Mai, Thailand. Int J Hygiene Environ Health 214:251–257

Wongsasuluk P, Chotpantarat S, Siriwong W, Robson M (2013) Heavy metal contamination and human health risk assessment in drinking water from shallow groundwater wells in an agricultural area in Ubon Ratchathani province, Thailand. Environ Geochem Health 36:169–182

WWF-Pak (2007) Report on National Surface Water Classification Criteria, Irrigation water quality guidelines for Pakistan. February-2007. World Wide Fund Pakistan

Xia X, Chen X, Liu R, Liu H (2011) Heavy metals in urban soils with various types of land use in Beijing. J Hazard Mater 186:2043–2050

Yousafzai AHK, Hashmi DR, Qaimkhani MI, Ahmad F, Siddiqui I (2001) Determination of heavy metals in vegetables and soils at sewerage farm in Sindh industrial trading estate (SITE), Karachi. J Chem Soc Pak, 23

Zafar A, Eqani SAMAS, Cincinelli NBA, Tahir F, Shah STA, Hussain A, Alamdar A, Huang Q, Peng S, Shen H (2014) Toxic metals signature in the human seminal plasma of Pakistani population and their potential role in male infertility. Environ Geochem Health. doi:10.1007/s10653-014-9666-8

Influence of Nanotoxicity on Human Health and Environment: The Alternative Strategies

Buddolla Viswanath and Sanghyo Kim

Contents

B. Viswanath
Department of Bionanotechnology, Gachon University, San 65, Bokjeong-Dong, Sujeong-Gu, Seongnam-Si, Gyeonggi-Do 461-701, Republic of Korea
e-mail: buddolla@gmail.com

S. Kim (✉)
Department of Bionanotechnology, Gachon University, San 65, Bokjeong-Dong, Sujeong-Gu, Seongnam-Si, Gyeonggi-Do 461-701, Republic of Korea

Gil Medical Center, Graduate Gachon Medical Research Institute, Incheon 405-760, Republic of Korea
e-mail: samkim@gachon.ac.kr

P. de Voogt (ed.), *Reviews of Environmental Contamination and Toxicology, Volume 242*, Reviews of Environmental Contamination and Toxicology 242,
DOI 10.1007/398_2016_12

1 Introduction

Nanotechnology allows the manipulation and application of engineered particles or systems that have at least one dimension less than 100 nm in length (Stander and Theodore 2011). New industries in collaboration with university research departments are being formed and are finding so many investors eager to back their ideas and products. There is no doubt that governments and major industrial companies are committing significant resources for research into the expansion of nanometre scale processes, materials and products. Growing exploration of nanotechnology has resulted in the discovery of many distinctive properties of nanomaterials such as superior catalytic, optical, magnetic, mechanical and electrical properties when compared to conventional formulations of the same material (Ferrari 2005; Pattan and Kaul 2014). Nanomaterials are progressively more being used for commercial purposes and in consumer products leading to increased direct and indirect exposure in humans (Pattan and Kaul 2014). In the field of medicine, nanoparticles are purposefully injected into the human body. For drug delivery and imaging nanomaterials are often deliberately coated with biomolecules such as protein, DNA and monoclonal antibodies to aim particular cells (Lewinski et al. 2008; Nalwa 2014). The novel physicochemical properties of these engineered nanomaterials may initiate new mechanisms of injury and toxicological effects due to destructive interactions of nanomaterials with biological systems and the environment (Ray et al. 2009; Yan et al. 2015). Even though less studied than human health, there is also basis for concern regarding environmental impacts and areas such as ecotoxicology, environmental chemistry, behaviour and fate are areas of concentrated current research (Zhu et al. 2008). Nanomaterials may potentially impact the environment in three possible ways: (1) direct effect on invertebrates, micro-organisms, fish and other species; (2) interaction with other pollutants, that may transform the bioavailability of toxic compounds and nutrients; and (3) changes to non-living environmental structures (Dhawan and Sharma 2010). Understanding the environmental and human health impacts of nanomaterials is a crucial stage for the responsible development of nanotechnology and to gain full benefit of its applications. As nanotechnology matures, questions are being aroused regarding whether the products or materials of nanotechnology will provide hazards to human health or the environment and whether the manufacturing of these materials will spawn new hazards or waste streams. Considerate the environmental and human health impact of nanomaterials is a fundamental stage for the liable development nanotechnology and to gain full advantage of its applications (Yadav et al. 2014).

Alternatively, green nanoscience has been described as the progress of clean technologies to reduce possible environmental and human health risks coupled with the production and use of nanomaterials and to promote replacement of existing products with new nano-products that are further environmentally friendly all over their life cycle (Hutchison 2008). Green nanoscience is the successful integration of green chemistry and nanoscience that brings the collective approaches and methods to build up greener products, processes, and applications (Iavicoli et al. 2014).

Recently green nanoscience is booming in the developing fields of nanoelectronics, nanocomposites and thermoelectric (Marconnet et al. 2011). Therefore, green chemistry and nanoscience are promising fields that take advantage of molecular-level design and have massive potential to overcome the adverse impacts of nanoparticles and nanomaterials on health and environment (Iavicoli et al. 2014). If the global research community can take advantage of green nanoscience then we can surely look forward to the advent of safe nanotechnologies. In this review we have summarized the adverse effects of different nanoparticles to human health and environment along with alternative strategies to overcome ill impacts.

2 Nanotoxicology

Nanotoxicology and nano-risk have been drawing increasing attention of toxicologists and regulatory scientists as the manufacturing of nanomaterials increases (Dhawan and Sharma 2010). As shown in Fig. 1, a number of hazardous exposure conditions are encountered by the workers occupied in nanotechnology activities. In fact, nanomaterials may have significant, still unknown, hazards properties that can pose risks for a broad range of workers: researchers, laboratory technicians, cleaners, production workers, transportation, storage and retail workers, employees in disposal and waste facilities and potentially, emergency responders who deal with spills and disasters of nanomaterials and may be differently exposed to these potential, innovative xenobiotics (Iavicoli et al. 2014). Thus the branch of nanotoxicology deals with the study relating to the toxicity of the nano materials, as it is essential to know the toxicity of nano material before using it for a variety of applications (Gao et al. 2015a; Donaldson and Poland 2013; Haynes 2010). In addition, the effects and impacts on human health also needs to be reviewed accordingly.

The International Council on Nanotechnology (ICON) has created a database of all the publications of several nano materials along with their impact on environmental health and safety. This highlights on the exciting trends associated with the field of nanotoxicology. In addition, it is involved in proposing reliable, robust, and data-assured test protocols for nanomaterials in human and environmental risk assessment (Donaldson and Poland 2013). The utmost challenge faced in the field of nanotoxicology now-a-days is the recognition as well as the estimation of the deleterious effects of a variety of engineered nanomaterials with their dissimilar physiochemical properties, which are regularly being manufactured and launched for versatile applications (Hutchison 2008). It is not very easy to discover the exact hazard denominations of nanoparticles due to different causes, for instance, we are not specific which physico-chemical property of the nanoparticles is influencing the toxicity (Di Bona et al. 2015). The extreme changes in biological, chemical and other properties that can formulate nanotechnology applications are so exciting, however, also may value examination to determine any effects on product safety, effectiveness, or other characteristics. Moreover, currently no clear processes and regulatory guidelines on the evaluation or testing of nanoparticulate materials are available.

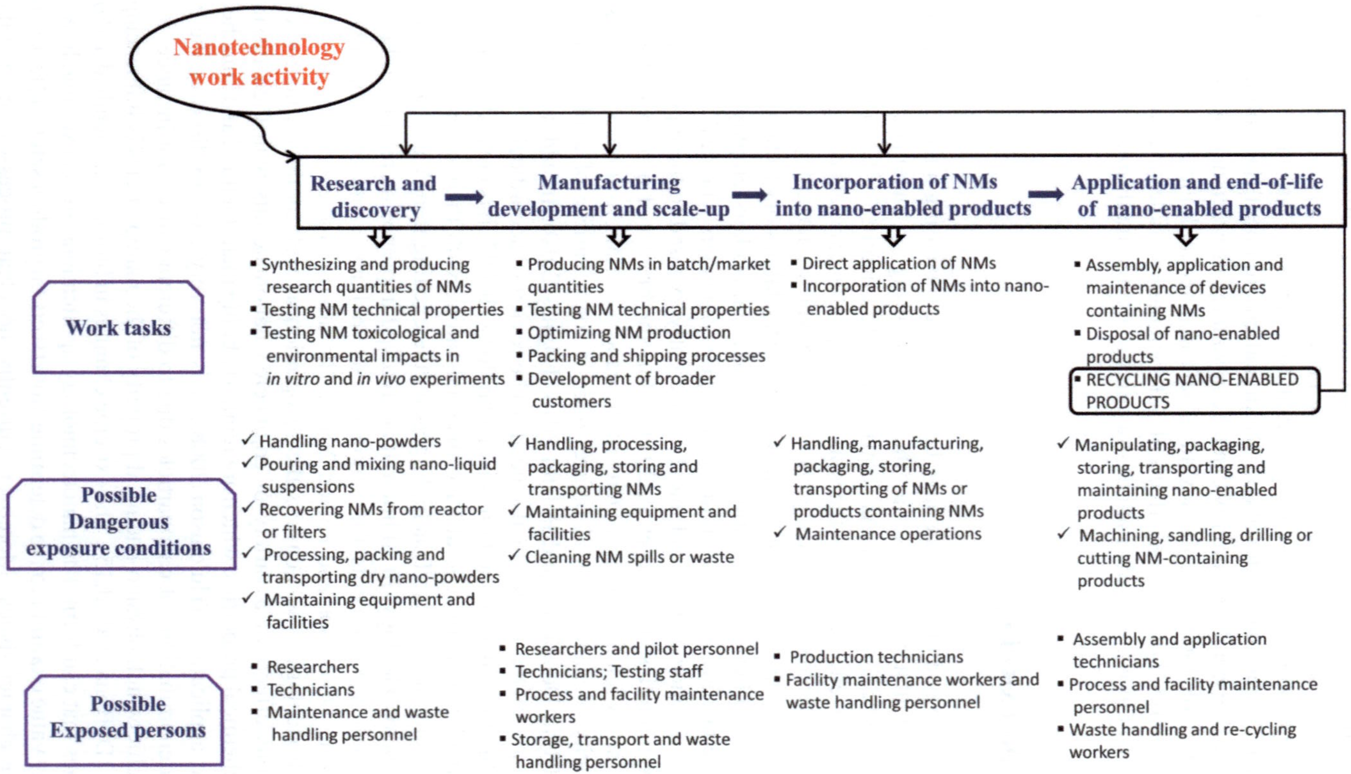

Fig. 1 Assessment of potentially hazardous exposure conditions for workers involved in nanotechnology activities (Source: Iavicoli et al. 2014)

2.1 Physicochemical Properties of Nanomaterial–Toxicity Relationship

As stated earlier, nanotechnology is a quick growing field that offers materials that have new dimensions, novel properties, and a broader array of applications but the basis of the toxicity of engineered nanoparticles is still unclear (Kumar et al. 2012). Till date, research results are more evocative than definitive and no quantitative structure–activity relationship models are existing. Physicochemical properties such as small size, specific surface area, structure, aspect ratio, morphology, solubility, chemical composition, reactivity, photochemistry, production of reactive species and surface properties (i.e. charge and coating) can be of prime importance (Yadav et al. 2014). In fact, the very similar properties that guide to the technical advantages of nanotechnology also lead to unique biological effects. For instance, size is the key feature determining the scope of uptake and toxicity of many nanomaterials, which have been shown to be size dependent (Oberdorster 2010). The dimension of nanoparticles may have vital impact on where nanoparticles end up in the human body and other organisms (Kumar et al. 2012). In humans, large inhalable nanomaterials of any composition with particle size above 2.5 μm have a tendency to deposit mostly in the nose and throat while smaller nanoparticles with particle size less than 2.5 μm can locate their way to the upper airways (Alfaro-Moreno et al. 2007). Shape and size of the nanoparticles has been shown to have a distinct effect on the biological activity. Pal and his group reported that silver nanoparticles undergo shape-dependent interaction with *E. coli* (Pal et al. 2007). In another study, Journeay et al. (2008) confirmed that water-soluble rosette nanotube structures exhibit low pulmonary toxicity due to their biologically stimulated design and self-assembled architecture. The smallest particles can make a way into deeper alveolar region and might penetrate to various parts of the respiratory tract (Kovacic and Somanathan 2009). Because of their small size, nanomaterials can possibly pass through the lungs into the bloodstream and to be taken up by cells, reaching potentially susceptible sites such as liver, spleen, kidney and heart (Sturm 2015). In case of anatase TiO_2 nanomaterial, it was revealed that modification to a fiber structure of greater than 15 μm formed a highly toxic particle that initiated an inflammatory response by alveolar macrophages and that length may be a vital determinant of nanomaterial biocompatibility (Hamilton et al. 2009). Physico-chemical characterization and their interaction with biological media are essential for widely used metal oxides and carbon nanomaterials (Landsiedel et al. 2010; Štengl et al. 2016). In one of the investigations by Schaeublin and his group, 1.5 nm sized gold nanoparticles exhibited their action on cellular processes like the charged NPs initiating cell death through apoptosis and neutral NPs leading to necrosis in HaCaT cells (Schaeublin et al. 2012). It was observed that surface charge was a key determinant of their action on cellular processes (Beddoes et al. 2015). Figure 2 illustrates a range of possible physicochemical properties of engineered nanomaterials leading to nanotoxicology.

Further, the size and shape of nanomaterials is the main factor in influencing their uptake across the gill membrane or the gastrointestinal tract (GI) of aquatic

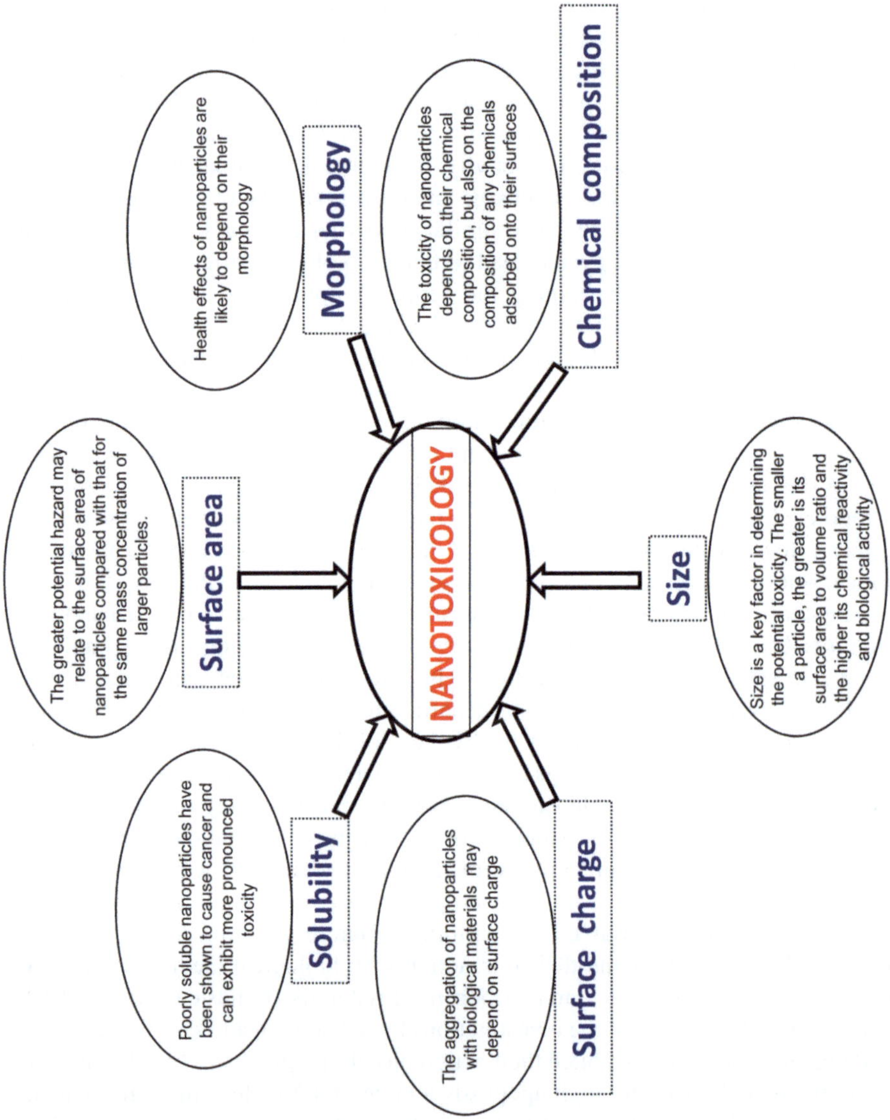

Fig. 2 Physicochemical properties of engineered nanomaterials leading to nanotoxicology

and terrestrial organisms. It has been observed that the absolute limit for passive diffusion through fish gills is about 1 nm (Batley et al. 2013). Surface charge and particle composition are also key factors for determining the uptake and toxicity of nanomaterials (Fan et al. 2015). Mineral particle stimulated apoptosis was dependent on particle size, while surface reactivity and composition were found to be most essential for the proinflammatory potential of the particles (Pattan and Kaul 2014). The aspect ratio of engineered nanoparticles along with biological persistence is likely to be an important factor for their toxicology. Fibres longer than 20 μm cannot be phagocytosed by alveolar macrophages, causing reduced clearance and accumulation. Poland and group have reported that dependent on structure, the carbon nanotubes (CNTs) perform in a similar manner to asbestos and may have an even superior biological activity and therefore is hazardous (Poland et al. 2008). Some of the reported morphologies of engineered nanoparticles have been summarized in Table 1. Therefore, nanomaterials are composed of primary

Table 1 Some of the reported morphologies of engineered nanoparticles

Morphologies of engineered nanoparticles	
 *Spherical* Generally in small crystalline particles and non-crystalline structures	 *Rod or wire* May have a series of cross sections like cubic, circular and pentagonal
 *Discotic* Thin flat plates regularly broader than 10 nm. May be a variety of shapes including hexagonal and irregular	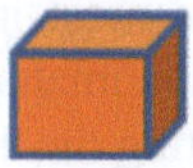 *Geometric solid* Usually in crystalline particles may be a array of shapes like cubic, tetrahydral and icosahedral
 *Dendritic* Comprises of nanoscale wires the dendrite may be larger than 100 nm	 *Tetrapod* Shaped by the enlargement of hexagonal phase rods from cubic seed crystals
 *Dumbell* Produced by the development of one materials only at the ends of the rod of another material	 *Tear drop* An extension of the spherical shape

and agglomerated particles that can vary in size, shape, charge, crystalline, chemical composition, surface properties and all these characteristics have been optional to influence the toxicity of nanomaterials. As a result, several efforts are made to increase the knowledge on the toxicity-determining characteristics of nanomaterials to categorize them into hazard groups in order to facilitate risk assessment (Braakhuis et al. 2014; Arts et al. 2015).

3 Potential Human Health Effects of Nanomaterials

Nanoparticles have the same dimensions as biological molecules such as proteins. Currently, exposure to nanoparticles is increased, and because of uncertainty regarding their toxic characteristics, concerns have arisen that such materials pretence novel health risks for consumers, workers, and the environment (Braakhuis et al. 2015). An extensive literature states that some of the nanoparticles may have an adverse effect and there is a cause for concern as to their effect on human health, which has largely developed out the ample literature on the health impacts of nanoparticles. This is mainly due to the several features that may contribute to the toxicity of nanomaterials. For instance, an organism's reaction to a particular nanomaterial may be related to the mass of the administered dose, or it may be related to other factors, together with the number of particles, shape, electrical charge, and coating, or a mixture of physiochemical properties (Buzea et al. 2007). Currently, there is no consensus within the scientific community on what characteristic may be the most important in elucidating this dose-response relationship for each type of nanomaterial. This lack of agreement is comprehensible, given that studies evaluating the health effects of nanomaterials show a range of findings and highlight the unfortunate generalization of responses across all types of nanomaterials.

Nanoparticles can interact with complex networks of immune cells located within and beneath epithelial surfaces and these NPs can act as allergens during the neonatal period, triggering the immune system to induce allergic inflammation in later life stages (Sly and Schüepp 2012). Detrimental cardiovascular consequences due to NPs exposure have been reported in epidemiological studies (Liou et al. 2012). Wang and their group reported that nano-cerium-element doped titanium dioxide induces apoptosis of Bel 7402 human hepatoma cells in the presence of visible light (Wang et al. 2007). Nanoparticles like zinc oxide and sunscreen titanium dioxide can cause oxidative damage to DNA in vitro and in cultured human fibroblasts (Dunford and Salinaro 1997). Inhaled ultrafine particles (UFPs) can increase access to the blood stream and can then be dispersed to other organs in the body; this has been shown for synthetically produced nanoparticles such as C^{60} fullerenes which accumulate in the liver (Hougaard et al. 2015). Even big particles outside the 'nano' range can enter the stratum corneum of human skin and get to the epidermis and occasionally the dermis and may be taken up into the lymphatic system (Smulders et al. 2015; Tinkle et al. 2003). There is a strong

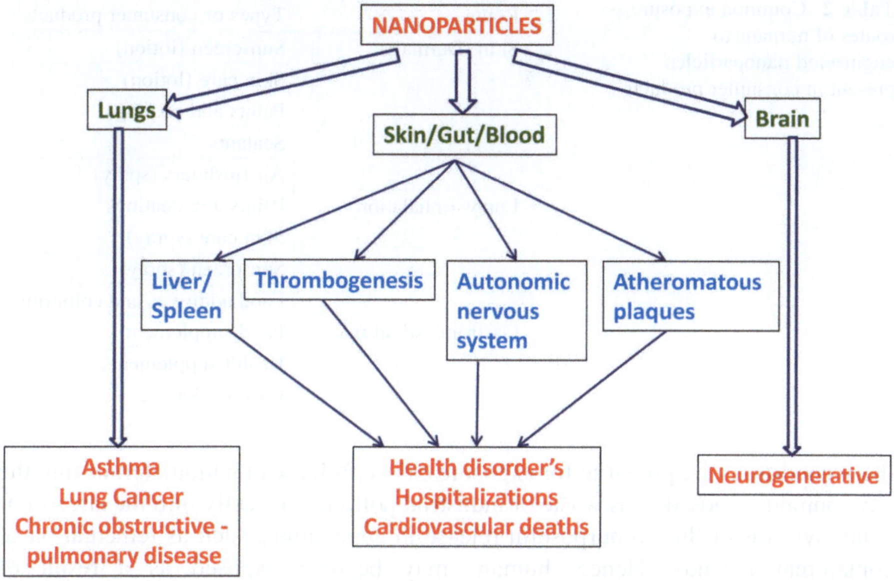

Fig. 3 Systematic health effects of nanoparticles on human body

possibility that nanoparticles can be assimilated into the body through the lungs, skin and gastrointestinal tract (Smulders et al. 2015). Additionally, several studies have been conducted in investigating the dissolution of nanoparticles in artificial body fluids such as fluids representing the environment of the stomach, blood and airways and found adverse influences (Stebounova et al. 2011; Leo et al. 2013). Figure 3 gives an illustrated picture of systematic health effects of nanoparticles on human body.

3.1 *Major Modes of Exposure*

There are numerous exposure scenarios depending on the particulars of manufacture, use and discarding. All through these scenarios, the levels of exposure, the population exposed, the duration of exposure and the nature of the material to which people are exposed are completely different (Civeira et al. 2015). In an industrial situation, exposure to nanomaterials can happen to workers at all phases of material life cycle. During the development of a new material, it is possible that material will be manufactured under strongly controlled conditions, usually in very little quantities. Once the material moves into commercial production, exposures can occur potentially during synthesis of the material or in downstream activities such as packaging, transport, recovery and storage (Leo et al. 2013). Nanomaterials may also be integrated into a composite material, which may be consequently re-engineered or reprocessed by cutting, finishing and sawing. Once again in

Table 2 Common exposure routes of humans to engineered nanoparticles present in consumer products

Route	Types of consumer products
Skin (Dermal)	Sunscreen (lotion)
	Skin care (lotion)
	Paints and coatings
	Sealants
	Air freshners (spray)
Lungs (inhalation)	Paints and coatings
	Skin care (spray)
	Sunscreen (spray)
	Food additives and colourings
Gastrointestinal tract	Food supplement
	Health supplements
	Food packaging

these conditions the potential for exposure exists. Release of nanomaterials into the environment is possible as waste or industrial pollution, directly into the air, soil or water systems or due to purposeful release in applications such as remediation of contaminated lands. Hence, humans may become exposed as a result of nanomaterial contamination in air, water or the food chain, or during the use of commercial products containing nanomaterials (Buzea et al. 2007). In view of human exposure for all of these scenarios, it is compulsory to think about the route of entry into the human body. Following are the some of the various human exposure routes in detail and Table 2 illustrates the common exposure routes of humans to engineered nanoparticles.

3.1.1 Skin

The significance of dermal exposure to dangerous substances continues to rise (Mackevica and Foss 2015). Detrimental effects arising from skin exposure may either happen locally within the skin or alternatively the substance may be absorbed through the skin and circulated via the bloodstream, probably causing systemic effects (Warheit and Donner 2015). Skin is the major primary defence organ in human body and directly comes into contact with many lethal agents. Skin exposure to engineered nanomaterials can also occur during the purposeful application of topical creams and other drug treatments (Hagens et al. 2007). Information on in vivo uptake of Ag nano particles due to dermal exposure are limited partly caused by the lack of appropriate analytical approaches for the determination of Ag in biological matrices, but strongly needed to enable risk assessment of skin exposure to nano silver containing products (Bianco et al. 2015; Völker et al. 2013). However, nanosilver-based dressings and surgical sutures have received approval for clinical application and good control of wound infection is attained; their dermal toxicity is still a matter of scientific dispute and worry. Ryman-Rasmussen et al. (2006) confirmed that quantum dots with various physicochemical properties could go through the intact stratum corneum barrier and get localized within the epidermal and dermal layers. Nano-titanium dioxide (TiO_2) is one of the most frequently used materials being synthesized and due to the extensive application of TiO_2 nanoparticles and their inclusion in many commercial products, the increased exposure of human beings to nanoparticles is possible. These TiO_2 particles could get through the human stratum corneum and get to epidermis and even dermis (Shakeel et al. 2016; Lademann et al. 1999). Rouse et al. (2007) demonstrated that fullerene-based peptides were capable of penetrating intact skin and mechanical stressors could make easy their traversal into the dermis. Monteiro-Riviere et al. (2005) proved that epidermal keratinocytes are able to phagocytising an array of engineered nanoparticles and setting off inflammatory responses. Intradermally taken quantum dots could penetrate subcutaneous lymphatics and regional lymph nodes (Gopee et al. 2007). It has been proved that engineered nanoparticles like quantum dots, single or multi-wall carbon nanotubes with nanoscale titania and surface coating have lethal effects on fibroblasts and epidermal keratinocytes and are competent of altering their gene or protein expression (Haliullin et al. 2015).

3.1.2 Respiratory Tract

The respiratory system provides a major doorway for ambient particulate materials and the diseases resulting from airborne particle materials like asbestos, quartz, and carbon have long been keenly investigated in clinical and environmental medicine (Pelclova et al. 2016). In recent times, the pathogenic effects and pathology of inhaled engineered nanoparticles are seized the attention of researchers. Nanoparticles deposition in the pulmonary system differs considerably according to the granulometry of nanoparticles and their airborne behaviour (Ma et al. 2015). Particle granulometry has a foremost impact on the pulmonary deposition site (Witschger and Fabries 2005; Ma et al. 2015). In many nanoparticle production processes, the granulometry can also vary significantly according to the stage of production. These differences in nanoparticle distribution in the lungs may have major consequences on the health effects of inhaled nanoparticle particles and the elimination mechanisms involved (Zhang et al. 2005; Ma et al. 2015). In the alveolar region, the macrophages will take up the insoluble particles by phagocytosis. However, the effectiveness of phagocytosis is badly dependent on nanoparticle shape and size. Several studies have proved that unagglomerated polyethylene nanoparticles of general nano scale size deposited in the alveolar region are not phagocyted efficiently by the macrophages (Keller et al. 2014). However, the macrophages are especially efficient for coarser particles in the one to three micrometre range. After the deposition of nanoparticles in the alveolar region they are absorbed across the lung epithelium, later they enter the blood and lymph to reach cells in the lymph nodes bone marrow, spleen and heart (Nurkiewicz et al. 2006). Further targets after translocation include the sensory nerve endings surrounded in the airway epithelia, followed by ganglia and the central nervous system by means of axons (Simkó and Mattsson 2014). Soto et al. (2007) demonstrated that aggregated silver nanoparticles and some other nanomaterials are cytotoxic to alveolar macrophage cells as well as epithelial lung cells. It was proved by both inhalation and instillation experiments that nanoparticles were taken up by alveolar macrophages to some extent and aggregated silver particles persevere there for up to 7 days. Results of in vitro studies indicate a dose-dependent programmed cell death included by oxidative stress as main possible pathway of toxicity by silver nanoparticles (Völker et al. 2013). In addition, silver nanoparticles may influence cellular enzymes by interference with free thiol groups and mimicry of endogenous ions. Cena et al. (2015) conducted a field study to estimate the amount of Cr, Mn, and Ni deposited in the respiratory system of welders in two amenities. Even though each worker wore a nanoparticle respiratory deposition (NRD) sampler during gas metal arc welding (GMAW) of mild and stainless steel and flux-cored arc welding (FCAW) of mild steel and several welders also wore side-by-side NRD samplers and closed-face filter cassettes for total particulate samples. The NRD sampler estimates the aerosol's nano-fraction deposited in the respiratory system. The obtained results proved that most of the Cr and more than half of the Ni and Mn in the fumes were in the fraction smaller than 300 nm.

3.1.3 Gastrointestinal Tract

Gastrointestinal tract acts as an organ system responsible for consuming and digesting foodstuffs, absorbing nutrients, as well as expelling waste. Nanomaterials can reach the gastrointestinal tract by ingesting directly in water, food, cosmetics, drugs and drug delivery devices or after mucociliary clearance from the respiratory tract via nasal region (Mann et al. 2012; Som et al. 2011). In addition, increased utilisation of nanoparticles may lead to increased environmental contamination and unintentional ingestion via water, food animals, or fish (Bergin and Witzmann 2013). It is measured that the exogenous sources of ingestion exposure mainly results from hand to mouth contact in the workplace. On the other hand, NPS can be ingested directly via food, water, drinking, drugs and drug delivery systems. Besides, NPs cleared from respiratory tract via the mucociliary escalator can then be ingested into the gastrointestinal tract. Thus, gastrointestinal tract is considered as vital target for nanoparticles exposure (Liu et al. 2013). Assessment of nanoparticles must taken into deliberation not only for absorption and extra-intestinal organ accumulation but also the potential for altered gut microbes and the effects of this perturbation on the host (Bergin and Witzmann 2013). Chen et al. (2006) reported the acute toxicity of copper particles (bulk) and nanocopper in mice and discovered that nanocopper was several folds lethal than bulk copper. Nanocopper was also reported to source pathological damage to kidney, liver and spleen. Russell-Jones (2000) proposed the usage of biodegradable nanoparticles in the delivery of oral vaccines for antigens known to be susceptible to proteolysis. In fact, studies on toxicity of nanomaterials after oral ingestion are very limited and very little work has been done till now on gastrointestinal tract exposure.

Gastrointestinal tract is a highly complex environment and understanding of the fate of ingested NPs requires consideration of multiple factors. Many features have been identified that are significant to interpretation of nanoparticles ingestion studies and these studies include physicochemical characterisation of nanoparticles and reporting of metadata from in vivo studies. Moreover, a characteristic somewhat exceptional to the ingestion route is the possible for toxic effects related to interactions with the gut microbiome. In the end, although doses higher than distinctive exposures are a standard and essential part of toxicity studies for establishing dose range parameters, these doses should be reasonably based and critically compared to probable exposure levels (Bergin and Witzmann 2013).

3.2 *Effects of Carbon Nano Tubes on Human Health*

Carbon nanotubes are a novel form of carbon molecule and are an important new class of technological materials that have numerous novel and useful properties (Donaldson et al. 2006, 2010). Surrounded by hexagonal complex of carbon atoms, these hollow cylinders can have diameter around 0.7 nm in size and reach some millimeters in length. They are single layered or multilayered and are chemically and thermally very stable. Their manufacturing normally involves the presence of metals, the ultimate content of which in the product will depend on the product's conditions of manufacturing and subsequent purification (Awasthi et al. 2013). The composite materials containing CNTs may have incredible strength, potentially sufficient to allow the building of spacecraft structures, space elevators, artificial muscles, combat jackets, membranes for gas separation and land and sea vehicles (Kumar et al. 2010). Wang et al. (2004) reported that Hydroxylated single-walled carbon nanotubes administered by gavage in mice are spread to most of the organs and tissues, apart from the brain. Pantarotto et al. (2004) investigated the intracellular transfer of functionalized single-walled carbon nanotubes (SWCNT), i.e., conjugated with lysine, on human and mouse fibroblasts in vitro. They reported that these carbon nanotubes could pass through the cellular membrane and gather up in the cell and end up in the cell nucleus. In an investigation by Monteiro-Riviere et al. (2005) multi-walled carbon nanotubes (MWCNT) were found in the cytoplasmic vacuoles of human epidermal keratocytes in vitro, leading a decline in cell viability and a considerable increase in an inflammation marker (interleukin-8). This reveals the capability of MWCNT to penetrate the cell membrane. Cui et al. (2005) confirmed that SWCNT can hold back cell proliferation, initiate apoptosis and reduce adherence of human embryonic kidney cells in vitro. Shvedova et al. (2003) proved through their investigation that the exposure to unrefined SWCNT can direct to an increase in cutaneous toxicity in exposed workers. In addition, CNTs are also reported to induce human fibroblast (Tian et al. (2006) and T-lymphocyte apoptosis (Bottini et al. 2006) and show cytotoxicity in alveolar macrophages as well (Jia et al. 2005). In differing to this Liu et al. (2009) reported absence of any obvious toxicity of properly functionalized and purified

CNTs. Engineered nanomaterials causes stress in the cellular environment which disturb the oxidative balance resulting in high concentrations of intracellular reactive oxygen species (ROS). Despite the potential impacts of carbon nanotubes (CNTs), on human health and the environment, they have been receiving more and more attention in the recent past, existing information on the toxicity of CNTs.

3.3 Effects of Inorganic Nanoparticles on Human Health

Among the various nanomaterials, inorganic nanoparticles are very significant in current technologies. Because they can be easily and cheaply synthesized and mass produced. In addition, they are more readily integrated into applications (Auffan et al. 2009). Insoluble inorganic nanoparticles can be made up of pure metals or a variety of inorganic products or alloys. Their nanometric size differentiates them from the same products usually found on a larger scale (Tee et al. 2015). These inorganic nanoparticles exhibit electrical, mechanical and other properties that do not survive when in larger dimensions (Sharma et al. 2015). Drug sensitization with various inorganic nanoparticles (NPs) has established to be a capable and a developing strategy in the field of nanomedicine. For instance, Rose Bengal (RB), a notable photosensitizer, triggers the formation of ROS under green light irradiation and as a result it induces cytotoxicity and cell death (Chaudhuri et al. 2015). Acticoat™ consists of a polyester or nylon mesh, trapping polyethylene and includes a silver nanocrystal coating. This product has been used for several years to speed up healing of injuries and decrease bacterial colonization. In the occurrence of moisture, the product liberates ions and silver radicals that would be accountable for antibacterial action. In an in vitro study of cultured human keratinocytes, Lam et al. (2004) observed a significant decrease in cell viability and conclude cytotoxicity of silver nanoparticles released by Acticoat™. Braydich-Stolle et al. (2009), demonstrated that 100 % anatase nanoparticles, in spite of of size, induce cell necrosis and membrane leakage, but they do not generate ROS. In contrast, the rutile nanoparticles initiate apoptosis through formation of ROS. Therefore, it seems that links between size and crystal structure may have a role in mediating nanoparticle toxicity. Lucarelli et al. (2004) conducted in vitro studies and reported that SiO_2 and cobalt (Co) nanoparticles displayed significant proinflammatory action for the activity of human marrow monocytes. Yao et al. (2015) determined the influence of particle size and concentration of gold nanoparticles (AuNPs) on their absorption, accumulation, and cytotoxicity in model intestinal epithelial cells. As the mean particle diameter of the AuNPs decreased (from 100 to 50 to 15 nm), their rate of absorption by the intestinal epithelium cells increased, but their cellular accumulation in the epithelial cells decreased. Moreover, accumulation of AuNPs caused cytotoxicity in the intestinal epithelial cells, which was evidenced by depolarization of mitochondria membranes and the results provide an important insights into the relationship between the dimensions of AuNPs and their gastrointestinal uptake and potential cytotoxicity (Yao et al. 2015).

Platinum drugs as anti-cancer therapeutics are held in particularly high in consideration. Regardless of their success, there are drawbacks associated with the use of platinum drugs. Their dose-limiting toxicity, their limited activity against an array of common cancers and patient resistance to Pt-based therapeutic regimes are some examples (Parker et al. 2016). Peters et al. (2004) examined the viability and behaviour of human endothelial cells in vivo and observed that PVC, TiO_2, SiO_2 and Co nanoparticles were incorporated into the cell vacuoles. Inorganic elements are an undeniable part of living organisms; therefore, the synthesis, stability, and toxicity of engineered metal nanoparticles (ENPs) have been broadly studied during the past two decades. But research on the formation, fate, and ecological effects of naturally-occurring nanoparticles (NNPs) has become a focus of attention only recently (Sharma et al. 2015).

Organic nanoparticles can be normally described as solid particles composed of organic compounds mainly lipids or polymeric compounds (Lambert et al. 2014). Over the past decades, this type of nanoparticles has met an immense development and intensive investigations due to their high potentialities in a wide spectrum of industrial areas ranging from electronic to photonic, conducting materials to sensors, medicine to biotechnology, and so forth (Grimsdale et al. 2009; Niu et al. 2009; Chen et al. 2015). In addition, organic nanoparticles can also be materials connecting with entrapment, encapsulation or surface adsorption of active biological substances (Kashi et al. 2012). The results are promising regarding the therapeutic aspects organic nanoparticles, mostly for polymeric nanoparticles, but the information is insufficient relating to their toxicity in the work environment because of the unusual route used and the limited information concerning their toxicological assessment. As per the literature available, very limited and incomplete data is present on human health hazards related organic nanoparticles. But when it comes to the animal studies there is sufficient data to prove the adverse effects of organic nanoparticles (Liu et al. 2014; Lehto et al. 2014; Kim et al. 2006; Lockman et al. 2004; Alvarez-Román et al. 2004).

3.4 Effects of Quantum Dots on Human Health

Quantum dots (QDs) are very tiny particules on the order of a nanometer in size and are composed of a hundred to a thousand atoms. They are a type of fluorescent label with applications in biological molecules, cells and in vivo imaging (Yuan et al. 2015). Although the applications of these QDs are rising, there are concerns about their potential ill effects to human health. But, the toxicity connected with these nanoparticles and the underlying mechanisms have not been systematically assessed (Nguyen et al. 2015). Kirchner et al. (2005) uncovered the cytotoxicity of ZnS and CdSe nanocrystal solutions for human fibroblasts and tumour cells. Cytotoxicity was higher if the nanocrystals were coated with mercaptopropionic acid, an unstable coating. Phosphosilicate, PEG-silica and polymer-coated inert gold nanoparticles also have a cytotoxic effect (Gambardella et al. 2015). The investigators concluded that the toxic effect may be associated to the straight effect of precipitated particles on the cells. Shiohara et al. (2004) demonstrated the in vitro cytotoxicity of ZnS and CdSe quantum dots layered with sheep serum albumin and mercaptoundecanoic acid. Human hepatocytes, primate kidney cells and cervical cancer cells were also exposed to these quantum dots and noticed a decrease in the viability of the cell lines. In a study, double-stranded DNA was incubated in a cadmium selenide solution encapsulated in zinc sulphite functionalized with surface biotin. Ultraviolet (UV) radiation was also applied. The outcome of this study showed that the quantum dots altered the DNA by releasing SO_2 free radicals, resulting from ZnS oxidation. The proportion of DNA alterations varied according to the application of UV (Green and Howman 2005). Lovric et al. (2005) recommended that the location of QDs could be relevant to the toxic and adverse cellular responses linked with these highly fluorescent NPs. In spite of this, an understanding of the actual cellular process of the uptake and sub-cellular localisation and distribution of QDs is limited. In a study by Nguyen et al. (2015) CdTe-QD caused disruption of mitochondrial membrane potential, increased intracellular calcium levels, impaired cellular respiration, and decreased adenosine triphosphate synthesis. Even with the apparent benefits proposed by QDs, sensitive concerns have been lifted as to their exposure to humans and what impact they might have upon human health.

3.5 *Effects of Nanospheres, Nanoshells and Nanocapsules on Human Health*

A wide variety of insoluble organic polymers are employed in the production of nanocapsules, nanospheres and nanoshells. Some of these structures are manufactured to allow integration with other substances, often medications. The exterior of these nanoparticles can also be altered to interact purposely with certain sites of the body. Because of their nanometric size, these nanoparticles can circulate in a living organism, serve as a drug vector or fix to specific cells. In a study by Favi et al. (2015) rod-shaped silver nanorods (AgNRs) and gold nanorods (AuNRs) were fabricated by electron beam physical vapor deposition (EBPVD), and their cytotoxicity toward human skin fibroblasts were assessed and the results showed the maximum toxicity with fibroblast cells by both types of nanorods. Torres-Lugo et al. (2002) showed the in vitro cytotoxicity and a reversible alteration of the electrical resistance of the epithelial cells by hydrogel nanospheres. Cahouet et al. (2002) intravenously introduced nanocapsules with a lipid core and a shell composed of 2-hydroxy-polyethylene glycol (PEG) stearate and lecithin in rats. The nanocapsules were stained with technetium-99 and iodine-125. The authors observed a longer persistence of the nanocapsules in the blood compartment than expected. The nanocapsules were dispersed in the animals' stomach liver, intestines, and penis, but there was no considerable cerebral distribution. Like this, cytotoxic effects of nanospheres, nanoshells and nanocapsules on cell membrane, mitochondrial function, prooxidant/antioxidant status, enzyme leakage, DNA, and other biochemical endpoints were elucidated by many researchers (El-Ansary et al. 2015; Perez et al. 2015; Jeannet et al. 2015; Liu et al. 2013). Researchers highlight the need for caution during the use and disposal of such manufactured nanomaterials to prevent unintended environmental impacts.

3.6 Effects of Fullerenes on Human Health

Fullerenes are carbon allotropes similar in structure to graphene but rolled up to form closed-cage, hollow spheres. The use of fullerenes have expanded significant consideration due to their anti-oxidant and radical scavenging characters and their current applications comprise targeted drug delivery, energy application, polymer adaptations and cosmetic products. The fabrication of fullerenes and their use in consumer products is expected to increase in future (Aschberger et al. 2010). Totsuka et al. (2009) manufactured nano/microparticles such as fullerenes (C60), carbon black (CB) and ceramic fiber that are being widely used because of their desirable properties in industrial, medical and cosmetic fields. However, these manufactured products, CB, C60 and kaolin, were shown to be genotoxic in in vitro assay systems. Functionalization of fullerenes with hydroxyl groups (fullerenols) can boost the solubility and potential for cellular interaction, but the health and safety effects of varying degrees of fullerene hydroxylation in biological systems is inadequately understood. Existing reports regarding the toxicity and inflammatory potential of fullerenols give contradictory conclusions. To further clarify the potential for toxicity of fullerenols, human epidermal keratinocytes (HEK) were exposed to fullerenols (low ($C_{60}(OH)_{20}$), medium ($C_{60}(OH)_{24}$), and high ($C_{60}(OH)_{32}$) at concentrations ranging from 0.000544 to 42.5 μg/mL for 24 and 48 h. The findings suggest that extrapolation across similar NP will be dependent upon surface chemistry and concentration which may affect the degree of agglomeration and thus leads to biological effects (Saathoff et al. 2011).

Santos et al. (2014) observed the balance between hydrophilicity and hydrophobicity resulting from the surface chemistry of fullerene nanoparticles, rather than the cluster size or the surface charge acquired by fullerenes in water, influences their membrane interactions and consequently their effects on mitochondrial bioenergetics. Sayes et al. (2004) demonstrated the cytotoxicity (CL50) of four water-soluble fullerenes on human cells like hepatic caricinoma cells and fibroblasts in vitro. They confirmed that toxicity differ with the nature of the functional group. Some studies report risks of oxidative damage and toxicity, particularly when cells were exposed to the photoexcited form of fullerene. The present knowledge of the lethal effects of fullerene nanoparticles is relatively inadequate. Toxic effects have already been recognized at the pulmonary, reproductive, cardiac, renal, cutaneous and cellular levels (Injac et al. 2013). The acknowledged toxic effects on humans and the physicochemical characteristics of fullerene nanoparticles give good evidence that through nanoparticles there is risk of occupational disease in researchers and students who develop these products and in workers who manufacture, transform or use nanoparticles.

4 Nanoecotoxicology-Potential Biological Effects

Little is identified about the fate and behaviour of synthetic nanomaterials in the environment, and appropriate methods to detect them in complex environmental media are only in the progress stage. Nanoecotoxicology is a sub-discipline of ecotoxicology and particularly aims to identify and predict effects drawn by nano sized materials on ecosystems. To attain this objective, nanoecotoxicology needs to take into consideration the entry routes and fate of nanomaterials in the abiotic and biotic environment to define exposure (Oughton et al. 2008; Kahru and Ivask 2013). Furthermore needs to recognize those interactions of nanomaterials with biota that alter the proper function of cells comprising an organism, thus impacting populations, which in turn can lead to changes in community structure and function. Collectively, this information can be used to evaluate the risk that nanomaterials in a given environment (Stone et al. 2010). Figure 4 gives a detailed sorting of

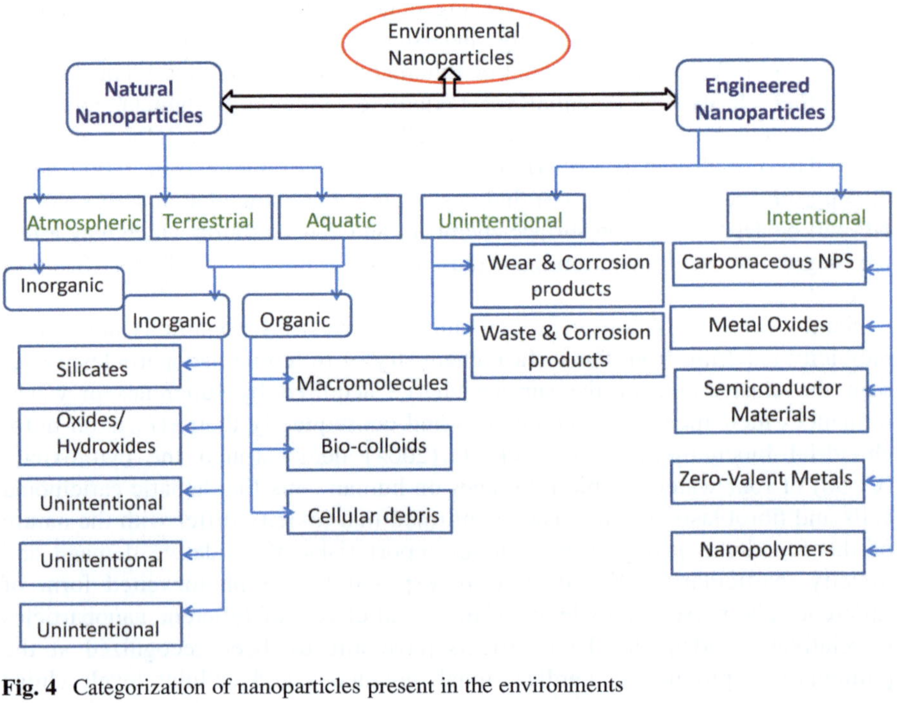

Fig. 4 Categorization of nanoparticles present in the environments

nanoparticles existing in the environments. Despite of a rising considerate that synthetic NPs should be evaluated for their probable environmental hazard prior their use in products and following inevitable release into the environment. There are currently few data on the toxicity of nanomaterials to environmentally relevant species, limiting the quantitative risk assessment of NPs (Kahru and Dubourguier 2010). As manufactured nanomaterials are used in many commercially available consumer products, production of nanomaterials is increasing and the environmental exposure to these materials is evident. It is clear that the physicochemical properties as well as the structure and morphology of nanomaterials have a high influence on toxicity (Sigg et al. 2014). Therefore, environmental risk assessments of engineered nanoparticles require thorough characterization of nanoparticles and their aggregates. In addition, quantitative analytical methods are required to assess environmental concentrations and enable both effect and exposure assessments. Many methods still need optimization and development, especially for new types of nanoparticles, but extensive experience can be gained from the fields of environmental chemistry of natural nanomaterials and from fundamental colloid chemistry (Hassellöv et al. 2008).

Based on the ecotoxicological data previously available and mitigating factors, hazard assessment is conducted. Further, chronic testing is compulsory to improve the risk assessment process if the substance is classified as Persistent, Bioaccummulative, Toxic (PBT). The greater part of present studies of the toxicity of nanomaterials have been carried out on a limited number of nanomaterials and aquatic species, generally at high concentration and over short exposure time. Fullerenol $C_{60}(OH)_{24}$ and Nanoclusters of C_{60} have been shown to create reactive oxygen species in water under polychromatic light and UV (Pickering and Wiesner 2005). Literature on potential toxicity of nanomaterials on terrestrial species like plants, micro-organisms and soil invertebrates is growing rapidly (Handy et al. 2008). Nanomaterials have the ability to unite substantial fractions of pollutants such as trace metals and organics. In the environment, contaminants adsorb to natural solid phases like nanoparticles. Nanomaterials have the capability to combine with substantial fractions of contaminants such as organics and trace metals (Behra and Krug 2008) Carbon nanotubes and Zero-valent iron oxides have been applied for the remediation of organic pollutants and used for the immobilization of organic compounds and metals. Nanoparticles in altered form may be more bioavailable and may be taken up through cell membranes more easily (Sigg et al. 2014; Rickerby and Morrison 2007).

Some of the biological effects of nanomaterials dispersed in the environment including the ecological and evolutionary effects of nanomaterials on terrestrial and aquatic ecosystems such as: species interactions, circulation of nanomaterials and their by-products within ecosystems, factors that influence bioaccumulation and biomagnification of nanomaterials in food network, biotic processes that influence the persistence and chemical transformations of nanomaterials in the environment, and the way and existence of effects on ecosystems (Rickerby and Morrison 2007; Kahru and Dubourguier 2010) Therefore, from the environment nanomaterials have the ability to interact with metabolic networks, cellular constituents and living

Table 3 In vivo observed effects induced by engineered nanoparticles that are supported by in vitro evidences

In vivo observed effects supported by in vitro evidences	
In vitro evidence	*In vivo observed evidences*
Increased cytotoxicity on exposed cell cultures	Chronic obstructive pulmonary disease (COPD)
Proliferative stimuli induced by extracts of DEP components	Hyperplasia
Macrophage-dendritic transepithelial cells network alterations in the Gap Junction Intercellular Communication (GJIC)	Particle translocation
Endothelial cell activation by direct contact with particles or indirectly induced in co-cultures where pneumocytes, macrophages, and other cell are exposed. Changes in the TEER values related to tight junctions	Systemic and endothelial dysfunction
ROS increases via NADPH-oxidase in lung epithelial cell exposed to PM	Oxidative stress
Secretion of IL-1b, IL-6, IL-8, TNFa, MCP-1, and so forth, by lung cells, macrophages, and cocultures	Local and systemic inflammation

tissues including interactions at the organ, cellular molecular and systemic levels. Engineered nanoparticles can effects on organism ontogeny and multi-generational life histories (Lv et al. 2015). In vivo observed effects induced by engineered nanoparticles are summarized in Table 3 and the multiple scenarios through which nanoparticles enter into environment and humans are exemplified in Fig. 5.

4.1 Environmental Issues

Nanotechnology is a transformative technology revolutionizing many areas including energy, security, information technology, agriculture, environmental protection, and healthcare (Gao et al. 2015b). Scientific activities related to the development of nanomaterials have been remarkable and the number of peer-reviewed papers related to the topic has shown an exponential growth over the last decade. Currently there are more than 60 countries that have already launched national nanotechnology programs. But definitely the victory or failure of nanotechnology may well depend on the skill to address environmental issues. Responsible Research and Innovation provides a framework for judging the ethical qualities of innovation processes, however guidance for researchers on how to implement such practices is limited. The propose of any research should be anticipatory, exploring in advance and expecting potential technological impacts; reflective, by examining objectives and purposes of technologies as well as the uncertainties in risk assessment; deliberative, the idea that public and diverse stakeholders' perspectives are actively considered during design processes and, lastly; responsive, the actual alteration and shaping of technological trajectories in response to deliberation and reflection (Spruit et al. 2015). Many researchers are

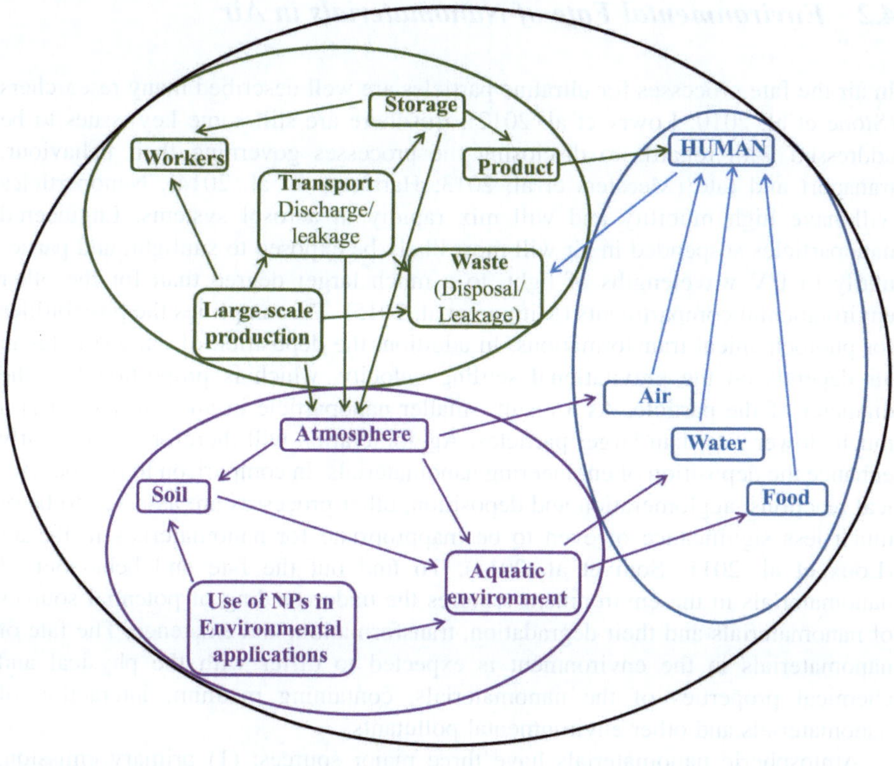

Fig. 5 Multiple scenarios through which nanoparticles enter into environment and humans

working hard to provide answers to many key environmental issues, including the following:

- How far industry and society can expect hazardous material to be delivered into the environment during either the production or use of nanoproducts?
- What are the possible environmental problems linked with this nanotechnology?
- Could nanoapplications show the way to environmental degradation, predominantly from bioaccumulation of nanoproducts in living tissue?
- What effects will regulations have on this nanotechnology?

Investigations on nanoparticles influence on ecology shows a certain hazard potential of some nanomaterials. Even though scientific uncertainties still exist, the precautionary principle should be applied in the sense of preventive risk minimization. Environmental inputs should be avoided to the extent possible. Ecotoxicological research should be progressively more focus on the environmental relevance of the materials and consider the complexity of natural systems. As the result of their minute size and elevated specific surface areas, chemical reactivities sensitive to exposed surface sites will likely be magnified with these products (Loux et al. 2011).

4.2 Environmental Fate of Nanomaterials in Air

In air the fate processes for ultrafine particles are well described many researchers (Stone et al. 2010; Lowry et al. 2012). But there are still some key issues to be addressed with regards to disclosing the processes governing their behaviour, transport and fate (Meesters et al. 2013; Hartmann et al. 2014). Nanoparticles will have high mobility and will mix rapidly in aerosol systems. Engineered nanoparticles suspended in air will most likely be exposed to sunlight, and particularly to UV wavelengths of light, to a much larger degree than for the other environmental compartments (Mitrano et al. 2015). This amplifies the possibilities for photochemical transformations. In addition, the deposition of nanoparticles in air depends on the gravitational settling velocity, which is proportional to the diameter of the particle. As a result, smaller nanoparticle in air will deposit at a much slower rate than larger particles. Agglomeration will therefore significantly enhance the deposition of engineering nanomaterials. In comparison to photochemical reactions, agglomeration and deposition, other processes are assessed to be of much less significance or even to be inappropriate for nanomaterials in the air (Loux et al. 2011; Soni et al. 2015). To find out the fate and behaviour of nanomaterials in the environment requires the understanding of potential sources of nanomaterials and their degradation, transformation, and existence. The fate of nanomaterials in the environment is expected to differ with the physical and chemical properties of the nanomaterials, containing medium, interaction of nanomaterials and other environmental pollutants.

Atmospheric nanomaterials have three major sources: (1) primary emission, refers to those that are openly released from road traffic exhaust and industrial combustion; (2) secondary emission, refers to those that are produced in the atmosphere from the compression of low volatility vapours produced from the oxidation of atmospheric gases; and (3) formation at the time of diesel exhaust dilution (Baalousha and Lead 2009). The lack of data about engineered nanomaterials in the atmosphere is due to the nonexistence of methods capable of discriminating engineered nanomaterials from the background concentration from other sources which is alike to the condition in aquatic and terrestrial environments (Baalousha and Lead 2009). From the literature it is known that fine, ultrafine particles, and nanomaterials can go through several processes in the atmosphere (Tiwari and Marr 2010; Gouin et al. 2011). Some nanomaterials can produce by condensation of low volatility compounds or reduce in size by evaporation of adsorbed water or other volatiles, resulting in the deviation in particle size distribution but not the overall numeral concentration. Atmospheric nanomaterials can combine resulting in an increase in particle dimension with a decrease in the numeral concentration (Gidhagen et al. 2004). Sometimes nanoparticles can also be missing from the atmosphere by dry and wet deposition, both of which are capable for extremely small particles of natural origin and so presumably also for engineered nanomaterials. This results in a decline in particle numeral concentration and a change in particle size distribution to larger sizes (Clarke et al. 2004).

4.3 Environmental Fate of Nanomaterials in Water

The fate of nanomaterials in the aquatic environment can be influenced by different processes, such as aggregation and disaggregation, diffusion, interaction between nanoparticles and natural water components, transformation, biotic and abiotic degradation and photoreaction (Vale et al. 2015). The fate and behaviour of engineered nanomaterials released into the aquatic environment can be understood with the help of existing literature on the fate and behaviour of natural colloidal particles. Currently, nanomaterials are extensively recommended for wastewater treatments due to their exceptional properties. Some studies report the various advantages of nanotechnology in the remediation of wastewaters, but inadequate research has been directed toward the fate and potential impacts of the solid residues produced after the application of such technologies (Nogueira et al. 2015). In a study, researchers examined the aggregation and sedimentation kinetics behaviors of citrate- (Cit-AgNPs) and polyvinylpyrrolidone-coated silver nanoparticles (PVP-AgNPs) in calcium chloride ($CaCl_2$) solutions, emphasizing the effects of particle size and type of coating material on both behaviors. As the ionic strength increased, Cit-AgNPs aggregated rapidly and settled down, while PVP-AgNPs did not aggregate, even at an ionic strength of 10 mM $CaCl_2$, due to likely steric hindrance effects of PVP coating. Interestingly, PVP-AgNPs sedimented without aggregating within week and this inclination seems to have relevance to the particle size. These results suggest that the particle size and type of coating material play important roles in determining nanoparticle fate in water (Jang et al. 2014). In addition, nanomaterials released into the water bodies may interact with aquatic organisms and induce adverse effects at different levels of biological organization. These potential ecotoxicological risks of ENMs to aquatic organisms have recently been reviewed (Rocha et al. 2015; Grillo et al. 2015; Baker et al. 2014; Corsi et al. 2014; Ma and Lin 2013; Matranga and Corsi 2012) but their mode of action and biological risk remain unclear.

Therefore, in aquatic environment nanoparticles may interact with natural organic matter, natural colloids and suspended particulate matter, resulting in aggregation and potentially sedimentation from solution. Sedimentation and aggregation may represent a pathway for the carrying of nanoparticles from the water column to benthic sediments. The nanoparticles in aquatic environment are bioaccumulated by deposit and filter feeding organisms. Research on such interactions have not been fully studied till date due to the fact that robust and sensitive analytical methods are not yet available for detecting and characterizing nanoparticles in complex environmental matrices such as natural waters and soils (Tiede et al. 2015) but may considerably affect nanoparticle fate and toxicity.

4.4 Environmental Fate of Nanomaterials in Soil

Soil is the matrix of a multilayer food web structure and it is a complex interface between gases-solid-water-organic/inorganic matters and organisms. Nanomaterials are small enough to pass through soil pores (Mukhopadhyay 2014). They can adhere to soil particles due to their high surface area and become immobilized (Michael et al. 2008). Large aggregates of nanomaterials can be immobilized by sedimentation, filtration, or straining in smaller pores (Mukhopadhyay 2014). At present few reports are available on the transport and fate of nanomaterials in the natural porous environment. It is reported that transport is moderately fast and influenced by the type of nanomaterials (Li et al. 2006; Boxall et al. 2007). While toxicity mechanisms have not yet been totally clarified for most nanoparticles, probable mechanisms include disruption of membranes or membrane potential, oxidation of proteins, genotoxicity, interruption of energy transduction, formation of reactive oxygen species, and release of toxic constituents (Zharov et al. 2006). These toxicity mechanisms may result from various factors such as: high surface area to volume ratio, surface charge, hydrophobic and lipophilic groups may permit them to interrelate with proteins and membranes, complementary effects of nanostructures which cause inhibition of enzyme activity, bioaccumulation and chemical composition which amplify their reactivity (Jafar and Hamzeh 2013). Polymers and surfactants enhance the transport of nanoparticles. Many researchers are investigating the role of natural organic matter in nanoparticle facilitated transport. The properties of the soil matrix may influence the diffusion and mobility of nanoparticles. The mobility of nanoparticles in soils depends on nanoparticle's physical–chemical properties; characteristics of the soil and environment; and the interaction of nanoparticles with natural colloidal material. List of some existing ENPs and their health and environmental effects are summarized in Table 4.

5 Global Strategies Designed to Address Human Health and/or Environmental Safety Aspects of Nanomaterials

The concern over the possible adverse influences of nanomaterials on living systems has given rise to 'nanotoxicology' (Dhawan et al. 2009; Schulte et al. 2016). However, nanotoxicology has lagged far behind nanotechnology due to a number of experimental challenges and problems faced in designing studies involving toxicological assessment of nanomaterials. Globally, many organizations initiated research programmes or strategies designed to address human health and/or environmental safety aspects of nanomaterials (Dobrovolskaia et al. 2016). A summary of global organizations and their objectives were illustrated in Table 4. The Organisation for Economic Co-operation and Development (OECD) is an intergovernmental organisation in which representatives of 34 industrialised nations in North and South America, Europe and the Asia and Pacific region, as

Table 4 List of some existing ENPs and their health and environmental effects

Nanoparticle	Environmental effects	Health effects
Carbon nanotubes	Cause indirect effects upon contact with the surface of the environmental organisms; environmental damage	Apoptosis; decreased cell viability; lung toxicity; oxidative stress; retarded cell growth; skin irritation etc.
Fullerenes	Effects on soil organisms and enzymes; aquatic ecosystems; binding of chemicals to fullerenes may affect the toxicity of other environmental contaminants	Retarded cell growth; decreased cell viability; oxidative stress and apoptosis etc.
Heterogeneous nanostructures	Toxicity depends on multiple physicochemical as well as environmental factors; adverse influence of ecosystem etc.	Arrest of cell growth and sometimes even cell death; chromatin condensation; free radical formation
Nanosilver	Undergoes several transformations when it is released into the environment and shows adverse effects	Alterations of the non-specific immune responses; altered cell signalling; apoptosis; necrosis of cells; oxidative stress etc.
Nanostructured flame retardants	Persistent and tend to accumulate in the environment; toxic to plants, wildlife etc.	Oxidative stress; fibrosis; cardiovascular effects; cytotoxicity; carcinogenic etc.
Polymeric nanoparticles	Potential hazardous factor for environmental exposure	Oxidative stress; inflammation; alteration in cellular morphology and functioning etc.
Silicon based nanoparticles	Potential hazardous factor for environmental exposure; adverse influence of ecosystem etc.	Cardiovascular effects; cytotoxicity; increase oxidative stress etc.
TiO_2 nanoparticles	Disrupt an aquatic ecosystem's carbon and nitrogen cycles; stress photosynthetic organisms	Excessive exposure in human may result in increased oxidative stress; retarded cell growth; slight changes in lungs etc.

well as the European Commission (EC), meet to coordinate and harmonise policies, discuss issues of mutual concern, and work together to respond to international problems (Rashidi et al. 2015). One of the objectives of OECD include the valuation of risk assessment approaches for manufactured nanomaterials through information exchange and the identification of opportunities to fortify and improve risk assessment capacity. Promoting policies and modern technologies for sustainable economic growth and employment, increasing standards of living, and trade liberalization is principal aims of the OECD. Drawing on facts and real-life experience, OECD recommend policies designed to improve the quality of people's lives (Rashidi et al. 2015). In addition, the OECD is also working for a stronger, cleaner, and fairer economy.

The National Institute for Occupational Safety and Health (NIOSH) established the Nanotechnology Research Center (NTRC) in 2004 to address the occupational safety and health concerns that might be associated with this engineered nanomaterials (Fischman et al. 2011). It is working in parallel with the development and implementation of commercial nanotechnology through conducting strategic planning and research, partnering with public and private-sector colleagues from the United States and abroad, and, making information widely available. The NIOSH goal is to provide national and world leadership for incorporating research outputs about the implications and applications of nanotechnology into good occupational safety and health practice for the advantage of all nanotechnology workers. NIOSH has developed a strategic plan for coordinating nanotechnology research and for use as a guide for the development of new research efforts (Kuempel et al. 2012; Brenner et al. 2016).

The NanoSafety Cluster, a cluster of projects funded by the EC, recognized the need for a computational infrastructure for toxicological data management of engineered nanomaterials (ENMs) (Jeliazkova et al. 2015). It is an EC programme to maximise the synergies between the past, ongoing and future FP7 nanosafety projects. Each of these projects addresses important aspects of nanosafety, including toxicology, ecotoxicology, exposure assessment, risk assessment, standardisation, and mechanisms of interaction. There are currently almost 40 such projects. In 2014, the NanoSafety Cluster conducted a Key Global NanoSafety Database Survey which is first in a biannual continuous effort and attempt to accumulate, organise and share up-to-date information about NanoSafety related databases worldwide. The resulting list of databases will continuously be updated with innovative valuable insight along its development (Oomen et al. 2014; Guadagnini et al. 2015). The Strategic Research Agenda (SRA) has been developed by members of the European NanoSafety Cluster, a forum for ongoing FP6 and FP7 projects covering all features of nanosafety. The implementation of the SRA is expected to afford a major step forward in the expansion of safe and sustainable nanomaterials (Kermanizadeh et al. 2016). In addition to these organizations, other organizations like Federal Institute for Materials Research and Testing (BAM), Federal Ministry of Education and Research (BMBF), Federal Environment Agency (Umweltbundesamt, UBA), Federal Institute for Risk Assessment (Bundesinstitut für Risikobewertung, BfR), Federal Institute of Occupational Safety and Health (BAuA), Federal Research Institute of Nutrition and Food (Max Rubner-Institut, MRI), MODENA COST etc. are contributing their efforts to address human health and/or environmental safety aspects of nanomaterials through different strategies as summarized in Table 5.

6 Green Nanoscience and Its Assurance

Many research and activist groups have expressed worries and concerns about nanotechnology toxicity to human and environment (Polshettiwar et al. 2012; Hutchison 2008). Conflicting research conclusions have increased concerns about nanomaterial safety. Ambiguous answers regarding nanomaterial safety and toxicity are available which make to think nanotechnology as a boon and or as a curse. There is an improved and intensified attention on addressing the Environmental Health and Safety (EHS) concerns interrelated to nanotechnologies by both

Table 5 Global strategies to address human health and/or environmental safety aspects of nanomaterials

Organization	Objectives
The Organisation for Economic Co-operation and Development (OECD)	• Evaluation of risk assessment approaches for manufactured nanomaterials through information exchange and the identification of opportunities to strengthen and enhance risk assessment capacity
National Institute for Occupational Safety and Health (NIOSH)	• Development and implementation of commercial nanotechnology through conducting strategic planning and research to provide national and world leadership for incorporating research findings about the implications and applications of nanotechnology into good occupational safety and health practice.
EU NanoSafetyCluster	• It works on the projects addressing all aspects of nanosafety including toxicology, ecotoxicology, exposure assessment, mechanisms of interaction, risk assessment and standardisation. • Conducts workshops and seminars to educate people particularly all nanotechnology workers
Federal Institute for Materials Research and Testing (BAM)	• BAM is involved in EU-funded (FP7) projects: – *NanoDefine* (FP7): The aim of NanoDefine is to support the governance challenges associated with the implementation of the nanomaterial legislation by addressing the issues on the availability of suitable measuring techniques, reference material, validated methods, acceptable for all stakeholders and delivering an integrated and interdisciplinary approach. – *NanoValid* (FP7): The main objective of NanoValid is the development of new reference methods and certified reference materials, including methods for characterization, detection/quantification, dispersion control and labeling, as well as hazard identification, exposure and risk assessment of ENs
Federal Ministry of Education and Research (BMBF)	• NanoCare—Safe handling of Manufactured Nanomaterials • Studying the effects of ENM on humans and the environment
Federal Environment Agency (Umweltbundesamt, UBA)	• NanoCare—Safe Handling of Manufactured Nanomaterials • Investigating Impacts on Health and the Environment

(continued)

Table 5 (continued)

Organization	Objectives
Federal Institute for Risk Assessment (Bundesinstitut für Risikobewertung, BfR)	• To demonstrate and establish new principles and ideas based on data from value chain implementation studies • To establish Safe-by-Design as a fundamental pillar in the validation of a novel manufactured nanomaterial • Establishing nanomaterial grouping/ classification strategies according to toxicity and biological effects for supporting risk assessment • Grouping of nanostructured materials for protection of workers, consumers, the environment and risk minimisation
Federal Institute of Occupational Safety and Health (BAuA)	• Grouping of nanostructured materials for protection of workers, consumers, the environment and risk minimization
Federal Research Institute of Nutrition and Food (Max Rubner-Institut, MRI)	• Detection and characterization of nanomaterials in complex matrices such as food etc. • Research on nano-sized carrier systems for bioactive compounds • Interaction of nanomaterials with compounds of the food matrix
Modena Cost	• The specific objectives are to study the synthesis of engineered nanomaterials (ENM) with controlled composition, size, area and nano-texture and • To develop of strategies to immobilise ENM in matrices, on substrates with minimum effect on the desired properties and surface reactivity and identify the relevant datasets for Quantitative Nanostructure-Toxicity Relationships (QNTR) modelling

scientists and policymakers (Hussain et al. 2015; Hutchison 2008). The focus on nanomaterial environmental health and safety depicts the requirement for a research agenda that can minimize nanomaterial hazards. As an ultimate answer to many of the concerns about nanotechnology toxicity and safety, green nanoscience has evolved as an approach to determine and implement the design rules for safer, greener nanomaterials for more efficient processes (Hussain et al. 2015). With green nanoscience we can design and process the materials that can eradicate hazards throughout the material's life cycle. Since green nanoscience offers significant promise as solutions to long-standing environmental, health and technological challenges (Hussain et al. 2015; Hutchison 2008; Nath and Banerjee 2013). Both the government and private industry are gearing up to invest billions of dollars in research and development of green nanoscience. Even though this green nano science is growing slowly, there are principal groups that have enunciated and illustrated the concepts of green nano with recent conferences and symposia highlighting work that is currently being conducted.

Green nanoscience is an integrated approach of green chemistry and nanotechnology that aims to generate and apply design rules proactively for greener nanomaterials and to develop potential synthetic strategies to manufacture nanomaterials reproducibly with definite structure, composition and purity. The following are the fundamental objectives of green nanoscience.

- To develop safer and greener substitute materials that can be used if a nanomaterial is found to be lethal or bioaccumulative in commercial or near-commercial phase.
- Discover the design rules for novel nanomaterial groups that have attractive properties and an elevated level of safety.
- Diminish the hazard and increases the potentiality of nanomaterials production.

Green nanoscience can provide further benefits. It can encourage innovation through the investigation of new materials with greener properties. Green nanoscience facilitates commercialization by reducing insecurity about material safety and providing more potential production approaches (Mckenzie and Hutchison 2004). It protects our investment in nanotechnology from the threats of public or consumer doubts about the unsure risks of the technology. Merging green chemistry and nanoscience will have noteworthy impact on the processes and products of the future. To maturely build up nanotechnology, we must evaluate, through new nanomaterials, appropriate metrics and nano manufacturing processes and relate these findings to direct this upcoming industry. The employment of nanomaterials in green chemistry to meet its goals will definitely offer new opportunities to develop superior safe products and green processes with low impact on the environment and human health. The integration of these two emerging fields provides even greater opportunity for reducing the impact of new technologies on the environment and human health. As asserted by many environmental scientists, if nanotechnology is to be the key to the future, then it should be developed with sustainability.

Green nanoscience employs the familiar twelve principles of green chemistry to the design, production, and application of nanomaterials (Hutchison 2008; Anastas and Warner 1998). Green nanoscience, similar to green chemistry, struggles to trim down or abolish hazards to human health and the environment through safe product design and process optimization. To practice those principles, it is essential to acquire the knowledge like mechanistic understanding, synthetic methods, characterization tools and strategies and bio or eco testing procedures. As a result we can acquire steps on time to find substitute for materials that are not safe to design new materials and to manufacture materials which are reliable and potential (Eckelman et al. 2008). The twelve principles of green chemistry which are applied in green nano science are; avoiding the generation of waste, atom economy, less hazardous chemical synthesis, designing safer materials, safer solvents and auxiliaries, design for energy efficiency, use of renewable feedstock's, eliminate the use of derivatives, catalysis which can improve the selectivity of reactions and reduces the amount of energy necessary to initiate a reaction, design for degradation, real-time analysis for pollution prevention, inherently safer chemistry which involves the process to evaluate the environmental issues associated with a product production. With these principles, green chemistry offers potential benefits in process development and manufacturing as well as product design of green nanoparticles. At present, numerous green nanoparticles with well-defined chemical composition, size, and morphology have been synthesized by different methods and their applications in many progressive technological areas have been explored (Hussain et al. 2015; Vinothkannan et al. 2015; Saini et al. 2015; Mashwani et al. 2015; Emmanuel et al. 2014).

Green chemistry has huge potential in navigating the responsible development of nanotechnology through the design of greener nanomaterials and the discovery of green nano producing methods. It provides potential benefits in process development and production as well as product design (Hutchison 2008). As several preparation processes for the molecular building blocks of nanotechnology engage high energy necessities or hazardous chemicals, the development of safe and greener processes for the manufacture of these materials is crucial. Several examples have been described where the risks of nanoparticle production have been drastically reduced by process redesign (Bazaka et al. 2016). Weare et al. (2000) developed an efficient and less hazardous production process for phosphine-stabilized gold nanoparticles that produces superior amounts of particles, in less time, in milder conditions, while using less hazardous reagents than the conventional preparation. The amino acid asparagine was used as a benign reducing and stabilizing agent for the production of monodisperse gold nanoparticles (AuNPs) using green chemistry principles. With an increasing concentration of amino acid asparagine (0.5–10 mM), the absorbance intensity at 525 nm increased; however, no effects on the color, size, or shape of the AuNPs were observed (Ghodake et al. 2015). The benefits of producing gold nanoparticles in supercritical CO_2 a greener solvent is safer and greener than is traditionally used solvent. Silver nanoparticles and Monodisperse gold have been manufactured through the use of living plants and in microorganisms (Ahmad et al. 2003). The application of micro reactors to manufacture nanoparticles in a rapid, continuous process results in improved energy efficiency, reduced waste, and increased control of product properties. Park et al. (2015) report a novel fluid-based fluorescent CD paints (C-paints) derived from polyethylene glycol and produced in quantum yields of up to ~14 %.

The sustainable synthesis of gold nanoparticles from gold ions was conducted with caffeic acid as a green reducing agent (Seo et al. 2015). In this study, newly synthesized gold nanoparticles exhibited catalytic activity toward the reduction of 4-nitrophenol to 4-aminophenol in the presence of sodium borohydride. The addition of metal oxide or metal into polymers make coatings that may sooner or later replace environmentally-harmful chromate coatings conventionally used for corrosion resistance (Štengl et al. 2016). Nanocomposite polymers based on biopolymers like chitosan and starch, intercalated with nanolayered clay has produced biodegradable polymers with the physical and mechanical characteristics that equal engineering plastics (Yong et al. 2015). In each of the above examples, green chemical strategies have offered opportunities for the improvement of synthetic methods that are more potential, reduce waste, and have enhanced health and environmental impacts (Geetha et al. 2015). Some of the green chemistry approaches in nanoproduction are evaluations of possible environmental and toxicological effects of nanoscale materials before they are included into technologies provide an opportunity to reduce negative consequences and encourage a sustainable nanotechnology industry. The improvement of low-waste methods and high-precision of nanoproduction will be crucial to commercialization.

Nanoscience provides new tools for green chemistry through novel design strategies and new nanostructured materials that are based on improved control of chemical and physical structure (Zhao et al. 2014). Devices with components of nanometer dimension are naturally lighter and faster, use less power, and often have improved performance. The novel chemical, optical, and electronic properties of these materials and the predictable high degree of control over these characteristics offer new scenario for greener materials and safe processes (Reddy et al. 2012). Green nanoscience grants various opportunities for better environmental technology through novel nanostructured materials with applications in remediation, sensing and pollution eradication. Nanoscale sensors based on nanowires or nanoparticles provide more rapid response time, give lower detection limits and make on-site, real-time detection possible. Several different types of inorganic nanoparticles have shown assurance as catalysts for photodegradation of pollutants in air, water, and soil. These nanomaterials are further the aims of green chemistry through superior advancement in evaluation and removal of hazards. Since green nanotechnology applications might involve a clean production process, such as synthesizing nanoparticles with sunlight or the recycling of industrial waste products into nanomaterials, such as turning diesel soot into carbon nanotubes, nanoscience applications in green chemistry can develop environmental remediation processes (Nath and Banerjee 2013). Another common function of green chemistry that is used in nanoscience is for increasing efficiency. Many nanostructured materials with particular preferred properties present more efficient material and energy consumption (Hussain et al. 2015). Catalysis by nanoscale metal complexes carries on to be an active area of research in both green chemistry and nanoscience (Kim et al. 2014). An additional tool that nanoscience offers green chemistry is dematerialization, a waste-reduction approach wherein nanoscale materials are used to replace macroscale counterparts. Considerable reductions in waste have been achieved through the use of molecular monolayers as an alternative of thick polymer films as masks for lithography and through nanoscale deposition of polymer electrolytes for non-lithographic patterning of metal oxide films. Nanoscience along with green chemistry has led to new potential catalysts where the catalyst is maintained in a nanoframework, which increases the, selectivity, catalytic activity and stability. Nanocomposite materials, which possess nano-scale components boost strength without adding weight, supply increased energy efficiency because of their lighter weight (Camargo et al. 2009). Thermoelectric devices made up of green nanoparticles are able to provide energy-efficient cooling or converting waste heat into electricity. Despite the benefits that nanomaterials are already providing in pollution prevention, remediation and well-organized use of resources, possibly the greatest contribution to green chemistry will be the novel production strategies available through nanoscience. Functional group assisted processes, such as self-assembly, can spectacularly decrease waste generation and energy requirements.

The nanomaterials of issue are those that are liberally dispersed or those that may become displaced from a bulk material during the usage of a material that contains embedded nanostructures. Although several hundreds of reports on nanomaterial hazards have been present in the literature, still it remains uncertain which characteristics of nanomaterials contribute to specific hazards (Winkler 2015; Bumbudsanpharoke et al. 2015). To develop the design rules for greener and safe nanomaterials; compositionally, structurally and well-defined nanomaterials are selected to study hypotheses regarding the influence of nano-structure on biological impact (Kumar et al. 2015). Systematic characterization and tests of purity are desirable to ensure that impacts can be interconnected to structural features (Ghodake et al. 2016). Biological testing is carried out to decide impacts and the mechanisms of action for exact endpoints. New hypotheses are proposed that contribute to materials design. Subsequent iterations of this process lead to enhanced understanding of the activity or structure relationships and to "design rules" for greener and safer nanomaterials (Balbus et al. 2007). To design new nanomaterials two approaches would be prudent; first avoid compositions which are toxic in nature; and second, thinking about the hazards of micrometer-scale and molecular materials of the targeted composition and keep away from compositions bracketed by smaller and larger materials with known hazards.

In terms of greener and safer nanomanfacturing, one would like to avoid use of toxic materials and reduce the production of hazardous by-product. Efficiency is also key factor to minimizing impact and improve manufacturability, that is, by increasing material efficiency, throughput, precise product control, and avoiding waste (Kaur et al. 2014). Nanoparticles with characteristics that depend particularly on each of their structural features like size, shape, composition surface coating etc., might be referred as fine nanomaterials (Kuhn et al. 2015). The recent methods of nanoparticle manufacturing of fine nanomaterials are usually depended upon the initial "discovery" routes. They regularly involve higher reactive and lethal reagents and have deprived efficiencies. The hazards and inefficiencies of these routes create significant risks during production and severely restrict production throughput. For these causes, the manufacture of nanoscale materials and devices is field where green chemistry principles can and should be readily applied to direct process improvement and innovation. Several examples are known that demonstrate the application of those principles and show how greener approaches often improve the manufacturability of the material by reducing costs, improving throughput and avoiding waste (Yan et al. 2016; Wang et al. 2015).

Therefore in consistence with universal efforts to minimize the generation of toxic waste and to develop energy-effective production, green chemistry and biochemical processes are increasingly integrating with contemporary developments in science and technology (Nath and Banerjee 2013). Green nanotechnology as a result aims to the application of green chemistry principles in designing nanoscale products, and the development of nanomaterial production methods with reduced toxic waste generation and safer applications. Hence, the definitive possible for green chemistry and nanoscience may lie in an integrated move towards the progress of 'green nanoscience'.

7 Green Nanotechnology-Challenges

As stated earlier, green nanotechnology is primarily focused on the application of the principles of green chemistry and sustainability. The highest levels of acceptance and benefits for society can be achieved if technological development of nanotechnology is attached with the evaluation of societal, environmental and economic impacts, posing the basis for complete sustainability evaluation of different NPs with the same functionality (Iavicoli et al. 2014). The health and safety concerns nanoparticles raised due to their unique physicochemical properties drew initial attention to their sustainability implications (Linkov et al. 2015). In addition to the recent immense progress, there are few challenges specific to green nanotechnology is that the science, the testing, the regulatory strategy, and even the processes required for commercial production are all being developed and deployed at the same time. There are still no clear guiding principles for researchers in preliminary discovery phases of green nanoscience. Many green nanomaterials require new commercial production techniques, which increases the need for basic research, engineering research, and coordination of the two between the industrial and research communities. Toxicology and analysis protocols need to be developed and constantly updated to reflect advances in the science. Regulatory uncertainty persists, and green nanotechnologies often face higher regulatory barriers than existing or conventional chemicals (Cinelli et al. 2016). The end-market demand is unclear, especially since there are only an inadequate number of commercial grade products that can be compared to conventional materials in terms of performance. A lot of research was then conducted to evaluate the impacts of NPs on sustainability, including life cycle assessments, economic and social impact assessments (Dobon et al. 2011) and broad development of criteria sets for a variety of risks, profits and sustainability implications.

8 Summary and Conclusion

Nanotechnology continues to offer new materials and applications that will benefit society. But, recently the stunning developments in nanotechnology have been with issues regarding to their potential effect on human health and the environment. There are no particular regulations on nanoparticles apart from existing regulations covering the similar material in bulk form. Difficulties abound in developing such regulations, further than self-imposed regulations by responsible companies, because of the likelihood of different properties exhibited by any one type of nanoparticle, which are tuneable by varying their size, shape and surface properties. Green nanoscience i.e green chemistry principles incorporated into nanotechnologies can answer many question posed by about whether the products or materials of nanotechnology will present hazards to human health or the environment and whether the production of these materials will generate new hazards. Merging green chemistry and nanoscience will have noteworthy impact on the products and processes of the future. To maturely develop nanotechnology, we must review, through appropriate metrics, new nanomaterials and nano-manufacturing processes and employ these findings to guide this promising industry. The use of nanomaterials to achieve the goals of green chemistry will provide new opportunity to build up superior products and processes with reduced impact on the environment and human health. The combination of these two emerging fields provides even superior opportunities for reducing the impact of new technologies on the environment and society.

Acknowledgments This research was supported by the R&D Program for Society of the National Research Foundation (NRF) funded by the Ministry of Science, ICT & Future Planning (2013M3C8A3078806 and 2015M3A9E2031372).

Competing interests The authors declare that they have no competing interests.

References

Ahmad A, Mukherjee P, Senapati S, Mandal D, Khan MI, Kumar R (2003) Extracellular biosynthesis of silver nanoparticles using the fungus *Fusarium oxysporum*. Colloids Surf 28:313–318

Alfaro-Moreno E, Nawrot TS, Nemmar A (2007) Particulate matter in the environment: pulmonary and cardiovascular effects. Curr Opin Pulm Med 13:98–106

Alvarez-Román R, Naik A, Kalia YN, Guy RH, Fessi H (2004) Skin penetration and distribution of polymeric nanoparticles. J Control Release 99:53–62

Anastas PT, Warner JC (1998) Green chemistry: theory and practice. Oxford University Press, Oxford England, New York

Arts JH, Hadi M, Irfan MA, Keene AM, Kreiling R, Lyon D, Maier M, Michel K, Petry T, Sauer UG, Warheit D, Wiench K, Wohlleben W, Landsiedel R (2015) A decision-making framework for the grouping and testing of nanomaterials (DF4nanoGrouping). Regul Toxicol Pharmacol 71(2 Suppl):S1–S27

Aschberger K, Johnston HJ, Stone V, Aitken RJ, Tran CL, Hankin SM, Peters SA, Christensen FM (2010) Review of fullerene toxicity and exposure—appraisal of a human health risk assessment, based on open literature. Regul Toxicol Pharmacol 58:455–473

Auffan M, Rose J, Bottero JY, Lowry GV, Jolivet JP, Wiesner MR (2009) Towards a definition of inorganic nanoparticles from an environmental, health and safety perspective. Nat Nanotechnol 4:634–641

Awasthi KK, John PJ, Awasthi A, Awasthi K (2013) Multi walled carbon nano tubes induced hepatotoxicity in Swiss albino mice. Micron 44:359–364

Baalousha M, Lead JR (2009) Overview of nanoscience in the environment. In: Emma S, Jamie RL (eds) Environmental and human health impacts of nanotechnology. Wiley-Blackwell Publishing Ltd, Hoboken, NJ, pp 1–25

Baker TJ, Tyler CR, Galloway TS (2014) Impacts of metal and metal oxide nanoparticles on marine organisms. Environ Pollut 186:257–271

Balbus JM, Maynard AD, Colvin VL (2007) Report: hazard assessment for nanoparticles-report from an interdisciplinary workshop. Environ Health Perspect 115:1654–1659

Batley GE, Kirby JK, McLaughlin MJ (2013) Fate and risks of nanomaterials in aquatic and terrestrial environments. Acc Chem Res 46:854–862

Bazaka K, Jacob MV, Ostrikov KK (2016) Sustainable life cycles of natural-precursor-derived nanocarbons. Chem Rev 116:163. doi:10.1021/acs.chemrev.5b00566

Beddoes CM, Case CP, Briscoe WH (2015) Understanding nanoparticle cellular entry: a physicochemical perspective. Adv Colloid Interface Sci 218:48–68

Behra R, Krug H (2008) Nanoecotoxicology: nanoparticles at large. Nat Nanotechnol 3:253–254

Bergin IL, Witzmann FA (2013) Nanoparticle toxicity by the gastrointestinal route: evidence and knowledge gaps. Int J Biomed Nanosci Nanotechnol 3:1–2

Bianco C, Kezic S, Visser MJ, Pluut O, Adami G, Krystek P (2015) Pilot study on the identification of silver in skin layers and urine after dermal exposure to a functionalized textile. Talanta 136:23–28

Bottini M, Bruckner S, Nika K, Bottini N, Bellucci S, Magrini A, Bergamaschi A, Mustelin T (2006) Multi-walled carbon nanotubes induce T lymphocyte apoptosis. Toxicol Lett 160:121–126

Boxall AB, Tiede K, Chaudhry Q (2007) Engineered nanomaterials in soils and water: how do they behave and could they pose a risk to human health? Nanomedicine (Lond) 2:919–927

Braakhuis HM, Gosens I, Krystek P, Boere JA, Cassee FR, Fokkens PH, Post JA, van Loveren H, Park MV (2014) Particle size dependent deposition and pulmonary inflammation after short-term inhalation of silver nanoparticles. Part Fibre Toxicol. doi:10.3109/17435390.2015.1012184

Braakhuis HM, Oomen AG, Cassee FR (2015) Grouping nanomaterials to predict their potential to induce pulmonary inflammation. Toxicol Appl Pharmacol. doi:10.1016/j.taap.2015.11.009

Braydich-Stolle LK, Schaeublin NM, Murdock RC, Jiang J, Biswas P, Schlager JJ, Hussain SM (2009) Crystal structure mediates mode of cell death in TiO2 nanotoxicity. J Nanopart Res 11:1361–1374

Brenner SA, Neu-Baker NM, Eastlake AC, Beaucham CC, Geraci CL (2016) NIOSH Field Studies Team assessment: worker exposure to aerosolized metal oxide nanoparticles in a semiconductor fabrication facility. J Occup Environ Hyg 12:1–31

Bumbudsanpharoke N, Choi J, Ko S (2015) Applications of nanomaterials in food packaging. J Nanosci Nanotechnol 15(9):6357–6372

Buzea C, Pacheco II, Robbie K (2007) Nanomaterials and nanoparticles: sources and toxicity. Biointerphases 2(4):17–71

Cahouet A, Denizot B, Hindre F (2002) Biodistribution of dual radiolabeled lipidic nanocapsules in the rat using scintigraphy and gamma counting. Int J Pharm 242:367–371

Camargo PHC, Satyanarayana KG, Wypych F (2009) Nanocomposites: synthesis, structure, properties and new application opportunities. Mater Res 12:1–39

Cena LG, Chisholm WP, Keane MJ, Chen BT (2015) A field study on the respiratory deposition of the nano-sized fraction of mild and stainless steel welding fume metals. J Occup Environ Hyg 12:721–728

Chaudhuri S, Sardar S, Bagchi D, Dutta S, Debnath S, Saha P, Lemmens P, Pal SK (2015) Photoinduced dynamics and toxicity of a cancer drug in proximity of inorganic nanoparticles under visible light. Chemphyschem. doi:10.1002/cphc.201500905

Chen Z, Meng H, Zing G (2006) Acute toxicological effects of copper nanoparticles *in vivo*. Toxicol Lett 163:109–120

Chen Y, Wang Q, Wang T (2015) Facile large-scale synthesis of brain-like mesoporous silica nanocomposites via a selective etching process. Nanoscale 7:16442–16450

Cinelli M, Coles SR, Sadik O, Karn B, Kirwan K (2016) A framework of criteria for the sustainability assessment of nanoproducts. J Clean Prod 126:277–287

Civeira MS, Pinheiro RN, Gredilla A, de Vallejuelo SF, Oliveira ML, Ramos CG, Taffarel SR, Kautzmann RM, Madariaga JM, Silva LF (2015) The properties of the nano-minerals and hazardous elements: potential environmental impacts of Brazilian coal waste fire. Sci Total Environ 544:892–900

Clarke AG, Robertson LA, Hamilton RS (2004) A Lagrangian model of the evolution of the particulate size distribution of vehicular emissions. Sci Total Environ 334:197–206

Corsi I, Cherr GN, Lenihan HS, Labille J, Hassellov M, Canesi L, Dondero F, Frenzilli G, Hristozov D, Puntes V, Della Torre C, Pinsino A, Libralato G, Marcomini A, Sabbioni E, Matranga V (2014) Common strategies and technologies for the ecosafety assessment and design of nanomaterials entering the marine environment. ACS Nano 8:9694–9709

Cui D, Tian F, Ozkan CS (2005) Effect of single wall carbon nanotubes on human HEK293 cells. Toxicol Lett 155:73–85

Dhawan A, Sharma V (2010) Toxicity assessment of nanomaterials: methods and challenges. Anal Bioanal Chem 398:589–605

Dhawan A, Sharma V, Parmar D (2009) Nanomaterials: a challenge for toxicologists. Nanotoxicology 3:1–9

Di Bona KR, Xu Y, Gray M, Fair D, Hayles H, Milad L, Montes A, Sherwood J, Bao Y, Rasco JF (2015) Short- and long-term effects of prenatal exposure to iron oxide nanoparticles: influence of surface charge and dose on developmental and reproductive toxicity. Int J Mol Sci 16:30251–30268

Dobrovolskaia MA, Shurin M, Shvedova A (2016) Current understanding of interactions between nanoparticles and the immune system. Toxicol Appl Pharmacol 5(299):78–89

Dobon A, Cordero P, Kreft F, Ostergaard SR, Antvorskov H, Robertsson M, Smolander M, Hortal M (2011) The sustainability of communicative packaging concepts in the food supply chain. A case study: part 2. Life cycle costing and sustainability assessment. Int J Life Cycle Assess 16:537

Donaldson K, Poland CA (2013) Nanotoxicity: challenging the myth of nano-specific toxicity. Curr Opi Biotechnol 24:724–734

Donaldson K, Aitken R, Tran L, Stone V, Duffin R, Forrest G, Alexander A (2006) Carbon nanotubes: a review of their properties in relation to pulmonary toxicology and workplace safety. Toxicol Sci 92:5–22

Donaldson K, Murphy FA, Duffin R, Poland CA (2010) Asbestos, carbon nanotubes and the pleural mesothelium: a review of the hypothesis regarding the role of long fibre retention in the parietal pleura, inflammation and mesothelioma. Part Fibre Toxicol 7:5. doi:10.1186/1743-8977-7-5

Dunford RA, Salinaro L (1997) Chemical oxidation and DNA damage catalysed by inorganic sunscreen ingredients. FEBS Lett 418:87–90

Eckelman MJ, Zimmerman JB, Anastas PT (2008) Towards green nano—E-factor analysis of several nanomaterials syntheses. J Industrial Ecol 12:316–328

El-Ansary A, Al-Daihan S, Bacha AB, Kotb M (2015) Toxicity of novel nanosized formulations used in medicine. Methods Mol Biol 1028:47–74

Emmanuel R, Karuppiah C, Chen SM, Palanisamy S, Padmavathy S, Prakash P (2014) Green synthesis of gold nanoparticles for trace level detection of a hazardous pollutant (nitrobenzene) causing methemoglobinaemia. J Hazard Mater 279:117–124

Fan J, Sun Y, Wang S, Li Y, Zeng X, Cao Z, Yang P, Song P, Wang Z, Xian Z, Gao H, Chen Q, Cui D, Ju D (2015) Inhibition of autophagy overcomes the nanotoxicity elicited by cadmium-based quantum dots. Biomaterials 78:102–114

Favi PM, Valencia MM, Elliott PR, Restrepo A, Gao M, Huang H, Pavon JJ, Webster TJ (2015) Shape and surface chemistry effects on the cytotoxicity and cellular uptake of metallic nanorods and nanospheres. J Biomed Mater Res A 103:3940–3955

Ferrari M (2005) Cancer nanotechnology: opportunities and challenges. Nat Rev Cancer 5:161–171

Fischman M, Storey E, McCunney RJ, Kosnett K (2011) National institute for occupational safety and health nanomaterials and worker health conference—medical surveillance session summary report. J Occup Environ Med 53:S35–S37

Gambardella C, Morgana S, Bari GD, Ramoino P, Bramini M, Diaspro A, Falugi C, Faimali M (2015) Multidisciplinary screening of toxicity induced by silica nanoparticles during sea urchin development. Chemosphere 139:486–495

Gao GY, Chen ML, Li MY, Yang ZB, Li ZP, Mei XG (2015a) Current status and prospect of translational medicine in nanotechnology. Yao Xue Xue Bao 50:919–922

Gao Y, Jin B, Shen W, Sinko P, Xie X, Zhang H, Jia L (2015b) China and the United States—global partners, competitors and collaborators in nanotechnology development. Nanomedicine 12(1):13–19, pii: S1549-9634(15)00181-1

Geetha P, Latha MS, Pillai SS, Koshy M (2015) Nanoalginate based biosorbent for the removal of lead ions from aqueous solutions: equilibrium and kinetic studies. Ecotoxicol Environ Saf 122:17–23

Ghodake G, Vassiliadis VS, Choi JH, Jang J, Lee DS (2015) Facile synthesis of gold nanoparticles by amino acid asparagine: selective sensing of arsenic. J Nanosci Nanotechnol 15:7235–7239

Ghodake G, Kim DY, Jo JH, Jang J, Lee DS (2016) One-step green synthesis of gold nanoparticles using casein hydrolytic peptides and their anti-cancer assessment using the DU145 cell line. J Ind Eng Chemist 33:1–6

Gidhagen L, Johansson C, Omstedt G (2004) Model simulations of NOx and ultrafine particles close to a Swedish highway. Environ Sci Tech 38:6730–6740

Gopee NV, Roberts DW, Webb P (2007) Migration of intradermally injected quantum dots to sentinel organs in mice. Toxicol Sci 98:249–257

Gouin T, Roche N, Lohmann R, Hodges G (2011) A thermodynamic approach for assessing the environmental exposure of chemicals absorbed to microplastic. Environ Sci Technol 45:1466–1472

Green M, Howman E (2005) Semiconductor quantum dots and free radical induced DNA nicking. Chem Commun (Camb) 1:121–123

Grillo R, Rosa AH, Fraceto LF (2015) Engineered nanoparticles and organic matter: a review of the state-of-the-art. Chemosphere 119C:608–619

Grimsdale AC, Chan KL, Martin RE, Jokisz PG, Holmes AB (2009) Synthesis of light emitting conjugated polymers for applications in electroluminescent devices. Chem Rev 109:897–1091

Guadagnini R, Halamoda Kenzaoui B, Walker L, Pojana G, Magdolenova Z, Bilanicova D, Saunders M, Juillerat-Jeanneret L, Marcomini A, Huk A, Dusinska M, Fjellsbø LM, Marano F, Boland S (2015) Toxicity screenings of nanomaterials: challenges due to interference with assay processes and components of classic in vitro tests. Nanotoxicology 9:13–24

Hagens WI, Oomen AG, de Jong WH (2007) What do we (need to) know about the kinetic properties of nanoparticles in the body? Regul Toxicol Pharmacol 49:217–219

Haliullin TO, Zalyalov RR, Shvedova AA, Tkachov AG (2015) Hygienic evaluation of multilayer carbon nanotubes. Med Tr Prom Ekol 7:37–42

Hamilton RF Jr, Wu NN, Porter D (2009) Particle length-dependent titanium dioxide nanomaterials toxicity and bioactivity. Part Fibre Toxicol 6:35–45

Handy RD, Vonder Kammer F, Lead JR (2008) The ecotoxicology and chemistry of manufactured nanoparticles. Ecotoxicology 4:257–264

Hartmann NIB, Skjolding LM, Hansen SF, Baun A, Kjølholt J, Gottschalk F (2014) Environmental fate and behaviour of nanomaterials: new knowledge on important transfomation processes. Copenhagen K, Danish Environmental Protection Agency, Environmental Project No. 1594

Hassellöv M, Readman JW, Ranville JF, Tiede K (2008) Nanoparticle analysis and characterization methodologies in environmental risk assessment of engineered nanoparticles. Ecotoxicology 17:344–361

Haynes CL (2010) The emerging field of nanotoxicology. Anal Bioanal Chem 398:587–588

Hougaard KS, Campagnolo L, Chavatte-Palmer P, Tarrade A, Rousseau-Ralliard D, Valentino S, Park MV, de Jong WH, Wolterink G, Piersma AH, Ross BL, Hutchison GR, Hansen JS, Vogel U, Jackson P, Slama R, Pietroiusti A, Cassee FR (2015) A perspective on the developmental toxicity of inhaled nanoparticles. Reprod Toxicol 56:118–140

Hussain I, Singh NB, Singh A, Singh H, Singh SC (2015) Green synthesis of nanoparticles and its potential application. Biotechnol Lett. doi:10.1007/s10529-015-2026-7
Hutchison JE (2008) Greener nanoscience: a proactive approach to advancing applications and reducing implications of nanotechnology. ACS Nano 2:395–402
Iavicoli I, Leso V, Ricciardi W, Hodson LL, Hoover MD (2014) Opportunities and challenges of nanotechnology in the green economy. Environ Health 13:78
Injac R, Prijatelj M, Strukelj B (2013) Fullerenol nanoparticles: toxicity and antioxidant activity. Methods Mol Biol 1028:75–100
Jafar G, Hamzeh G (2013) Ecotoxicity of nanomaterials in soil. Ann Biol Res 4:86–92
Jang MH, Bae SJ, Lee SK, Lee YJ, Hwang YS (2014) Effect of material properties on stability of silver nanoparticles in water. J Nanosci Nanotechnol 14:9665–9669
Jeannet N, Fierz M, Schneider S, Künzi L, Baumlin N, Salathe M, Burtscher H, Geiser M (2015) Acute toxicity of silver and carbon nanoaerosols to normal and cystic fibrosis human bronchial epithelial cells. Nanotoxicology 26:1–13
Jeliazkova N, Chomenidis C, Doganis P, Fadeel B, Grafström R, Hardy B, Hastings J, Hegi M, Jeliazkov V, Kochev N, Kohonen P, Munteanu CR, Sarimveis H, Smeets B, Sopasakis P, Tsiliki G, Vorgrimmler D, Willighagen E (2015) The eNanoMapper database for nanomaterial safety information. Beilstein J Nanotechnol 6:1609–1634
Jia G, Wang H, Yan L, Wang X, Pei R, Yan T, Zhao Y, Guo X (2005) Cytotoxicity of carbon nanomaterials: single-wall nanotube multi-wall nanotube, and fullerene. Environ Sci Technol 39:1378–1383
Journeay WS, Suri SS, Moralez JG (2008) Rosette nanotubes show low acute pulmonary toxicity *in vivo*. Int J Nanomed 3:373–383
Kahru A, Dubourguier HC (2010) From ecotoxicology to nanoecotoxicology. Toxicology 269:105–119
Kahru A, Ivask A (2013) Mapping the dawn of nanoecotoxicological research. Acc Chem Res 46:823–833
Kashi TS, Eskandarion S, Esfandyari-Manesh M, Marashi SM, Samadi N, Fatemi SM, Atyabi F, Eshraghi S, Dinarvand R (2012) Improved drug loading and antibacterial activity of minocycline-loaded PLGA nanoparticles prepared by solid/oil/water ion pairing method. Int J Nanomed 7:221–234
Kaur IP, Kakkar V, Deol PK, Yadav M, Singh M, Sharma I (2014) Issues and concerns in nanotech product development and its commercialization. J Control Release 193:51–62
Keller J, Wohlleben W, Ma-Hock L, Strauss V, Gröters S, Küttler K, Wiench K, Herden C, Oberdörster G, van Ravenzwaay B, Landsiedel R (2014) Time course of lung retention and toxicity of inhaled particles: short-term exposure to nano-Ceria. Arch Toxicol 88:2033–2059
Kermanizadeh A, Gosens I, MacCalman L, Johnston H, Danielsen PH, Jacobsen NR, Lenz AG, Fernandes T, Schins RP, Cassee FR, Wallin H, Kreyling W, Stoeger T, Loft S, Møller P, Tran L, Stone V (2016) A multilaboratory toxicological assessment of a panel of 10 engineered nanomaterials to human health—ENPRA project—the highlights, limitations, and current and future challenges. J Toxicol Environ Health B Crit Rev 19:1–28
Kim JS, Yoon T-J, Yu KN, Kim BG, Park SJ, Kim HW, Lee KH, Park SB, Lee J-K, Cho MH (2006) Toxicity and tissue distribution of magnetic nanoparticles in mice. Toxicol Sci 89:338–347
Kim JH, Nam DH, Park CB (2014) Nanobiocatalytic assemblies for artificial photosynthesis. Curr Opin Biotechnol 28:1–9
Kirchner C, Liedl T, Kudera S (2005) Cytotoxicity of colloidal CdSe and CdSe/ZnS nanoparticles. Nano Lett 5:331–338
Kovacic P, Somanathan R (2009) Pulmonary toxicity and environmental contamination: radicals, electron transfer, and protection by antioxidants. Rev Environ Contam Toxicol 201:41–69
Kuempel ED, Geraci CL, Schulte PA (2012) Risk assessment and risk management of nanomaterials in the workplace: translating research to practice. Ann Occup Hyg 56:491–505
Kuhn M, Ivleva NP, Klitzke S, Niessner R, Baumann T (2015) Investigation of coatings of natural organic matter on silver nanoparticles under environmentally relevant conditions by surface-enhanced Raman scattering. Sci Total Environ 535:122–130

Kumar S, Sharma A, Tripathi B, Srivastava S, Agrawal S, Singh M, Awasthi K, Vijay YK (2010) Enhancement of hydrogen gas permeability in electrically aligned MWCNT-PMMA composite membranes. Micron 41:909–1014

Kumar V, Kumari A, Guleria P, Yadav SK (2012) Evaluating the toxicity of selected types of nanochemicals. Rev Environ Contam Toxicol 215:39–121

Kumar S, Lather V, Pandita D (2015) Green synthesis of therapeutic nanoparticles: an expanding horizon. Nanomedicine (Lond) 10:2451–2471

Lademann J, Weigmann H, Rickmeyer C (1999) Penetration of titanium dioxide microparticles in a sunscreen formulation into the horny layer and the follicular orifice. Skin Pharmacol Appl Skin Physiol 12:247–256

Lam CW, James JT, Mc Cluskey R (2004) Pulmonary toxicity of single-wall nanotubes in mice 7 and 90 days after intratracheal instillation. Toxicol Sci 77:126–134

Lambert S, Sinclair C, Boxall A (2014) Occurrence, degradation, and effect of polymer-based materials in the environment. Rev Environ Contam Toxicol 227:1–53

Landsiedel R, Ma-Hock L, Kroll A (2010) Testing metal-oxide nanomaterials for human safety. Adv Mater 22:2601–2627

Lehto M, Karilainen T, Róg T, Cramariuc O, Vanhala E, Tornaeus J, Taberman H, Jänis J, Alenius H, Vattulainen I, Laine O (2014) Co-exposure with fullerene may strengthen health effects of organic industrial chemicals. PLoS One 9(12), e114490

Leo BF, Chen S, Kyo Y, Herpoldt KL, Terrill NJ, Dunlop IE, McPhail DS, Shaffer MS, Schwander S, Gow A, Zhang J, Chung KF, Tetley TD, Porter AE, Ryan MP (2013) The stability of silver nanoparticles in a model of pulmonary surfactant. Environ Sci Technol 47:11232–11240

Lewinski N, Colvin V, Drezek R (2008) Cytotoxicity of nanoparticles. Small 4:26–49

Li XQ, Elliott DW, Zhang WX (2006) Zero-valent iron nanoparticles for abatement of environmental pollutants: materials and engineering aspects. Crit Rev Solid State Mater Sci 31:111–122

Linkov I, Kurth MH, Hristozov D, Keisler JM (2015) Nanotechnology: promoting innovation through analysis and governance. Environ Syst Decis 35:22–23

Liou SH, Tsou TC, Wang SL, Li LA, Chiang HC, Li WF et al (2012) Epidemiological study of health hazards among workers handling engineered nanomaterials. J Nanopart 14:878

Liu Z, Tabakman S, Welsher K, Dai HJ (2009) Carbon nanotubes in biology and medicine in vitro and *in vivo* detection, imaging and drug delivery. Nano Res 2:85–120

Liu H, Liu T, Wang H, Li L, Tan L, Fu C, Nie G, Chen D, Tang F (2013) Impact of PEGylation on the biological effects and light heat conversion efficiency of gold nanoshells on silica nanorattles. Biomaterials 34:6967–6975

Liu Y, Deng H, Xiao C, Xie C, Zhou X (2014) Cytotoxicity of calcium rectorite micro/nanoparticles before and after organic modification. Chem Res Toxicol 27:1401–1410

Lockman PR, Koziara JM, Mumper RJ, Allen DD (2004) Nanoparticle surface charges alter blood-brain barrier integrity and permeability. J Drug Target 12:635–641

Loux NT, Su YS, Hassan SM (2011) Issues in assessing environmental exposures to manufactured nanomaterials. Int J Environ Res Public Health 8:3562–3578

Lovric J, Bazzi HS, Cuie Y, Fortin GRA, Winnik FM, Maysinger D (2005) Differences in subcellular distribution and toxicity of green and red emitting CdTe quantum dots. J Mol Med 83:377–385

Lowry GV, Gregory KB, Apte SC, Lead JR (2012) Transformations of nanomaterials in the environment. Environ Sci Technol 46:6893–6899

Lucarelli M, Gatti AM, Savarino G (2004) Innate defence functions of macrophages can be biased by nano-sized ceramic and metallic particles. Mast cell activation and its relation to proinflammatory cytokine production in the rheumatoid lesion. Eur Cytokine Netw 15:339–346

Lv M, Huang W, Chen Z, Jiang H, Chen J, Tian Y, Zhang Z, Xu F (2015) Metabolomics techniques for nanotoxicity investigations. Bioanalysis 7:1527–1544

Ma S, Lin D (2013) The biophysicochemical interactions at the interfaces between nanoparticles and aquatic organisms: adsorption and internalization. Environ Sci Process Impacts 15:145
Ma J, Mercer RR, Barger M, Schwegler-Berry D, Cohen JM, Demokritou P, Castranova V (2015) Effects of amorphous silica coating on cerium oxide nanoparticles induced pulmonary responses. Toxicol Appl Pharmacol 288:63–73
Mackevica A, Foss HS (2015) Release of nanomaterials from solid nanocomposites and consumer exposure assessment—a forward-looking review. Nanotoxicology 14:1–50
Mann EE, Thompson LC, Shannahan JH, Wingard CJ (2012) Changes in cardiopulmonary function induced by nanoparticles. Wiley Interdiscip Rev Nanomed Nanobiotechnol 4:691–702
Marconnet AM, Yamamoto N, Panzer MA, Wardle BL, Goodson KE (2011) Thermal conduction in aligned carbon nanotube-polymer nanocomposites with high packing density. ACS Nano 5:4818–4825
Mashwani ZU, Khan T, Khan MA, Nadhman A (2015) Synthesis in plants and plant extracts of silver nanoparticles with potent antimicrobial properties: current status and future prospects. Appl Microbiol Biotechnol 99:9923–9934
Matranga V, Corsi I (2012) Toxic effects of engineered nanoparticles in the marine environment: model organisms and molecular approaches. Mar Environ Res 76:32–40
Mckenzie LC, Hutchison JE (2004) Green nanoscience: an integrated approach to greener products, processes, and applications. Chem Today 2004:25–28
Meesters JA, Veltman K, Hendriks AJ, van de Meent D (2013) Environmental exposure assessment of engineered nanoparticles: why REACH needs adjustment. Integr Environ Assess Manag 9(3):e15–e26
Michael AW, Nguyen HT, Adrian SM, Kannangara GSK, Volk H, Lu GQM (2008) Nanomaterials in soils. Geoderma 146(1–2):291–302
Mitrano DM, Motellier S, Clavaguera S, Nowack B (2015) Review of nanomaterial aging and transformations through the life cycle of nano-enhanced products. Environ Int 77:132–147
Monteiro-Riviere NA, Nemanich RJ, Inman AO (2005) Multi-walled carbon nanotube interactions with human epidermal keratinocytes. Toxicol Lett 155:377–384
Mukhopadhyay SS (2014) Nanotechnology in agriculture: prospects and constraints. Nanotechnol Sci Appl 7:63–71
Nalwa HS (2014) A special issue on reviews in nanomedicine, drug delivery and vaccine development. J Biomed Nanotechnol 10:1635–1640
Nath D, Banerjee P (2013) Green nanotechnology—a new hope for medical biology. Environ Toxicol Pharmacol 36:997–1014
Nguyen KC, Rippstein P, Tayabali AF, Willmore WG (2015) Mitochondrial toxicity of cadmium telluride quantum dot nanoparticles in mammalian hepatocytes. Toxicol Sci 146:31–42
Niu X, Zou W, Liu C, Zhang N, Fu C (2009) Modified nanoprecipitation method to fabricate DNA-loaded PLGA nanoparticles. Drug Dev Ind Pharm 35:1375–1383
Nogueira V, Lopes I, Rocha-Santos T, Gonçalves F, Pereira R (2015) Toxicity of solid residues resulting from wastewater treatment with nanomaterials. Aquat Toxicol 165:172–178
Nurkiewicz TR, Porter DW, Barger M (2006) Systemic microvascular dysfunction and inflammation after pulmonary particulate matter exposure. Environ Health Perspect 114:412–419
Oberdorster G (2010) Safety assessment for nanotechnology and nanomedicine. J Intern Med 267:89–105
Oomen AG, Bos PM, Fernandes TF, Hund-Rinke K, Boraschi D, Byrne HJ, Aschberger K, Gottardo S, von der Kammer F, Kühnel D, Hristozov D, Marcomini A, Migliore L, Scott-Fordsmand J, Wick P, Landsiedel R (2014) Concern-driven integrated approaches to nanomaterial testing and assessment—report of the NanoSafety Cluster Working Group 10. Nanotoxicology 8:334–348
Oughton DH, Hertel-Aas T, Pellicer E, Mendoza E, Joner EJ (2008) Neutron activation of engineered nanoparticles as a tool for tracing their environmental fate and uptake in organisms. Environ Toxicol Chem 27:1883–1887

Pal S, Tak YK, Song JM (2007) Does the antibacterial activity of silver nanoparticles depend on the shape of the nanoparticle? A study of the Gram-negative bacterium *Escherichia coli*. Appl Environ Microbiol 73:1712–1720

Pantarotto D, Briand JP, Prato M (2004) Translocation of bioactive peptides across cell membranes by carbon nanotubes. Chem Commun (Camb) 26:16–17

Park SY, Lee HU, Lee YC, Choi S, Cho DH, Kim HS, Bang S, Seo S, Lee SC, Won J, Son BC, Yang M, Lee J (2015) Eco-friendly carbon-nanodot-based fluorescent paints for advanced photocatalytic systems. Sci Rep 5:12420. doi:10.1038/srep12420

Parker JP, Ude Z, Marmion CJ (2016) Exploiting developments in nanotechnology for the preferential delivery of platinum-based anti-cancer agents to tumours: targeting some of the hallmarks of cancer. Metallomics 8(1):43–60

Pattan G, Kaul P (2014) Health hazards associated with nanomaterials. Toxicol Ind Health 30:499–519

Pelclova D, Zdimal V, Fenclova Z, Vlckova S, Turci F, Corazzari I, Kacer P, Schwarz J, Zikova N, Makes O, Syslova K, Komarc M, Belacek J, Navratil T, Machajova M, Zakharov S (2016) Markers of oxidative damage of nucleic acids and proteins among workers exposed to TiO_2 (nano)particles. Occup Environ Med 73(2):110–118, pii: oemed-2015-103161

Perez JE, Contreras MF, Vilanova E, Felix LP, Margineanu MB, Luongo G, Porter AE, Dunlop IE, Ravasi T, Kosel J (2015) Cytotoxicity and intracellular dissolution of nickel nanowires. Nanotoxicology 22:1–38

Peters K, Unger RE, Kirkpatrick CJ (2004) Effects of nano-scaled particles on endothelial cell function *in vitro*: studies on viability, proliferation and inflammation. J Mater Sci Mater Med 15:321–325

Pickering KD, Wiesner MR (2005) Fullerol, sensitized production of reactive oxygen species in aqueous solution. Environ Sci Technol 39:1359–1365

Poland CA, Duffin R, Kinloch I, Maynard A, Wallace WAH, Seaton A (2008) Carbon nanotubes introduced into the abdominal cavity of mice show asbestos-like pathogenicity in a pilot study. Nat Nanotechnol 3:423–428

Polshettiwar V, Basset JM, Astruc D (2012) Nanoscience makes catalysis greener. ChemSusChem 5:6–8

Rashidi K, Shabani A, Saen RF (2015) Using data envelopment analysis for estimating energy saving and undesirable output abatement: a case study in the Organization for Economic Co-Operation and Development (OECD) countries. J Clean Product 105:241–252

Ray PC, Yu H, Peter PF (2009) Toxicity and environmental risks of nanomaterials: challenges and future needs. J Environ Sci Health C Environ Carcinog Ecotoxicol Rev 27:1–35

Reddy VBA, Reddy GK, Madhavi V (2012) Degradation of chlorpyrifos in aqueous solutions with chitosan—stablilized FeO nanoparticles. Int J Sci Innov Discov 2:106–112

Rickerby DG, Morrison M (2007) Nanotechnology and the environment: a European perspective. Sci Technol Adv Mat 8(1-2):19–24

Rocha TL, Gomes T, Sousa VS, Mestre NC, Bebianno MJ (2015) Ecotoxicological impact of engineered nanomaterials in bivalve molluscs: an overview. Mar Environ Res 111:74–88

Rouse JG, Yang J, Ryman-Rasmussen JP (2007) Effects of mechanical flexion on the penetration of fullerene amino acid derivatized peptide nanoparticles through skin. Nano Lett 7:155–160

Russell-Jones GJ (2000) Oral vaccine delivery. J Control Release 65:49–54

Ryman-Rasmussen JP, Riviere JE, Monteiro-Riviere NA (2006) Penetration of intact skin by quantum dots with diverse physicochemical properties. Toxicol Sci 91:159–165

Saathoff JG, Inman AO, Xia XR, Riviere JE, Monteiro-Riviere NA (2011) *In vitro* toxicity assessment of three hydroxylated fullerenes in human skin cells. Toxicol In Vitro 25:2105–2112

Saini P, Saha SK, Roy P, Chowdhury P, Sinha Babu SP (2015) Evidence of reactive oxygen species (ROS) mediated apoptosis in Setaria cervi induced by green silver nanoparticles from *Acacia auriculiformis* at a very low dose. Exp Parasitol 160:39–48

Santos SM, Dinis AM, Peixoto F, Ferreira L, Jurado AS, Videira RA (2014) Interaction of fullerene nanoparticles with biomembranes: from the partition in lipid membranes to effects on mitochondrial bioenergetics. Toxicol Sci 138:117–129

Sayes C, Fortner J, Guo W (2004) The differential cytoxicity of water-solute fullerenes. Nano Lett 4:1881–1887

Schaeublin NM, Braydich-Stolle LK, Schrand AM (2012) Surface charge of gold nanoparticles mediates mechanism of toxicity. Nanoscale 3:410–420

Schulte PA, Iavicoli I, Rantanen JH, Dahmann D, Iavicoli S, Pipke R, Guseva Canu I, Boccuni F, Ricci M, Polci ML, Sabbioni E, Pietroiusti A, Mantovani E (2016) Assessing the protection of the nanomaterial workforce. Nanotoxicology. doi:10.3109/17435390.2015.1132347

Seo YS, Cha SH, Yoon HR, Kang YH, Park Y (2015) Caffeic acid: potential applications in nanotechnology as a green reducing agent for sustainable synthesis of gold nanoparticles. Nat Prod Commun 10:627–630

Shakeel M, Jabeen F, Shabbir S, Asghar MS, Khan MS, Chaudhry AS (2016) Toxicity of nano-titanium dioxide (TiO2-NP) through various routes of exposure: a review. Biol Trace Elem Res 172(1):1–36

Sharma VK, Filip J, Zboril R, Varma RS (2015) Natural inorganic nanoparticles—formation, fate, and toxicity in the environment. Chem Soc Rev 44:8410–8423

Shiohara A, Hoshino A, Hanaki K (2004) On the cytotoxicity caused by quantum dots. Microbiol Immunol 48:669–675

Shvedova AA, Castranova V, Kisin E (2003) Exposure to carbon nanotube material: assessment of nanotube cytotoxicity using human keratinocyte cells. J Toxicol Environ Health A 66:1909–1926

Sigg L, Behra R, Groh K, Isaacson C, Odzak N, Piccapietra F, Röhder L, Schug H, Yue Y, Schirmer K (2014) Chemical aspects of nanoparticle ecotoxicology. Chimia (Aarau) 68:806–811

Simkó M, Mattsson MO (2014) Interactions between nanosized materials and the brain. Curr Med Chem 21:4200–4214

Sly PD, Schüepp K (2012) Nanoparticles and children's lungs: is there a need for caution? Paediatr Respir Rev 13:71–72

Smulders S, Larue C, Sarret G, Castillo-Michel H, Vanoirbeek J, Hoet PH (2015) Lung distribution, quantification, co-localization and speciation of silver nanoparticles after lung exposure in mice. Toxicol Lett 238:1–6

Som C, Wick P, Krug H, Nowack B (2011) Environmental and health effects of nanomaterials in nanotextiles and façade coatings. Environ Int 37:1131–1142

Soni D, Naoghare PK, Saravanadevi S, Pandey RA (2015) Release, transport and toxicity of engineered nanoparticles. Rev Environ Contam Toxicol 234:1–47

Soto K, Garza KM, Murr LE (2007) Cytotoxic effects of aggregated nanomaterials. Acta Biomater 3:351–358

Spruit SL, Hoople GD, Rolfe DA (2015) Just a cog in the machine? The individual responsibility of researchers in nanotechnology is a duty to collectivize. Sci Eng Ethics. doi:10.1007/s11948-015-9718-1

Stander L, Theodore L (2011) Environmental implications of nanotechnology—an update. Int J Environ Res Public Health 8:470–479

Stebounova LV, Guio E, Grassian VH (2011) Silver nanoparticles in simulated biological media: a study of aggregation, sedimentation, and dissolution. J Nanopart Res 13:233–244

Štengl V, Henych J, Janoš P, Skoumal M (2016) Nanostructured metal oxides for stoichiometric degradation of chemical warfare agents. Rev Environ Contam Toxicol 236:239–258

Stone V, Nowack B, Baun A, van den Brink N, Fv K, Dusinska M, Handy R, Hankin S, Hassellöv M, Joner E, Fernandes TF (2010) Nanomaterials for environmental studies: classification, reference material issues, and strategies for physico-chemical characterization. Sci Total Environ 408:1745–1754

Sturm R (2015) A computer model for the simulation of nanoparticle deposition in the alveolar structures of the human lungs. Ann Transl Med 3:281. doi:10.3978/j.issn.2305-5839.2015.11.01

Tee JK, Ong CN, Bay BH, Ho HK, Leong DT (2015) Oxidative stress by inorganic nanoparticles. Wiley Interdiscip Rev Nanomed Nanobiotechnol. doi:10.1002/wnan.1374

Tian F, Cui D, Schwarz H, Estrada GG, Kobayashi H (2006) Cytotoxicity of single-wall carbon nanotubes on human fibroblasts. Toxicol In Vitro 20:1202–1212

Tiede K, Hanssen SF, Westerhoff P, Fern GJ, Hankin SM, Aitken RJ, Chaudhry Q, Boxall AB (2015) How important is drinking water exposure for the risks of engineered nanoparticles to consumers? Nanotoxicology 12:1–9

Tinkle SS, Antonini JM, Rich BA (2003) Skin as a route of exposure and sensitization in chronic beryllium disease. Environ Health Perspect 111:1202–1208

Tiwari AJ, Marr LC (2010) The role of atmospheric transformations in determining environmental impacts of carbonaceous nanoparticles. J Environ Qual 39:1883–1895

Torres-Lugo M, Garcia M, Record R (2002) Physicochemical behavior and cytotoxic effects of p (methacrylic acid-g-ethylene glycol) nanospheres for oral delivery of proteins. J Control Release 80:197–205

Totsuka Y, Higuchi T, Imai T, Nishikawa A, Nohmi T, Kato T, Masuda S, Kinae N, Hiyoshi K, Ogo S, Kawanishi M, Yagi T, Ichinose T, Fukumori N, Watanabe M, Sugimura T, Wakabayashi K (2009) Genotoxicity of nano/microparticles in in vitro micronuclei, *in vivo* comet and mutation assay systems. Part Fibre Toxicol 6:23

Vale G, Mehennaoui K, Cambier S, Libralato G, Jomini S, Domingos RF (2015) Manufactured nanoparticles in the aquatic environment-biochemical responses on freshwater organisms: a critical overview. Aquat Toxicol 170:162–174

Vinothkannan M, Karthikeyan C, Gnana kumar G, Kim AR, Yoo DJ (2015) One-pot green synthesis of reduced graphene oxide (RGO)/Fe3O4 nanocomposites and its catalytic activity toward methylene blue dye degradation. Spectrochim Acta A Mol Biomol Spectrosc 136 Pt B:256–264

Völker C, Oetken M, Oehlmann J (2013) The biological effects and possible modes of action of nanosilver. Rev Environ Contam Toxicol 223:81–106

Wang H, Wang J, Deng X (2004) Biodistribution of carbon single wall carbon nanotubes in mice. J Nanosci Nanotechnol 4:1019–1024

Wang L, Mao J, Zhang GH (2007) Nano-cerium-element-doped titanium dioxide induces apoptosis of Bel 7402 human hepatoma cells in the presence of visible light. World J Gastroenterol 13:4011–4014

Wang Z, Xu C, Li X, Liu Z (2015) *In situ* green synthesis of Ag nanoparticles on tea polyphenols-modified graphene and their catalytic reduction activity of 4-nitrophenol. Col Surf A Physicochem Eng Asp 485:102–110

Warheit DB, Donner EM (2015) Risk assessment strategies for nanoscale and fine-sized titanium dioxide particles: recognizing hazard and exposure issues. Food Chem Toxicol 85:138–147

Weare W, Scott MR, Warner MG, Hutchison JE (2000) Improved Synthesis of Small (dCORE $\approx$ 1.5 nm) phosphine-stabilized gold nanoparticles. J Am Chem Soc 122:12890–12891

Winkler D (2015) Recent advances, and unresolved issues, in the application of computational modelling to the prediction of the biological effects of nanomaterials. Toxicol Appl Pharmacol. doi:10.1016/j.taap.2015.12.016

Witschger O, Fabries JF (2005) Particules ultra-fines etsante au travail 1- caracteristiques et effets potentiels sur la santé. INRS—Hygiene et securite du travail—Cahiers de notes documentaires—2e trimester 199:21–35

Yadav T, Mungray AA, Mungray AK (2014) Fabricated nanoparticles: current status and potential phytotoxic threats. Rev Environ Contam Toxicol 230:83–110

Yan L, Feng M, Liu J, Wang L, Wang Z (2015) Antioxidant defenses and histological changes in *Carassius auratus* after combined exposure to zinc and three multi-walled carbon nanotubes. Ecotoxicol Environ Saf 125:61–71

Yan W, Chen C, Wang L, Zhang D, Li AJ, Yao Z, Shi LY (2016) Facile and green synthesis of cellulose nanocrystal-supported gold nanoparticles with superior catalytic. Carbohydr Polym 140:66–73

Yao M, He L, McClements DJ, Xiao H (2015) Uptake of gold nanoparticles by intestinal epithelial cells: impact of particle size on their absorption, accumulation, and toxicity. J Agric Food Chem 63:8044–8049

Yong SK, Shrivastava M, Srivastava P, Kunhikrishnan A, Bolan N (2015) Environmental applications of chitosan and its derivatives. Rev Environ Contam Toxicol 233:1–43

Yuan R, Yu WM, Cheng F, Zhang XB, Ruan Y, Cao ZX, Larré S (2015) Effect of quantum dots on the biological behavior of the EJ human bladder urothelial carcinoma cell line. Mol Med Rep 12:6157–6163

Zhang Z, Kleinstreuer C, Donohue JF (2005) Comparison of micro- and nano-size particle depositions in a human upper airway model. J Aerosol Sci 36:123–129

Zhao X, Cui H, Chen W, Wang Y, Cui B, Sun C, Meng Z, Liu G (2014) Morphology, structure and function characterization of PEI modified magnetic nanoparticles gene delivery system. PLoS One 9(6), e98919

Zharov VP, Mercer KE, Galitovskaya EN, Smeltzer MS (2006) Photothermal nanotherapeutics and nanodiagnostics for selective killing of bacteria targeted with gold nanoparticles. Biophys J 90:619–627

Zhu X, Zhu L, Li Y (2008) Comparative toxicity of several metal oxide nano-particle aqueous suspensions to zebrafish (*Danio rerio*) early developmental stage. J Environ Sci Health A Tox Hazard Subst Environ Eng 43:278–284

Degradation Processes of Pesticides Used in Potato Cultivations

M. Kurek, H. Barchańska, and M. Turek

Contents

M. Kurek (✉) • H. Barchańska • M. Turek
Department of Inorganic, Analytical Chemistry and Electrochemistry, Faculty of Chemistry, Silesian University of Technology, B. Krzywoustego 6, 44-100 Gliwice, Poland
e-mail: Magdalena.Beata.Kurek@polsl.pl; Hanna.Barchanska@polsl.pl; Marian.Turek@polsl.pl

P. de Voogt (ed.), *Reviews of Environmental Contamination and Toxicology, Volume 242*, Reviews of Environmental Contamination and Toxicology 242,
DOI 10.1007/398_2016_13

1 Introduction

Potato comes from the Andes in South America, where it was called *batata* in indigenous language. In the sixteenth century Spanish conquistadors brought it to Europe. They called the tubers *patatas* which evolved into the English word *potato*. During the next centuries, the consumption of this vegetable has been steadily increasing. In the world, potato is the fourth most commonly grown crop, after rice, wheat and maize (Barceloux 2008; Chang et al. 2012; Chiipanthenga et al. 2012).

High fiber, carbohydrates, as well as water-soluble vitamin concentration and low-fat content make potato tuber, the edible part of potato, a valuable comestible (Lazzarini et al. 2005; Webb and Smith 2011).

The chemical composition of a tuber is presented in Fig. 1 (Source: The potato: Tuber).

The green parts of potato plant (especially sprouts, stems and green skin) contain toxic glycoalkaloids: α-solanine and α-chaconine. The content of glycoalkaloids varies considerably between different cultivars depending on the strain and postharvest conditions (light, mechanical injury, temperature storage). In vitro studies indicate that both α-solanine and α-chaconine are reversible inhibitors of human plasma cholinesterase. Clinical features of α-solanine poisoning include gastrointestinal and neurologic symptoms, particularly vomiting, headache, and flush (Barceloux 2008).

Numerous potato cultivars are susceptible to viral diseases, and are frequently infected by various bacteria and fungi. They are also hosts for various insects. Moreover, during the rather long vegetative period weeds disturb the growth of the potatoes (Kryczyński 2010).

Therefore, to increase the efficiency of the farm production and the quality of this crop, the pesticides are used worldwide.

Some examples of potato diseases caused by different pathogens were listed in Table 1.

Potato cultivation is attacked by different pathogens, such as viruses, bacteria, fungi, and is destroyed by weeds and insects. It is necessary to use agrochemicals, which are a mixture of pesticides characterized by different chemical properties.

Fig. 1 The chemical composition of potato tuber

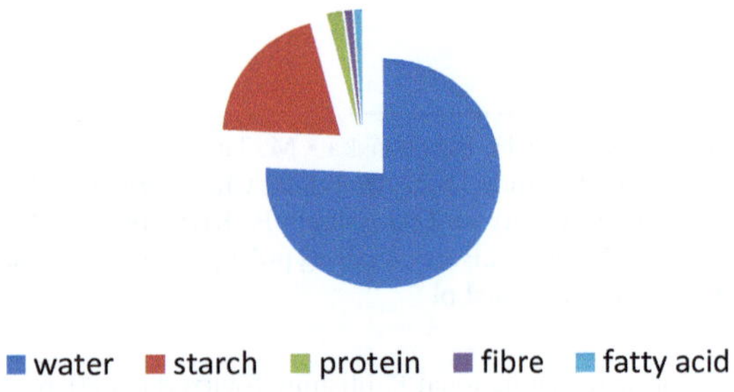

Table 1 Potato diseases (Sources: Hooker 1981; International year of potato 2008; University of Kentucky, College of Agriculture, Food and Environment)

Disease	Origin	Characteristic
Potato blackleg	Bacterial	Causes tubers to rot in the ground and in storage
Brown rot		Wilting and yellowing of the foliage
Ring rot		Witling of leaves and stem, vascular ring of tuber and stem
Pink eye		Pink areas which turn brown and it may transform in cavities
Late blight	Fungal	Destroys leaves, stems and tubers
Powdery snab		Forming raised lesions
Wart		Developing of warty outgrowthing at the stem
Skin spot		Light brown lesions on roots and stems
Leak		Causes swollen of potato's tuber
Pink rot		Wilting of leaves
Powdery middle		Light stipples on stems (initially powdery white)
Septoria leaf spot		Lesions on leaves rounded oval rings
Potato tuber moth	Pest	The most damaging pest of planted and stored potatoes in warm, dry areas
Melon thrips		Losing of potato vigour
Colorado potato beetle		Destroying of leaves and stems of potatoes
Aphids		
Potato leafhopper		

It is also important to control pesticides degradation in plants, soil and water, because their derivatives can be more toxic and stable in environment than parent compounds. Therefore, this article is dedicated to the degradation processes of selected pesticides used in potato cultivation, their toxicity and influence on human.

2 Pesticides

Pesticides (from Latin *pestis*—pest, *ceadere*—to kill) are a group of chemical compounds characterized by high toxicity. According to the definition of the Polish Environmental Protection Agency, pesticides are substances or mixture of substances which can destroy undesirable forms of plants, animals, and microorganisms or inhibit growth of pests (Seńczuk 2005).

Already in the Antique pesticides were used as plant protection products, e.g. arsenic—to control insects on plants; in the Middle Ages—an insecticidal powder—pyrethrum, while in the eighteenth and nineteenth century tobacco infusion was used against aphids (Seńczuk 2005).

The first synthetic pesticide, potassium dinitroorthocrezolate, was introduced in 1892, but the discovery of pesticidal properties of DDT (dichlorodiphenyltrichloroethane) and its use during World War II marked the beginning of the development, synthesis and use of pesticides on a large scale (Sikorska and Wędzisz 2009).

Application of pesticides has advantages and disadvantages. The advantages include: an increase in the crop yield, reduction in the frequency of animal, bird and human epidemics, the persistence of industrial products, e.g. paper or textiles. Main disadvantages include: contamination of water reservoirs and soil as a result of the transfer by wind or rain, bioaccumulation and toxic effects on non-target organisms including humans (Seńczuk 2005).

Insecticides are a class of pesticides of both natural and synthetic origin that are used to destroy insects or disrupt their development processes. The most important classes of insecticides are: organophosphorous, carbamates, organochlorines, polychlorinated biphenyls, neonicotinoids and pyrethroids (Seńczuk 2005).

Organophosphorous are used mainly in agriculture as insecticides, although they show properties of fungicides and herbicides. They cause damage to the cuticle, gastrointestinal irritation or disrupt the functioning of the respiratory system of insects. They constitute a tri-phosphoric acids and esters triphosphates. They can be degraded by hydrolysis, but at $pH > 8$ rate of this type of decomposition is reduced. Products of hydrolysis are diesters which can be also decomposed to monoesters, even to phosphorus. They decompose easily inside or on the surface of the plants and in soil. Organophosphorous insecticides act directly or non-directly on cholinesterase, which degraded neurotransmitter, acetylcholine (Seńczuk 2005).

Carbamates are esters of N-alkyl carbamic acid. N-aryl derivatives are used as herbicides. Some of them possess also fungicidal properties. These compounds can be quickly degraded in soil but their products of degradation have low toxicity, similar to parent compounds. They can be easily hydrolyzed in alkaline environment (Seńczuk 2005).

Organochlorines include three chemical groups: dichlorodiphenyltrichloroethane (DDT) analogs, hexachlorobenzene (HCB) isomers and cyclodiene compounds. They have low water solubility, but they have high solubility in organic solvents, oils and fats. They are resistant to external factors such as humidity, pH, temperature and light. Initially, these properties were recognized as advantages, but their use and production must be limited due to the potential for bioaccumulation (Seńczuk 2005).

Polychlorinated biphenyls are chemical compounds, which consist of two benzene rings and chlorine atoms attached to biphenyl. They have a wide range of applications e.g. in addition to insecticides. However, their main use in lacquers or lubricating oils caused environment contamination and consequently condensation in the food chain, which was dangerous for humans and animals. Therefore, its production was banned by the United States Congress in 1979 and by the Stockholm Convention on Persistent Organic Pollutants in 2001 (Seńczuk 2005).

Neonicotinoids are nicotine related insecticides, since they originate from nicotine that was initially applied as a natural plant protection product. They possess either a nitromethylene, nitroimine or cyanoamine group, which provide the insecticidal activity. They act as agonists of insect synaptic and extrasynaptic nicotinic acetylcholine receptors. An essential enzyme in insect central nervous system—acetylcholinesterase—is responsible for breaking the neurotransmitter acetylcholine. Electro-physiological studies show that neonicotinoids, like thiamethoxam (TMX), act on nicotinic acetylcholine receptors by reducing the effect of acetylcholine, which affects nervous system of insects (Mickaël et al. 2012).

Pyrethroids are synthetic analogues of pyrethrins. These compounds have improved physical and chemical properties, and biological activity compared to their natural analogues (Lee et al. 2002).

Natural pyrethrins are esters consisting of an "acid" component, which has a cyclopropane core and "alcohol" component. They are natural toxins produced by the flower of *Chrysanthemum*. Already in Antique and in the Middle Ages the dried flowers were used as insecticides. In nineteenth century, in Europe appeared "Persian dust", which was produced from dried flowers of *Chrysanthemum*. The powder of pyrethrum flower was also used as a plant protection product in Kenya, Tanzania and Ecuador (Elliott and Janes 1978; He et al. 2008).

The first synthetic pyrethroid was produced in twentieth century by alcohol substitutions in naturally occurring pyrethrins. Toxicity of pyrethroids is highly dependent on stereochemical structure (He et al. 2008; Koprucu and Aydm 2004).

Pyrethroids are the most lipophilic compounds among insecticides and they are characterized by high insecticidal and low mammal toxicity; they are biodegradable (Tabaeran and Narahashi 1998).

Classification of pyrethroids:

- Type I—insecticidal esters of chrysanthemic acid
- Type II—insecticidal esters of pyrethric acid

The pyrethroid insecticide affects the nervous system of insects. Type I pyrethroids (e.g., permethrin) only affect sodium channels. Type II compounds, such as lambda-cyhalothrin (LC) and deltamethrin (DM), have α-cyano group at the α-benzylic position (the α-carbon of the 3-phenoxybenzyl alcohol) and they also affect chloride and calcium channels, which are important for correction of nerve functions (He et al. 2008).

The different physiological effects are explained by the fact that the persistence of modified sodium currents by compounds of the first type lasts only tens or hundreds of milliseconds, while those of type II lasts for several seconds or longer (Elliott and Janes 1978).

Herbicides are a group of chemical compounds, both organic and inorganic, used for weed control. Organic herbicides are mainly derivatives of phenoxy acids, dinitrophenols, triazines and urea derivatives (Biziuk 2001; Seńczuk 2005).

Phenoxy acids include phenoxypropionic and phenoxybutyric derivatives. They are water-soluble. They destroy weeds by stimulating the endocrine system, what results in uncontrolled plants growth and their destruction. These compounds undergo photolysis, hydrolysis and biodegradation in environment. They do not pose a serious risk for human and animal health (Seńczuk 2005).

Dinitrophenols are compounds, which can be also used as insecticides, acaricides and ovicides. The mechanism of action is based on oxidative phosphorylation decoupling. They have low water solubilities and are characterized by high solubility in organic solvents. They are very toxic to mammals, beneficial insects and fish (Seńczuk 2005).

Triazines are highly toxic to mono- and dicotyledonous plants. They exhibit stability under soil conditions. They are photosynthesis inhibitors (photosystem II). The modes of their degradation are grouped into three types of reaction: dealkylation, oxidation, and dechlorination. Studies show, that chemical hydrolysis is the dominant degradation pathway for triazines (Seńczuk 2005).

Sulfonylureas became quickly accepted and popular because of low application doses, and favorable environmental and toxicological properties. They inhibit acetolactate synthase, an enzyme involved in the branched chain amino acid biosynthesis in plants, which inhibits weed cell growth. The soil half-life depends on soil pH and temperature. They decompose faster in warm and low pH soil conditions. Their water solubility is high and also depends on water pH. They can be degraded mainly via two pathways: hydrolysis and microbial activity. They do not pose a serious risk for environment and humans (Scrano et al. 1999; Nageswara Rao et al. 2012).

Fungicides are a large group of inorganic and organic compounds. They can act fungicidal (retain the growth and reproduction) or fungistatic (inhibition of the growth). They are used preventively. The most toxic to living organisms among all fungicides are inorganic mercury compounds, which, after long-term exposure, cause permanent damage of the nervous system, mental health and fatal poisoning, and therefore have been withdrawn from use (Seńczuk 2005).

Dithiocarbamates include three main groups: salts of dimethyl dithiocarbamates, ethylenebisdithiocarbamates and sulfides. They can be used in medicine to treat diseases caused by fungi and in agriculture both as fungicides as well as insecticides. Depending on their chemical structure, they can be characterized by low or high solubility in water. Heavy metals dithiocarbamates are characterized by low, but alkali metals are characterized by high solubility in water (Seńczuk 2005).

Phenylamides are selective fungicides used to control soilborne fungi *Pythium* and *Phytophtora* and foliar fungi that cause plant diseases. They belong to a group of systemic fungicides, which are taken up and distributed within plants and inhibit fungal growth by disrupting RNA synthesis. They are stable over a wide range of pH, temperature and light (Seńczuk 2005).

3 Degradation of Pesticides

Degradation of organic pollutants is a natural process of their decomposition in the environment. The degradation process is complete when the organic compound is decomposed to naturally occurring compounds; usually carbon dioxide and water (Dąbrowska et al. 2002).

Pesticides degradation in the environment is the result of biochemical processes, chemical reactions and photodegradation. Moreover, their decomposition may be a result of physical processes, e.g. evaporation or sublimation. Generally, the rate of polar pesticides degradation is higher, in comparison to nonpolar pesticides, due to the higher water solubility (Biziuk 2001). The main pesticides degradation processes, that occur in environment are: hydrolysis, photodegradation, and biodegradation.

Hydrolysis is a chemical process of chemical bond breaking in the presence of water, resulting in the conversion of the parent compound into its degradation product(s). Usually, it includes additional reactions: reduction, oxidation, and complexation. It may include also biodegradation and photodegradation processes, which occur in aqueous solutions (Katagi 2002).

Hydrolysis is also observed as one of the possible pathways of pesticides transformation in soil and living organisms. This process depends on pH and temperature, as well as metal ion presence, and dissolved organic matter (humic acids), which influence hydrolysis rate. Pesticides hydrolysis usually fits to first-order kinetics and the pesticide's disappearance rate is proportional to its initial concentration (Eq. 1) (Katagi 2002).

$$-d[P]/dt = k_{obs.}[P] \quad (1)$$

[P]—pesticide concentration
k_{obs}—observed hydrolysis rate concentration

Equation 1 allows calculating the pesticide half-life that does not depend on pesticide concentration, pH and temperature (Eq. 2).

$$T_{1/2} = 0.693/k_{obs.} \quad (2)$$

$T_{1/2}$—pesticide half-life

In the pH conditions of natural surface waters (5–8), most of pesticides are stable, e.g. pyrethroid insecticides or sulfonylurea herbicides. A higher rate of their degradation by hydrolysis is observed in alkaline and acidic conditions, respectively (Katagi 2002).

Photodegradation is a chemical process of organic compounds, e.g. pesticide, transformation, that occur under exposure to light. It occurs in the atmosphere, water and in upper soil layer. It can be classified into four groups: direct and indirect photodegradation, photosensitized degradation, photocatalyzed degradation (Burrows et al. 2002; Dąbrowska et al. 2002).

Direct photodegradation is based on photon absorption by molecules and they are transferred to the state of higher energy by homolysis, heterolysis or photolysis (Fig. 2). As a result of these processes chemical bonds are broken and parent compounds are transformed into various derivatives (Burrows et al. 2002; Dąbrowska et al. 2002).

Indirect photodegradation it is reaction with photo-oxidizing agents: OH-radicals, O_3 or NO_3, which naturally occur in the atmosphere. The mechanism

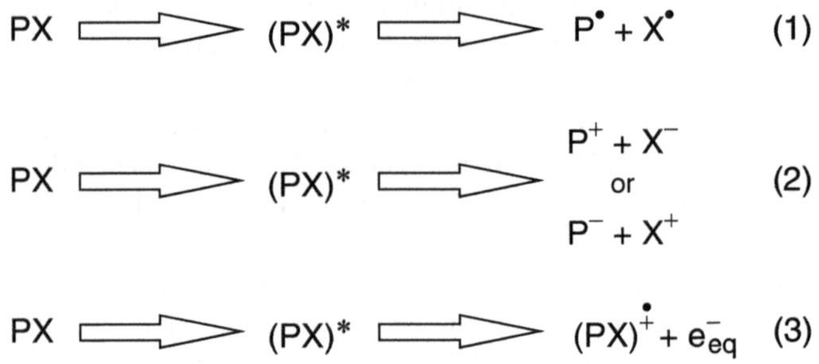

Fig. 2 Chemical reactions under direct photodegradation: (*1*) homolysis, (*2*) heterolysis, (*3*) photoionization

of this process consists of a few stages. Initially, a molecule absorbs the photon and chemical bond is broken. Then very reactive radical forms, e.g. OH are generated (Burrows et al. 2002; Dąbrowska et al. 2002; Chen and Liu 2008).

This indirect degradation process can be achieved by different ways:

- hydrogen peroxide addition

$$H_2O_2 + h\nu \rightarrow 2HO^{\bullet}$$

- ozone photolysis and singlet oxygen atoms generation and subsequently generation of $HO^{\bullet}$

$$O_3 + h\nu \rightarrow O_2 + O\left({}^1D\right)$$
$$O\left({}^1D\right) + H_2O \rightarrow 2HO^{\bullet}$$

- ozone photolysis and singlet oxygen atoms generation, which subsequently generate hydrogen peroxide

$$O_3 + H_2O + h\nu \rightarrow H_2O_2 + O_2$$

- Fe^{2+} oxidation by hydrogen peroxide

$$Fe^{2+} + H_2O_2 \rightarrow Fe^{3+} + OH^- + HO^{\bullet}$$
$$Fe^{3+} + H_2O + h\nu \rightarrow Fe^{2+} + HO^{\bullet} + H^+$$

- water hydrolysis

$$H_2O \rightarrow HO^{\bullet} + H^+ + e^-_{aq}$$

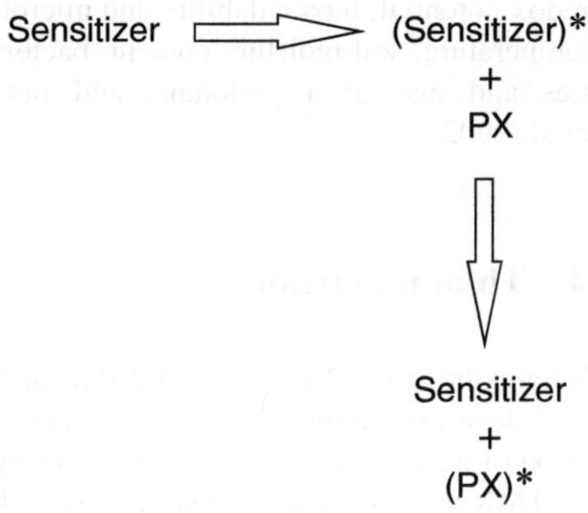

Fig. 3 Chemical reactions under photosensitized photodegradation

Photosensitized degradation is similar to direct photodegradation based on photon absorption by a molecule. The energy from its excited state (Sensitizer)* can be transferred to pesticide (PX) (see Fig. 3) (Burrows et al. 2002).

Subsequently, this excited compound can undergo homolysis, heterolysis and photolysis, even as a result of direct photodegradation. It also involves redox processes, e.g. the photo-Fenton reaction (an example is described in Sect. 5.1.2).

Photocatalytic degradation is a degradation process of pesticides, which involves catalysts and is conducted under light. It allows conducting continuous degradation process. Literature reports inform, that the following compounds are used as photocatalysts: Ag_2O/Ag_2CO_3 (Yu et al. 2014), Ag_3PO_4/TiO_2 (Yao et al. 2012), Fe_2O_3/TiO_2 (Palanisamy et al. 2013), TiO_2 (Guo et al. 2012; Liu et al. 2012; Kondrakow et al. 2014; Patel et al. 2014; Zheng et al. 2014), ZnO/CuO nanocomposite (Saravanan et al. 2013) and ZnO nanoparticles (Rahman et al. 2013).

Biodegradation is a multi-step process of biological decomposition of organic compounds involving microorganisms, in which their enzymes are used as catalysts. This degradation process can be defined as mineralization, when the organic compound is transformed into inorganic compound(s). Most of pesticides are transformed to organic compounds, which are characterized by different properties in comparison to the parent compounds. Microbial transformation includes: hydrolysis, hydroxylation, dehalogenation, ester or ether cleavage, deamination, nitro-reduction, methylation or demethylation (dealkylation) (Dąbrowska et al. 2002).

The following conditions are necessary to microbial activity:

- the microorganism must be present in environment containing the pesticide
- the pesticide must be accessible for the microorganism
- microorganism biomass should allow its proliferation

Biodegradation in the aquatic environment depends on the particulate matter presence, inorganic and organic nutrients, temperature, oxygen concentration, redox potential, bioavailability and microbial adaptation. In soil, it depends on soil temperature, soil moisture content, bacterial composition and activity, the properties and age of a pollutants and nutritive ingredients content (Dąbrowska et al. 2002).

4 Thiamethoxam

Thiamethoxam, 3-(2-chloro-1,3-thiazol-5-ylmethyl)-5-methyl-1,3,5-oxadiazinan-4-ylidene(nitro)amine, belongs to insecticides. It is a member of the second generation of neonicotinoids, in the thianicotinyl subclass (Maienfisch 2001, 2006).

TMX was first synthesized in 1991 and entered the market in 1998 by Syngenta Crop Protection. Like other neonicotinoids, it has been introduced to the market in order to replace organophosphates, a widely used group of insecticides. It is used as a soil, foliar and seed treatment agent for the control aphids, whiteflies and certain beetles (Maienfisch 2001; Hem et al. 2010).

The molecule of TMX contains two rings and five nitrogen atoms. N-nitroguanidine and the 2-chlorothiazole moieties are essential for biological activity of the compound. A replacement of this pharmacophore by other groups, e.g. N(3)-C(N)=NNO_2, causes the loss of pesticidal activity of the compound. The pharmacophore "backbone", a 1,3,5-oxadiazinane ring is characterized by stronger biological activity than any other heterocyclic ring, where instead of oxygen are =N-CH_3, =S, -CH_2. Introduction of a methyl group at N(5) leads to an increase in the insecticidal activity, but introduction other groups, e.g. hydrogen, ethyl, n-propyl or ester group, results in a complete loss of activity (Maienfisch 2001; Taillebois et al. 2014).

TMX is a moderately persistent compound in soil environment. The half-life in laboratory conditions is about 12 days, but in field conditions its degradation is slower, with a half-life of about 39 days. Degradation under field conditions is faster due to the presence of microorganisms and exposure to light. In aquatic environment its decomposition by light is moderately fast, but it is resistant to hydrolysis. Half-life of TMX in aqueous solution (laboratory conditions), at room temperature at pH 5 is more than 1 year, but at pH 7 is estimated at 200–300 days. It is more labile at pH 9 and with half-life a few days. In sediment systems it is degraded moderately fast, therefore does not cause contamination of water (Maienfisch 2001; Hem et al. 2010).

TMX is relatively non-toxic to birds, fish and mammals but it exhibits high toxicity to honeybees. Decrease of their population can negatively impact the natural crop pollination process (Mickaël et al. 2012).

Physical and chemical properties of TMX, including its half-lives in soil and water are listed in Table 2 (Source: The PPDB).

Table 2 Physical and chemical properties of pesticides, including their half lives in soil and water (Source: The PPDB)

Property		Thiamethoxam	Rimsulfuron	Metalaxyl	Lambda-cyhalothrin	Deltamethrin
Pesticide type		Insecticide	Herbicide	Fungicide	Insecticide	Insecticide
Substance group		Neonicotinoid	Sulfonylourea	Phenylamide	Pyrethroid	Pyrethroid
Substance origin		Synthetic	Synthetic	Synthetic	Synthetic	Synthetic
Chiral molecule		No	No	Yes	No	Yes
Chemical formula		$C_8H_{10}ClN_5O_3S$	$C_{14}H_{17}N_5O_7S_2$	$C_{15}H_{21}NO_4$	$C_{23}H_{19}ClF_3NO_3$	$C_{22}H_{19}Br_2NO_3$
Molecular mass (g mol^{-1})		291.71	431.44	279.33	449.85	505.2
Solubility in water at 20 °C (mg l^{-1})		4100	7300	8400	0.005	0.0002
Log P		−0.13	−1.46	1.65	5.5	4.6
Vapour pressure at 25 °C (mPa)		6.6×10^{-6}	8.9×10^{-4}	0.75	0.0002	0.0000124
Soil degradation (aerobic) (days)	DT50 (typical)	50	24.3	36	175	13
	DT50 (laboratory conditions at 20 °C)	121	24.3	36	175	26
	DT50 (field conditions)	39	10.8	38.7	26.9	21
Foliar DT50 (days)		–	3	30	5	3
Aqueous photolysis DT50 (days) at pH = 7		27	11.7	Stable	40	48
Aqueous hydrolysis DT50 (days) at 20 °C and pH = 7		Stable	7.2	106	Stable	Stable
Water sediments DT50 (days)		40	6	56	15.1	65
Water phase only DT50 (days)		3.6	4	56	0.24	17
Risk classification		Xn—harmful, N—dangerous for the environment	N—dangerous for the environment	Xn—harmful, N—dangerous for the environment	T+—very toxic, T—toxic, Xn—harmful, N—dangerous for the environment	T—toxic, N—dangerous for the environment

4.1 Abiotic Degradation of Thiamethoxam

4.1.1 Hydrolysis of Thiamethoxam

In details, the influence of pH on the persistence of four neonicotinoid insecticides: acetamiprid, thiacloprid, TMX and imidacloprid in water was described in Guzsvany et al. (2006). Among all investigated compounds TMX was the most stable in acidic conditions during 4 months. After 5 weeks of experiment conducted in water at pH = 7, in ^{1}H NMR (proton nuclear magnetic resonance) spectrum two signals were observed: one corresponded to TMX, and the second one belonged to TMX degradation product. The TMX hydrolysis was influenced by the pH of medium: the higher pH of medium, the hydrolysis was faster. The lowest persistence of TMX was observed at pH = 9 (Guzsvany et al. 2006; Maienfisch 2006). This conclusion was in agreement with the findings obtained by Zheng and Liu (1999) and Liquing and coworkers (2006). Moreover, IR (infrared spectroscopy) and GC-MS (gas chromatography–mass spectrometry) spectra analysis revealed, that there was only one main product of TMX hydrolysis. In comparison to parent compound, this derivative was stable in alkaline conditions.

According to above-mentioned studies, the main product of TMX hydrolysis was 3-(2-chloro-thiazolyl-5-ylmethyl)-5-methyl-4-oxooxadiazine. This pathway of TMX hydrolysis is presented in Fig. 4a.

The mechanism of TMX hydrolysis results from the character of its nitro group. This strong electron acceptor induces a small positive charge on the carbon of the cyano group in the oxidiazine ring. As a result it is easily attacked by hydroxyl anion (Liquing et al. 2006).

Another studies, which were conducted under laboratory conditions, on TMX degradation indicated that there is the second pathway of thiamethoxam hydrolysis.

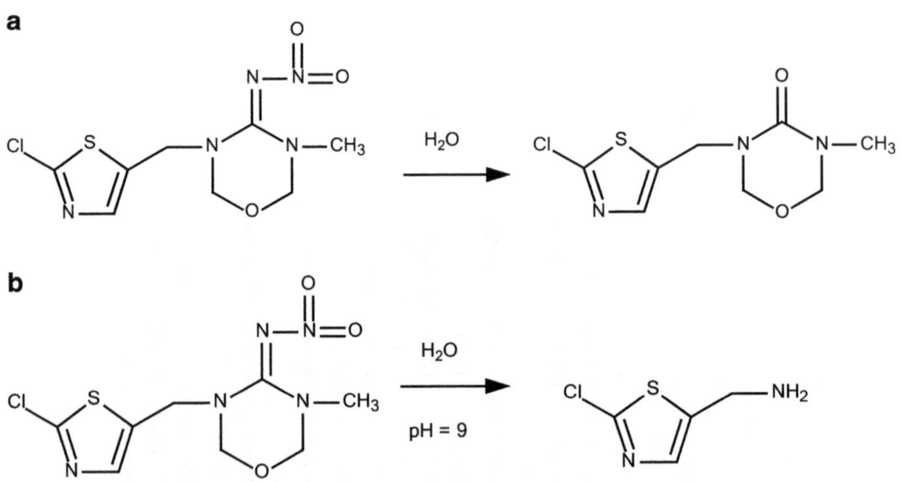

Fig. 4 Hydrolysis of TMX: (**a**) the first pathway (**b**) the second pathway

It leads to the opening of N-nitro-urea ring (at pH range in the range 5–9) and to the formation of 2-chloro-5-aminomethyl-thiazole, as presented in Fig. 4b (Maienfisch 2006).

Maienfisch (2006) observed, that harsh degradation conditions (concentrated HCl, 80 °C) gave clothianidin (that is also used as pesticide) as a TMX degradation product. It was an important observation for better understanding of its degradation process occurring in plants and insects, since these organisms converted TMX to clothianidin (Nauen et al. 2003).

4.1.2 Photodegradation of Thiamethoxam

Depending on the conditions, there are many pathways of TMX photodegradation (de Urzedo et al. 2007).

The first one, observed after exposure of TMX to artificial sunlight of photoperiods of 12 h over 12 days, is shown in Fig. 5.

Photodegradation of TMX was verified by de Urzedo and coworkers (2007) in aqueous solutions under UV radiation. Nucleophilic attack of H_2O on the nitro group in TMX (A), caused the formation of imine (B) and release of HNO_3. The conversion of compound B to the intermediate compound C occurred by nucleophilic substitution of the Cl atom by the OH group.

The same experiment showed that under these conditions TMX half-life was 10 min and this pesticide was completely degraded after 30 min. Simultaneously, the control experiment confirmed high stability of TMX in aqueous solution,

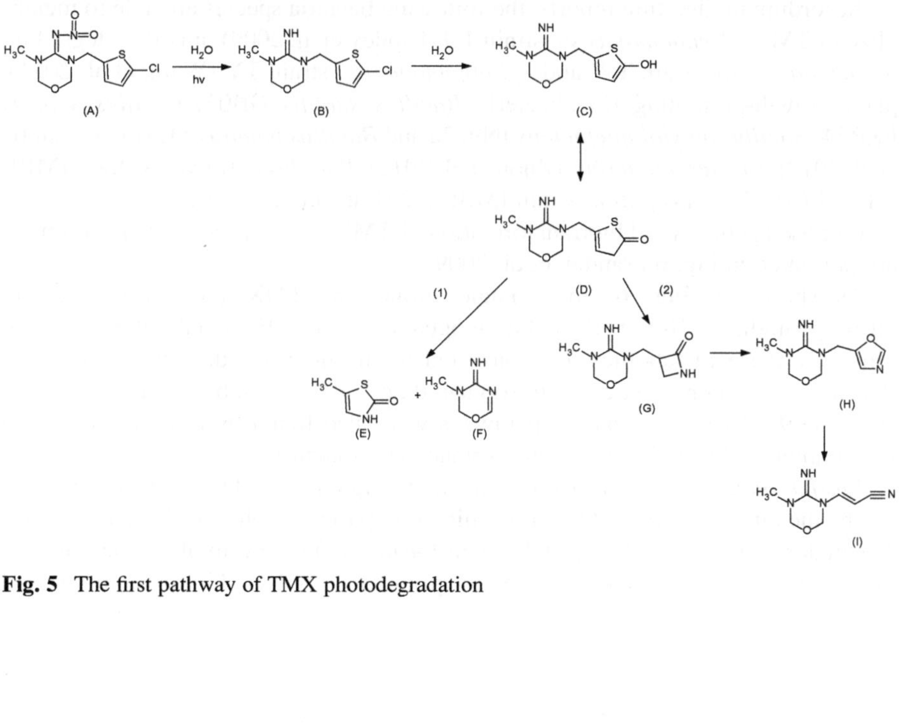

Fig. 5 The first pathway of TMX photodegradation

protected from UV light (de Urzedo et al. 2007). Mechanism of this reaction is similar to hydrolysis, which is presented in Sect. 4.1.1.

The product of the first pathway of TMX degradation, which is marked as compound D in Fig. 5, can be further degraded according to different pathways (Fig. 5).

Photolytic decomposition of metabolite D led to the formation of two degradation products: 5-methyl-2(3H)thiazolone (E) and compound F. The extrusion of the sulphur atom from compound D (second reaction) can generate azetidinone (G), which can be transformed to the oxazine (H). The final product of degradation (I) was formed by the hydrolysis of the imine group and the breakdown of the oxazol ring of compound H (third reaction) (de Urzedo et al. 2007). It is also possible, that photolytic degradation of TMX in aqueous medium may lead to formation of a carboxyl group instead of the imine (Schwartz et al. 2000).

As it is shown in Fig. 5, there are many pathways of TMX photodegradation, resulting from its complex structure—two rings and five nitrogen atoms. Moreover, the nitroguanidine group which is a chromophore, may decrease photolysis efficiency.

4.2 Biodegradation of Thiamethoxam

Metabolism of TMX by microorganisms in soil is the main pathway of its degradation. Because of TMX "magic nitro" group, which is responsible for its insecticidal activity and high selectivity to nAChR (nicotinic acetylcholine receptor), this moiety was thoroughly studied in biotransformation processes (Zhou et al. 2012).

According to literature reports, the following bacteria species are able to metabolize of TMX: *Pseudomonas sp.* strain 1G (Pandey et al. 2009), nicotine-degrading *Acinetobacter sp.* strain TW and *Sphingomonas sp.* strain TY (Wang et al. 2011), plant growth-promoting rhizobacteria *Bacillus subtilis* GB03, *Bacillus subtilis* FZB24, *Bacillus amyloliquefaciens* IN937a and *Bacillus pumilus* SE34 (Myresiotis et al. 2012), *Ensifer adhaerens* (Zhou et al. 2012), *Bacillus aeromonas* strain IMBL 4.1 and *Pseudomonas putida* strain IMBL 5.2 (Rana et al. 2015).

Proposed pathways of biotransformation of TMX by *Pseudomonas sp.* strain 1G are presented in Fig. 6 (Pandey et al. 2009).

As shown in Fig. 6, the "magic group" of TMX (A), is reduced to nitrosoguanidine =N-NO (B) and aminoguanidine –N-NH_2 metabolites (C). Subsequently these metabolites are converted to the nontoxic urea metabolite =O (E) and the more toxic desnitro/guanidine =NH metabolite (D) (Pandey et al. 2009). These degradation pathways were also found by Zhou et al. (2012) in mammals, mice and some plants (spinach and tomato).

Ensifer adhaerens is nitrogen-fixing microorganism, able to transform TMX (A) by the nitro-reduction of the nitroimino group to nitrosoimino (D) and urea (F). The major metabolic pathway of TMX in *Ensifer adhaerens* involved the transformation of its =N-NO_2 (N—nitroimino group) to =N-NO (N-nitrosoimino) and

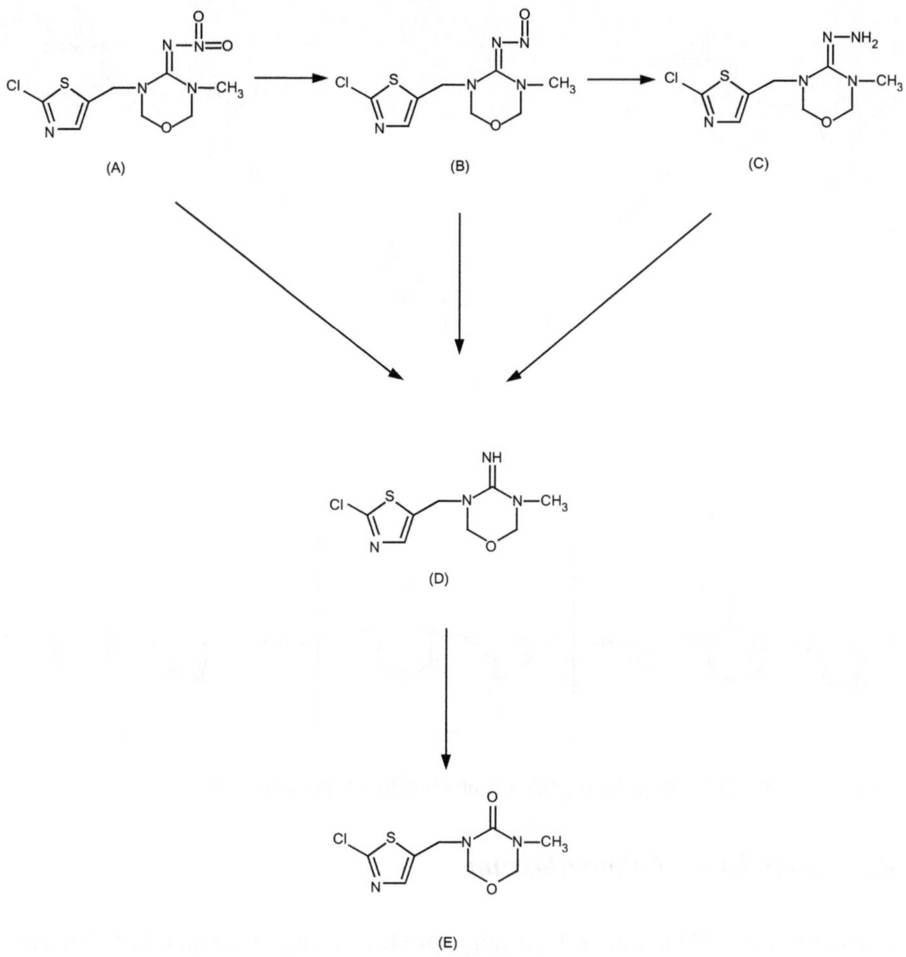

Fig. 6 Biotransformation of TMX by *Pseudomonas sp.* strain 1G

C=O (urea) metabolites. These compounds were also detected in mouse, insect and few plant (cotton, spinach and tomato) tissues exposed to TMX. This way of TMX degradation is presented in Fig. 7 (Zhou et al. 2012).

Moreover, in spinach, tomatoes, insects and mice organisms, desmethyl metabolite (B) is formed by N-desmethylation of TMX. Through the cleavage of oxidiazine ring clothianidin (C) is formed. This compound was also formed as a result of TMX hydrolysis (Zabar et al. 2012; Zhou et al. 2012).

Pathways of TMX degradation by microorganisms: *Pseudomonas sp* and *Ensifer adhaerens* were similar, as evidenced by the same intermediate degradation products (Figs. 6 and 7). Moreover, the final product of biodegradation was the same as a result of hydrolysis: 3-(-2-chlorothiazolyl-5-methyl-)-5-methy-4-oxadiazine. It suggested that degradation of TMX by these bacteria was based on hydrolysis (Pandey et al. 2009; Zabar et al. 2012; Zhou et al. 2012).

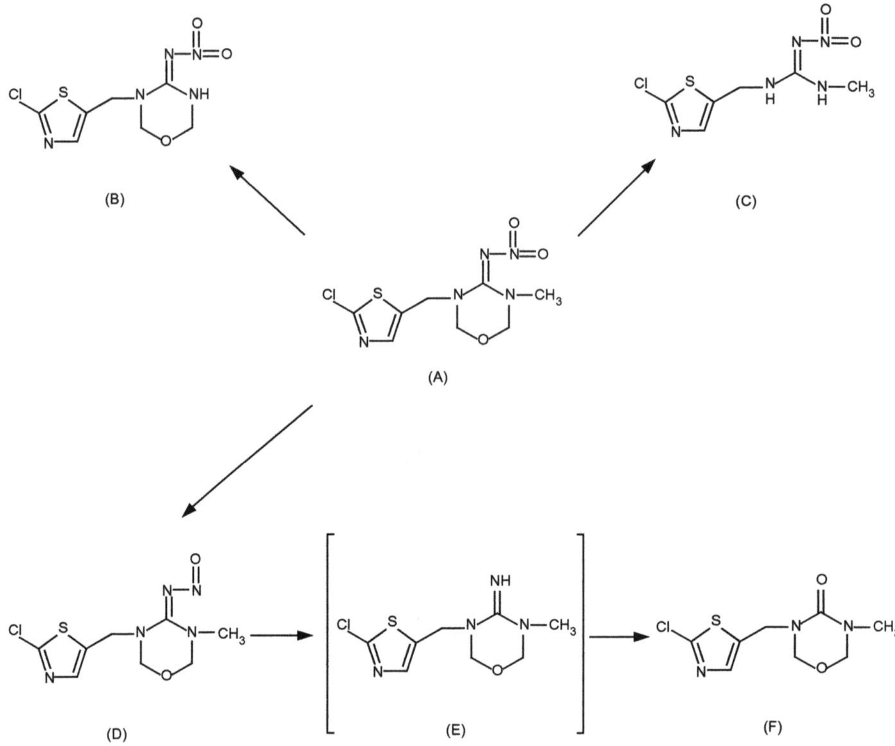

Fig. 7 Possible pathways of biodegradation of TMX by *Ensifer adhaerens*

4.3 Toxicity of Thiamethoxam

The influence of TMX and its degradation products on algae, zooplankton, bacteria, terrestrial plants, invertebrate species, and insect feeding birds was the objective of many studies, since TMX in one of the most spread insecticide in natural environment (over 25 % of total global insecticide applications (Bass et al. 2015)). Since its stability in the environment is, depending on conditions, in the range of 3.6–1000 days (Table 2; Bonmatin et al. 2015)) there is strong evidence that soils and waterways are contaminated with variable levels of TMX and its degradation products.

Šojić and coworkers (2012) investigated the ecotoxic effects of TMX and their UV-photolysis degradation intermediates (5-methyl-2(3H)thiazolone and imine) towards algae *Pseudokirchneriella subcapitata*, zooplankton *Daphnia magna* and bacteria *Vibrio fischeri*. The toxicity of TMX towards the algae culture was expressed by the normalized inhibition values of the solutions after 72 h incubation at 21 °C. The parent compound itself inhibited slightly the growth of *P. subcapitata*. However, at the end of the treatment (when the parent molecules were completely transformed) the toxicity of the solutions was higher than the normalized inhibition values of the initial solutions. The likely reason for this was the formation of degradation intermediates that were more toxic than the parent compound.

The toxicity of TMX towards *Daphnia magna* was expressed by the mortality rate of the zooplankton after 48 h of incubation at 21 °C. From the results it could be concluded that *Daphnia magna* was more sensitive to TMX degradation by-products and less sensitive to the parent compound than *P. subcapitata*. The mortality rate increased with the treatment time, suggesting the presence of toxic degradation products. Moreover, the investigations of mutagenicity and DNA breaking revealed that some very harmful degradation intermediates of TMX were produced in the photolytic processes (Šojić et al. 2012). The higher toxicity of TMX degradation products (the experiments were conducted on lettuce, *Lactuca sativa*) was confirmed by Laberge and Rollinson (2013).

Mir et al. (2013) conducted photocatalyzed degradation of TMX in aqueous suspension of TiO_2. Depending on the conditions of this reaction, a mixture of up to eight by-products was obtained. Most of them were different than naturally occurring TMX degradation products. Human erythrocytes as a model system under in vitro conditions were applied to access the toxicity of TMX and the mixture of its photocatalytic by-products.

On the basis of the obtained results, it was concluded that photocatalytic degradation of TMX led to reduction in toxicity implying that degradation intermediates were less toxic than the original substrate itself. Thus, the photocatalytic degradation of the organic compounds like TMX is a promising method for eliminating these compounds from water. However, as a result of the cleavage of oxidiazine cycle, clothianidin is formed, which is toxic for bees (*Apis mellifera*) (EFSA 2005; Zhou et al. 2012).

According to many studies (Beketov and Liess 2008; Blacquiere et al. 2012; Van den Brink et al. 2016), TMX as well as other neonicotinoids is very toxic to invertebrate, especially bees. Neonicotinoids concentrations present in pollen and nectar is the principal determinant of toxicity to individual bees or their colonies (Carreck and Ratnieks 2014; Bonmatin et al. 2015).

Since both TMX, classified as Xn—harmful, and N—dangerous for the environment (Table 2) as well as its degradation products exhibit negative effects on non-target organisms, there is a moratorium on neonicotinoid use as seed treatments, or as granules, on certain "bee attractive crops" such as maize, sunflower and oilseed rape in the European Union (EU) due to "high acute risks" to bees (Europa 2013; Balfour et al. 2015).

5 Lambda-Cyhalothrin and Deltamethrin

Lambda-cyhalothrin (LC) ((*R* + *S*)-α-cyano-3-(phenoxybenzyl(1*S* + 1*R*)-*cis*-3-(*Z*-2-chloro-3,3,3,-trifluoroprop-1-enyl)-2,2-dimethyl-cyclopropane-carboxylate) and deltamethrin (DM) (cyano(3-phenoxyphenyl)methyl-3-(2,2-dibromoethenyl)-2,2-dimethyl-cyclopropane-1-carboxylate), are pyrethroids insecticides.

LC was firstly reported in 1984 by Robson and Crosby and 1 year later it was introduced by ICI Agrochemical (now Syngenta) (He et al. 2008).

DM, is fourth generation synthetic pyrethroid pesticide and belongs to an α-cyano, type II pyrethroids. It was obtained by Elliott in 1974 and has been commercialized since 1982 (Pavan et al. 1999; Koprucu and Aydm 2004).

LC and DM are used in agriculture against a broad spectrum of insect pests of economically important crops. They are also used for the control of household insect pests such as mosquitoes, cockroaches, flies, termites, and fleas. Both of them exhibit low mammal toxicity (Tabaeran and Narahashi 1998).

Despite the similar structures of LC and DM, the first one is resistant to degradation in soil, while the second is not. Both of them are photostable and very resistant to hydrolysis in aquatic system. It results from their lipophilic properties. In water sediments system they are decomposed fast. They are biodegradable (Tabaeran and Narahashi 1998).

Physical and chemical properties of LC and DM, including their half-lives in soil and water are listed in Table 2 (Source: The PPDB).

5.1 Abiotic Degradation of Lambda-Cyhalothrin

5.1.1 Hydrolysis of Lambda-Cyhalothrin

LC is stable in acidic solution, in alkaline conditions it is hydrolyzed since it is a lipophilic compound. It results from presence of α-cyano benzyl moiety (II pyrethroid class), which causes alkaline hydrolysis acceleration (He et al. 2008; Katagi 2002).

The mechanism of LC hydrolysis is based on nucleophilic attack of the hydroxyl ion (Fig. 8) (He et al. 2008).

As a result, a cyanohydrin derivative (B) was formed, which was unstable and transformed into acidic derivative (D) and hydrocyanic acid (E) (He et al. 2008).

Because of LC lipophilic properties, its strong adsorption to soil and low risk of water contamination, most studies focused on photodegradation and biodegradation this compound in soil.

5.1.2 Photodegradation of Lambda-Cyhalothrin

LC exposure to UV light (254 nm) resulted in nearly complete degradation with losses of more than 95 % of the initial amount. According to He et al. (2008) the pathways of LC photodegradation included decarboxylation, reductive dehalogenation and ester or other bond cleavage. These processes are presented in Fig. 9.

Compounds: B (3-(2-chloro-3,3,3,-trifluoroprop-1-en-1-yl)-2,2 dimethylcyclopropanecarboxylic acid) and D ((3-phenoxyphenyl)acetonitrile) were products of LC ester bond cleavage (A). Moreover, compound D can be also

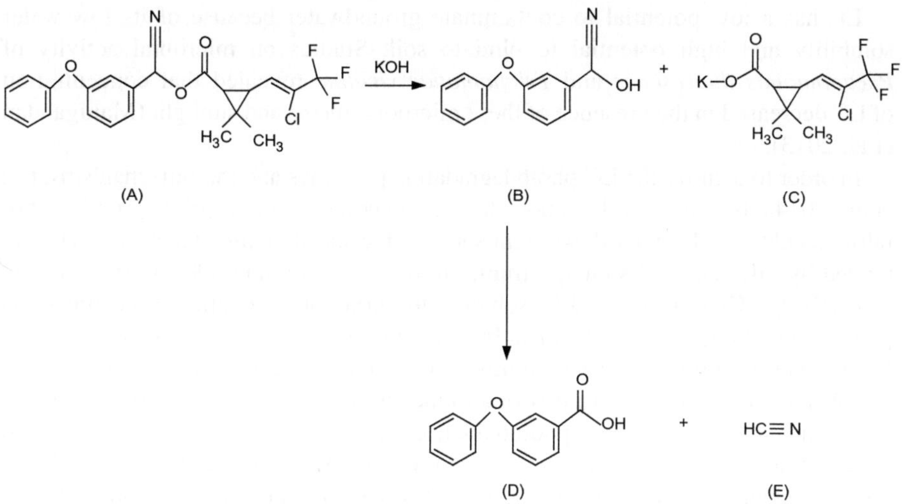

Fig. 8 Hydrolysis of LC in alkaline conditions

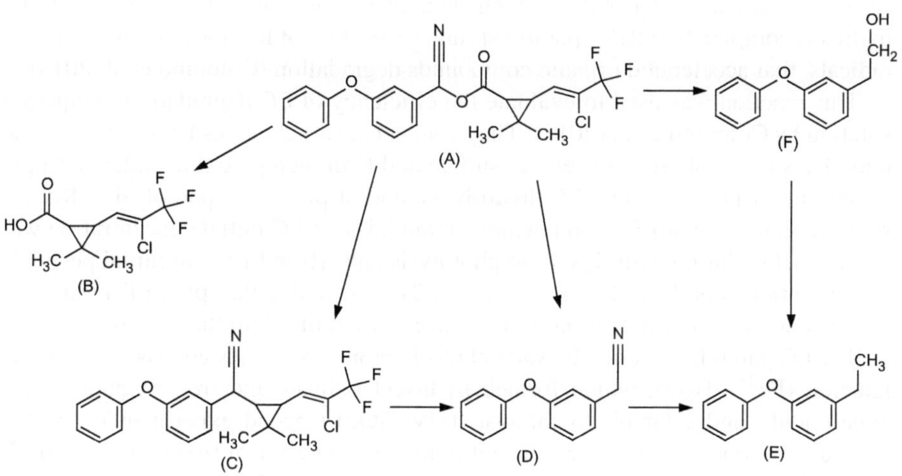

Fig. 9 Photodegradation of LC

generated by cleavage of C-C bond of compound C (decarboxyhalothrin), which was formed through the decarboxylation pathway. Photoproduct F ((3-phenoxyphenyl)methanol) was formed by the ester cleavage and subsequent loss of the cyano group of LC and photoproduct E (1-methyl-3-phenoxybenzene) was formed by hydroxyl or cyano group loss from compounds D or F (He et al. 2008).

In soil surfaces and in aqueous solutions at pH = 5, LC was degraded in the presence of sunlight with a half-life of 30 days. The degradation in soil occurred through hydroxylation followed by cleavage of the ester linkage resulting in two main degradation products that were finally degraded to carbon dioxide.

LC has a low potential to contaminate groundwater because of its low water solubility and high potential to bind to soil. Studies on microbial activity of *Pseudomonas fluorescens* and *Trichomonas viridiae* revealed that concentration of LC decreased in the presence of these microorganisms and sunlight (Manigandan et al. 2013).

In order to identify the LC photodegradation pathways and their mechanisms, Liu et al. (2014a, b) examined LC photochemical decomposition in solution under laboratory conditions. They used two light sources: Hg and Xe lamp, which were characterized by different emission spectrum: ultraviolet region and 800–1000 nm range, respectively. Exposition of LC solution to ultraviolet radiation accelerated its photodegradation, because chemical bond energy of LC, like other pyrethroids, was below energy corresponding to ultraviolet region. Moreover, it was proven that LC photodegradation fitted to first-order kinetics and were performed through ester, C-C and C-O cleavage, photooxidation (mostly), photoisomerization and decyanation. The major products of these processes were: 3-phenoxybenzaldehyde, 3-phenoxybenzoic acid methyl ester, 4-phenoxybenzoic acid and m-phenoxybenzoic acid, and two other of products, which were not identified.

In order to remove organic compounds from water, advanced oxidation processes (AOP) are used, which include photo-Fenton reaction. It is based on the ferric hydroxyl complex $Fe(OH)^{2+}$ photolysis and generation of ferrous ions and hydroxyl radicals, that accelerated organic compounds degradation (Colombo et al. 2013).

This reaction was used to evaluate the efficiency of LC degradation in aqueous solution by Colombo et al. (2013). They conducted simultaneous LC hydrolysis and photolysis in the absence of ferrous sulfate and hydrogen peroxide under UV light irradiation at pH = 6.2 and 2.5 (hydrolysis) and at pH = 2.5 (photolysis). Results showed, that for photo-Fenton reaction, about 85 % of LC initial concentration was degraded, but during hydrolysis and photolysis only 10 and 17 % of initial pesticide concentration was degraded, respectively. They revealed that photo-Fenton reaction enabled efficient reduction in the concentration of LC in the environment.

The LC photodegradation by solar photo-Fenton was also tested, based on Fenton reaction ($Fe^{2+} + H_2O$), which formed hydroxyl radicals and iron oxidation (first experiment), and solar photo catalysis (UV/TiO2) (second experiment). The LC degradation process was more efficient in the first experiment (removal of 80.65 % of initial concentration of LC) than in the second one (63.7 %) (Alalm et al. 2015).

5.2 *Biodegradation of Lambda-Cyhalothrin*

Due to the fact that LC is an ester compound and consists of acid and alcohol moiety, it is susceptible to esterase activity. Therefore, it esterase gene pytY was isolated from the genomic library of *Ochrobactrum anthropi* YZ-1 and its activity was tested. An enzyme assay described by Ruan and coworkers (2013) showed, that this enzyme was able to degrade various pyrethroids, including LC. This process is influenced by pH and temperature, since these parameters effect esterase activity.

PytY showed the highest activity at pH and temperature in the range of 5.0–9.0 and 15–45 °C respectively. Its activity decreased when pH exceeded 7.5 and temperature was higher than 45 °C (Ruan et al. 2013).

Some research on LC biodegradation indicated that it is used as carbon and nitrogen source by microorganisms in soil (Peackock et al. 2014; Chen et al. 2015).

Soil rhizosphere microorganisms are capable to use LC as a carbon source, as was shown by Peackock et al. (2014), who applied a redox gradostat reactor for measurement of microbial activity of LC.

Shen and coworkers (2015a, b) conducted studies on LC degradation in sterile and non-sterile soils under laboratory conditions. Results revealed longer LC half-life in sterile soil than in presence of soil microorganisms. In sterile soil the half-life was around 161 days, in non-sterile soil the average half-life was around 120 days. They also observed longer LC half-life at low pH and reduced exposure to light. Under field conditions, LC half-lives were also longer than under laboratory conditions, what suggested the dependence of LC degradation in soil on pH, temperature, exposure to light and microbial activity (Wang et al. 1997; Laabs et al. 2000).

Chen et al. (2015) performed a detailed study of LC degradation pathways by *Bacillus thuringiensis* strain ZS-19. It was proven, that this process was highly effective since the bacterium used this pesticide as a source of carbon and nitrogen, similar to rhizosphere microorganisms.

Pathway of LC biodegradation is presented in Fig. 10.

Degradation is initiated by the cleavage of the carboxylester linkage through hydrolysis, which is the main pathway of LC (A) degradation. Metabolites are: α-hydroxy-3-phenoxy-benzeneacetonitrile (B) and (1R,3R)-trans-2,2-dimethyl-3-(2-methyl-1-propenyl)cyclopropane-1-carboxylic acid (C) (Chen et al. 2015).

Compound B is also formed in alkaline hydrolysis LC (Sect. 5.1.1, Fig. 8), which indicates that in these microorganisms hydrolysis also occurs.

As shown in Fig. 11 main metabolites are produced by cleavage of the diaryl bond and aromatic ring of compound B. These metabolites are: 3-phenoxy-

Fig. 10 Pathway of LC degradation by *Bacillus thuringiensis* strain ZS-19

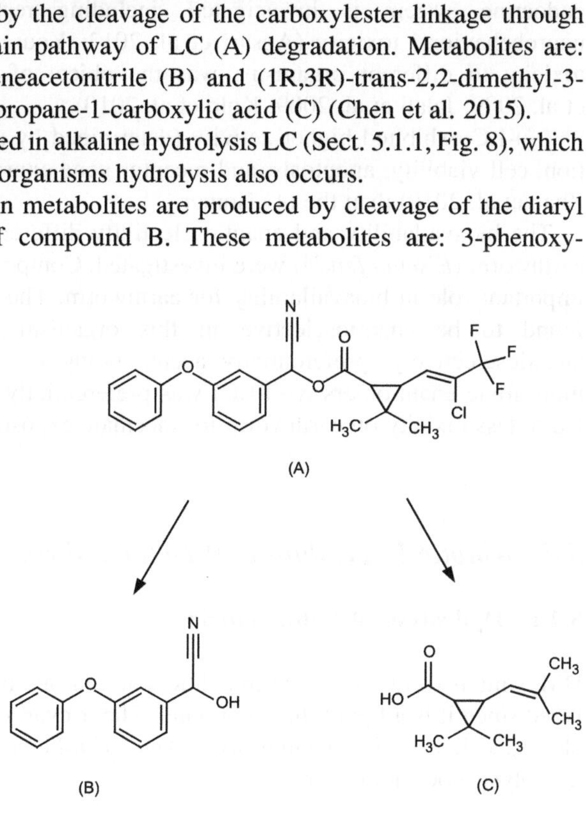

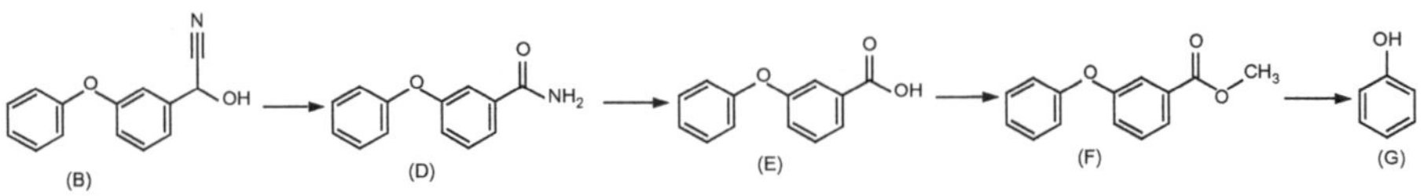

Fig. 11 Transformation of α-hydroxy-3-phenoxy-benzeneacetonitrile, one of LC metabolite

benzamide (D), 3-phenoxybenzoic acid (E), 3-phenoxybenzoate (F) and phenol (G) (Chen et al. 2015).

This pathway confirms that cyanohydrin derivative was unstable, as mentioned in Sect. 5.1.1. Its hydrolysis and biodegradation products were similar, which suggested that *Bacillus thuringiensis* strain ZS-19 could also perform hydrolysis (Chen et al. 2015).

5.3 Toxicity of Lambda-Cyhalothrin

Since LC is a chiral compound, the particular attention of researchers is drawn to determining the toxicity of its enantiomers. Since synthetic pyrethroids have endocrine disruptive properties, their possible influence on the functioning of the endocrine system is investigated, including reproductive as well as its neurobehavioral toxicity (Ansari et al. 2012; Yousef 2010). Assessment of LC toxicity to different organisms was the object of many investigators (Fetoui et al. 2009; Jalali et al. 2009; Kolo et al. 2010).

(−)-LC exhibited higher activity (determined by measuring the cell proliferation, cell viability, apoptosis, and receptor gene expression) (Zhao and Liu 2009; Zhao et al. 2010) then the (+) isomer.

The bioavailability and enantioselectivity differences between LC isomers in earthworm (*Eisenia fetida*) were investigated. Compound sorption on soil plays an important role in bioavailability for earthworm. The bioaccumulation of LC was found to be enantioselective in this organism. Earthworms took up LC enantioselectively, preferentially accumulating the less toxic enantiomer. The more toxic enantiomers ((−)-LC) was preferentially degraded in earthworm and led to less toxicity on earthworm for racemate exposure (Chang et al. 2016).

5.4 Abiotic Degradation of Deltamethrin

5.4.1 Hydrolysis of Deltamethrin

DM, similar to LC, is stable in acidic solution, in alkaline conditions it is hydrolyzed since it is a lipophilic compound. The α-cyano benzyl moiety (II pyrethroid class) accelerates alkaline hydrolysis. Due to DM low water solubility (Table 2), the hydrolysis does not occur.

5.4.2 Photodegradation of Deltamethrin

DM, like other pyrethroids, contains double bonds that are unstable after exposure to light, therefore these compounds are susceptible for photodegradation in the environment. Among the main degradation processes: hydrolysis, biodegradation and photodegradation, the latter one is prevailing. Possible pathways for the DM photodegradation under laboratory conditions were proposed by Liu et al. (2009). Photodegradation of DM is presented in Fig. 12.

According to Liu and coworkers (2009) DM (A) is degraded by ester cleavage; as a result photoproduct B is formed. Subsequently decyanation occurs, resulting in a formation of 3-phenoxybenzalcohol (compound C). This photoproduct is afterwards degraded in the photooxidation process to photoproducts E and F.

Similar results were obtained by Nahri-Niknafs and Ahmadi (2013). They investigated the degradation of pyrethroid insecticides: DM and fenvalerate in natural

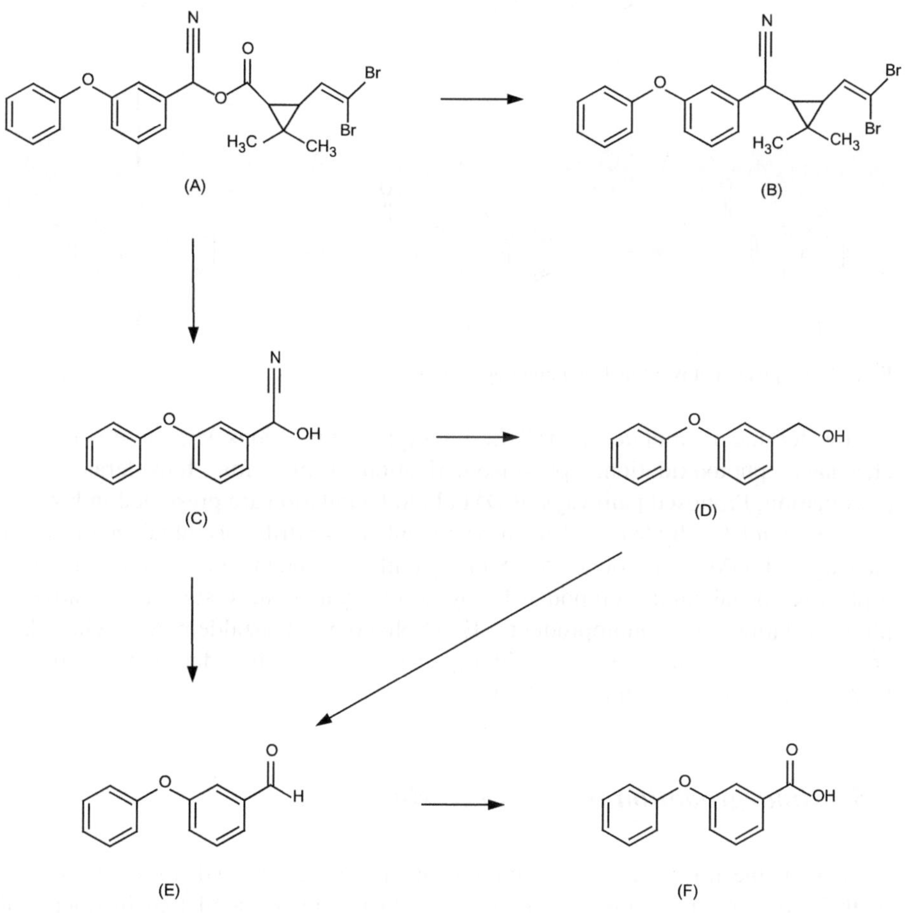

Fig. 12 Photodegradation of DM

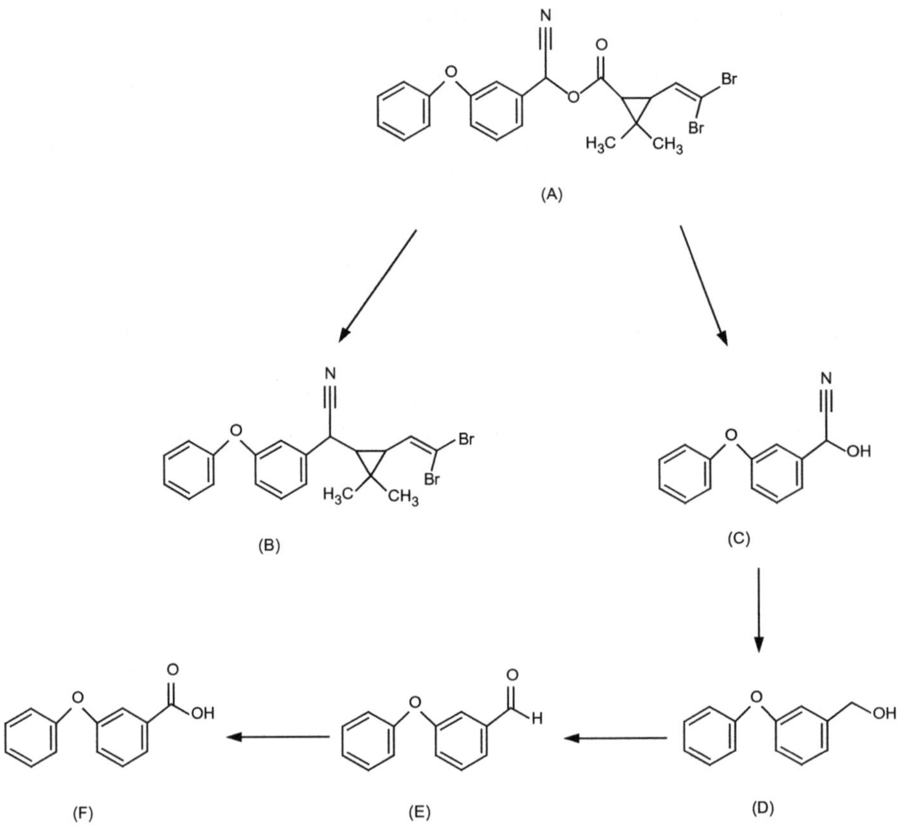

Fig. 13 Proposed pathway of DM photodegradation

river water under simulated solar light. Four types of reactions were observed: ester cleavage, photooxidation, photoisomerization (only for fenvalerate) and decyanation. Proposed pathways of DM photodegradation are presented in Fig. 13.

Compound C—hydroxyl 3-phenoxyphenyl acetonitrile—is obtained by ester cleavage of DM (compound A). Subsequently compound C is transformed to 3-phenoxy benalcohol (compound D) by DM decyanation. Result of derivative D photooxidation is photoproduct E—3-phenoxy benzaldehyde. The last compound F, 3-phenoxybenzoic acid, is generated by photooxidation of compound E (Nahri-Niknafs and Ahmadi 2013).

5.5 Biodegradation of Deltamethrin

Studies on the microbial transformation of the five pyrethroid insecticides: permethrin, deltamethrin, fastac, fenvalerate, fluvalinate revealed that in microbial culture, composed of *Bacillus cereus*, *Pseudomonas fluorescens* and

Achromobacter sp., DM was the most persistent among all of the investigated pesticides (Maloney et al. 1988).

The first DM-degrading bacterium strain was isolated in 1988, but it was not able to completely decompose this pesticide, because of toxic effects of the major product of this process to this microorganism (Chen et al. 2011).

Devisri and Iyer (2013) isolated from Koovar river water microorganisms: *Bacillus sp*, *Klebsiella sp*, *Pseudomonas sp* and *Staphylococcus sp*, which were able to biotransform DM.

Because of DM's lipophilic properties, it is characterized by strong adsorption to soil organic matter, and therefore low mobility in soil. Nevertheless, DM can degrade by soil microorganisms and photolysis. These processes mainly depend on soil type and oxygen availability.

According to studies described in (Grant et al. 2002; Cycoń et al. 2014) a bacterium *Serratia marcescens* strain DeI-1 and DeII 2 uses DM as a source of carbon and energy. Strain DeI-1 was more effective than DeII-2, which results from the higher sensitivity of the first strain to DM degradation product—3-phenoxybenzaldehyde.

Zhang et al. (2016) proved, that DM biodegradation depended on pH and temperature. Results indicate faster degradation in neutral and alkaline conditions because of lower enzyme activity in acidic conditions. The rate of biodegradation decreases at high temperature (40 °C), while enzyme activity decreases at low temperature (15 °C), as the bacterial growth is slower. Higher degradation rate is observed in soils containing *Bacillus cereus* strain Y1, in comparison to sterilized soil. It was also suggested that this bacterium strain could be applied in bioremediation processes.

Another bacterium, which is capable of complete degradation of DM and other pyrethroids, e.g. cypermethrin or permethrin is *Streptomyces aureus* strain HP-S-01. Degradation of DM was the most efficient in 18–38 °C temperature range and at pH in the range of 5–10 (Chen et al. 2011). Moreover, this pyrethroid pesticide could be also transformed by the following bacterium: *Micrococcus sp* (Tallur et al. 2008), *Brevibacterium aureum* (Chen et al. 2013), *Pseudomonas aeruginosa* strain JQ-41 (Song et al. 2015) and *Bacillus subtilis* strain BSF01 (Xiao et al. 2015).

As presented in Fig. 14, DM (A) is degraded by hydrolysis of carboxylester linkage to α-hydroxy-3-phenoxy-benzeneacetonitrile (B). This product is unstable and quickly transforms into 3-phenoxybenzaldehyde (C), which is subsequently degraded to 2-hydroxy-4-methoxybenzophenone (D). The major product of DM biodegradation is 3-phenoxybenzaldehyde (Chen et al. 2011).

5.6 *Toxicity of Deltamethrin*

DM is a chiral compound. Its toxicity depends on its stereochemical structure. Investigations show, that this pyrethroid insecticide is highly toxic to aquatic organism because it is easily washed from land into surface water. The WHO

Fig. 14 Pathway of DM biodegradation by *Streptomyces aureus* strain HP-S-01

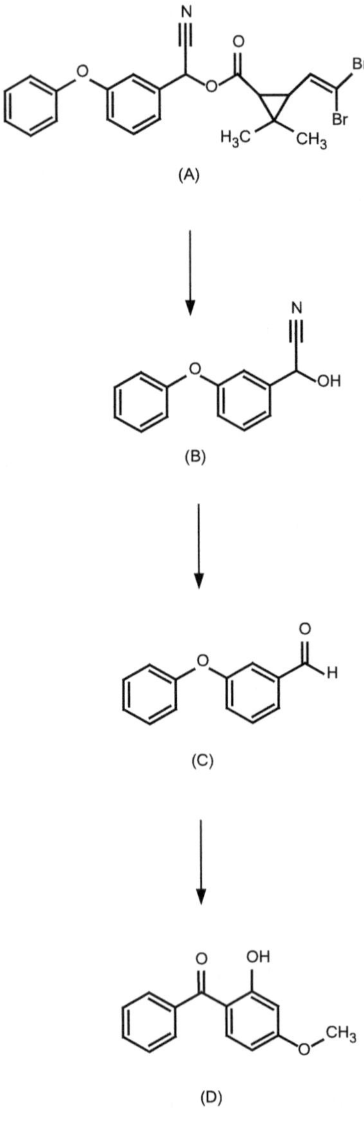

(World Health Organization) issued a report about the high toxicity of DM to fish (Ural and Saglam 2005; Tuzmen et al. 2008).

DM shows an adverse effect on the fecundity of the females of *Lysiphlebus fabarum Marshall* (Maradani et al. 2016). DM induced a significant rise in levels of oxidative stress in earthworms (*Eisenia foetida*), increasing the production of peroxidation products (malondialdehyde and protein carbonyls) and increasing the comet assay tail DNA% (determined by single-cell gel electrophoresis) (Shen et al. 2015a, b).

Abdel-Daim and coworkers (2016) examined the potential hepatoprotective, nephroprotective, neuroprotective and antioxidant effects of DM toxicity in male mice. Oral exposition of mice to DM revealed a significant increase in serum hepatic and renal injury biomarkers as well as TNF-a level and AChE activity. Moreover, liver, kidney and brain lipid peroxidation and oxidative stress markers were altered due to DM toxicity.

The teratogenicity of a DM solution and Decis® (an insecticide formulation containing DM) in chick embryos was investigated by Bhaskar and coworkers (2012, 2015).

Eggs of *Gallus domesticus* were immersed in aqueous emulsions of DM as well as Decis solution and incubated. After incubation the recovered embryos were evaluated for mortality rate, wet body weight, gross morphological and skeletal malformations. The result revealed that embryonic mortality markedly increased after administration of DM. The significant decrease in wet body weight and significant increase in percentage of abnormal survivors was observed in a dose dependent manner. A spectrum of external and skeletal malformations was observed in DM treated embryos. Among biochemical changes, total glycogen and RNA contents were significantly decreased in embryos exposed to DM. Similarly, a significant alteration was observed in alanine transaminase activity. These findings suggests that DM exhibits embryotoxic and teratogenic effects in the developing chick embryos.

6 Rimsulfuron

Rimsulfuron (RIM), N-((4,6-dimethoxypyrimidin-2-yl)aminocarbonyl)-3-(ethylsulfonyl)-2-pyridine-sulfonamide, is a herbicide of the sulfonylurea class. It was commercialized in 90s by Du Pont de Nemours (Martins et al. 2001).

It controls growth of grasses and some broadleaf weeds. This sulfonylurea herbicide is applied on corn crops at the 2–6 leaves stages. It can be also applied alone or in mixture with metribuzin (selective herbicide of the same chemical class as triazines) in potato crops. It is absorbed by the leaves and roots of the plants (Rouchaud et al. 1997).

RIM, like other sulfonylureas, quickly became popular because of low application doses and favorable environmental and toxicological properties (Nageswara Rao et al. 2012).

Rimsulfuron is a weak acid (see pKa in Table 2). Its molecule contains an aryl group, a sulfonylurea bridge and a heterocycle derived from triazine (Martins et al. 2001). Introduction of chlorine in the herbicide molecule enhances the stability of these compounds and prolongs their activity in the environment.

Its stability in the environment is low; it rapidly degrades in water and soil and it does not pose a risk to the environment. However, it is important to monitor pathways of its degradation, because derivatives can be more toxic than the parent compound. In soil under field conditions its decomposition is faster than under laboratory conditions. It results, like for TMX, from higher microbial activity and additional exposure to light. It is moderately fast degraded in aquatic system by photolysis. RIM has polar properties (see solubility in water at 20 °C and logP in Table 2) and it is not resistant to hydrolysis. It is also rapidly decomposed in water sediment systems and does not cause contamination of water (Martins et al. 2001; Seńczuk 2005). It has little or no toxological effects on mammals (Scrano et al. 1999).

Physical and chemical properties of RIM, including its half-lives in soil and water are listed in Table 2 (Source: The PPDB).

6.1 Abiotic Degradation of Rimsulfuron

6.1.1 Hydrolysis of Rimsulfuron

Hydrolysis of sulfonylurea herbicides like RIM in aqueous solutions proceeds according to first-order kinetics, and depends on pH and temperature. Hydrolysis under acidic conditions is based on cleavage of the sulfonylurea bridge. The following degradation products were identified: sulfonamide, the heterocyclic amine and carbon dioxide. Since these compounds are weak acids, in neutral and alkaline conditions they are much less likely to undergo hydrolysis (more persistent to hydrolysis). The rate of hydrolysis at pH above 10 increases due to base-catalyzed reaction mechanism. It was also proven, that with increases of temperature, stability of this compound to hydrolysis decreases. These studies were conducted under laboratory conditions, but the same results were obtained under field conditions (Kearney and Kaufman 1969).

6.1.2 Photodegradation of Rimsulfuron

Schneider and coworkers (1993) showed that RIM degradation in aqueous neutral solutions and in soil does not depend on irradiation. In acidic conditions (pH = 5) they observed photodegradation of RIM. The same conclusions were reached by Scrano and coworkers (1999). They also observed that exposure to light did not affect the rate of degradation under neutral and alkaline conditions. Simultaneously, it was shown that hydrolysis conducted under sunlight is faster in acidic or alkaline conditions, than neutral.

Under simulated sunlight in water, at pH 5 and 9, the photodegradation half-life was in the range of 1–9 days. The hydrolysis rate was as high as the photolysis rate and both decrease with increasing pH. The pathway of RIM degradation in neutral and alkaline conditions is presented in Fig. 15 (Scrano et al. 1999).

RIM was degraded to N-(4,6-dimethoxy-2-pyrimydinyl)-N-[(3-(ethylsulfonyl)-2-pyridynyl)]urea (B) and N-[(3-ethylsulfonyl)-2-pyridynyl]-4,6-dimethoxy-2-pyridineamine (C) under neutral and alkaline conditions. The major metabolite

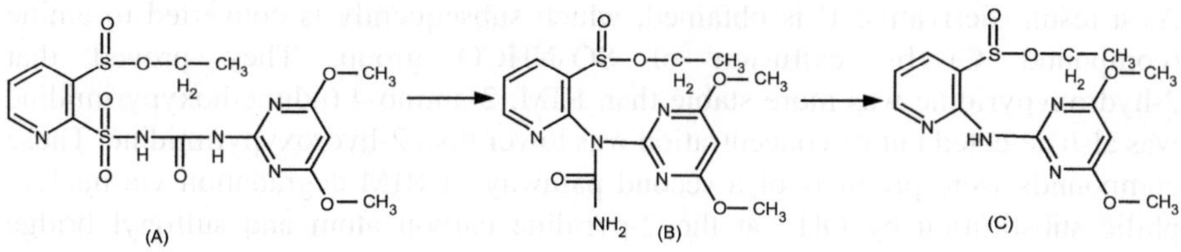

Fig. 15 Abiotic degradation of RIM in neutral and alkaline conditions

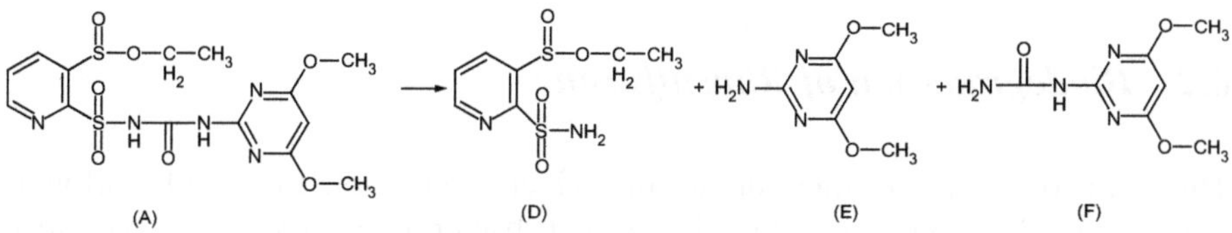

Fig. 16 Abiotic degradation of RIM in acidic conditions

under these conditions was compound C, which was more stable than compound B (Scrano et al. 1999).

Hydrolysis in acidic solutions (pH < 7) was also observed. The main pathway of RIM abiotic degradation under acidic condition is presented in Fig. 16 (Scrano et al. 1999).

The major metabolites of RIM (A) under acidic conditions were: 3-(ethylsulfonyl)-2-pyridinesulfonamide (D), 2-amino-4,6-dimethoxypyrimidine (E) and N-(4,6)-dimethoxy-2-pyrimidynyl)urea (F) (Scrano et al. 1999).

Martins and Marmoud (1999) conducted experiments on RIM degradation in aqueous solutions and soil suspensions. They observed, similar to above-mentioned scientists, that this herbicide was rapidly degraded with the same rate in both environments and its degradation rate depended on pH. They revealed that derivative B (Fig. 15) is also persistent in soil.

In distilled water containing KOH (pH = 10) Martins and coworkers (2001) also observed that RIM is hydrolyzed to compound C (Fig. 15). These studies confirmed the similar hydrolysis sensitivity of metabolite B to RIM and the higher persistence of compound C than the parent compound, which could pose a risk for groundwater contamination. They observed a higher rate of RIM degradation in aqueous soil suspensions under field conditions than in water. Moreover, it was shown that the initial concentration of RIM also influences the degradation processes of its metabolites. These investigations revealed a higher rate of metabolite C degradation for the low concentration of parent compound, while it was not observed for metabolite B (Fig. 15). It was explained by inhibition of microbial activity at high concentrations.

Metabolism of RIM in soil was also examined by Rouchaud and coworkers (1997). RIM transformation undergoes intramolecular nucleophilic substitution.

As a result, derivative B is obtained, which subsequently is converted to amine (compound C) by extrusion of SO_2NHCO group. They proved that 2-hydroxypyridine was more stable than RIM. 2-amino-4,6-dimethoxypyrimidine was also detected but its concentration was lower than 2-hydroxypyrimidine. These compounds were products of a second pathway of RIM degradation via nucleophilic substitution by OH^- at the 2-pyridine carbon atom and sulfonyl bridge breakage. These results were partially confirmed by Schneider et al. (1993), who observed only compounds B and C.

6.2 Biodegradation of Rimsulfuron

There are some types of microorganisms, which are capable to degrade sulfonylurea herbicides: actinomycetes, fungi and bacteria. *Streptomyces griseolus*, belongs to acinetomycetes, degraded chlorsulfuron via conversion of the methoxy to hydroxyl group. *Aspergillus niger* and *Penicillium sp.* (fungi) were able to hydrolyze the sulfonylurea bridge. Sulfonamide and heterocyclic compounds were the products of this reaction. Two degradation pathways of RIM (Fig. 17) were proposed, similarly to other sulfonylurea herbicides (Kearney and Kaufman 1969).

According to Song et al. (2013) RIM, as well as others sulfonylurea herbicides: tribenuron methyl, chlorsulfuron, bensulfuron methyl, methametsulfuron methyl, cinosulfuron and nicosulfuron, were degraded by *Talaromyces flavus*, strain LZM1. These fungi probably used RIM as a nitrogen source, since the pathways of RIM's degradation occurred via cleavages of N-containing bonds.

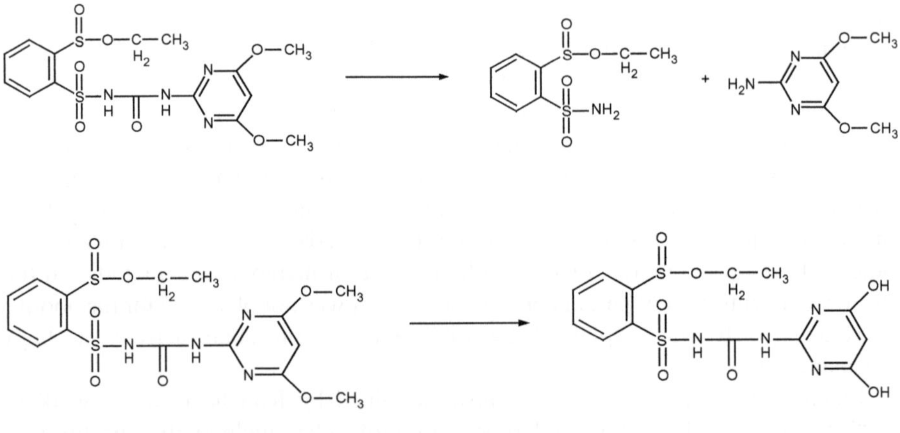

Fig. 17 The possible biodegradation pathway of RIM

6.3 *Toxicity of Rimsulfuron*

In the environment RIM decomposes to five degradation products, the major two of them are: [N-(4,6-dimethoxypyrimidin-2-yl)-N-((3-ethylsulfonyl)-2-pyridinyl) urea] and [N-((3-ethylsulfonyl)-2-pyridinyl)-4,6-dimethoxy-2-pyrimidineamine], both of which are more stable in the environment than the parent compound (380–1100 days, 101–214 days for metabolites, respectively and 6–40 days for parent compound) (EFSA 2005; Rosenbom et al. 2010).

The herbicidal activity of RIM is specific to plants, however, its degradation products are not herbicides and they may be less toxic to algae than the parent compound, therefore the toxicity test was conducted with *D. manga* and *V. fischeri*. The assessment of the toxicity of RIM and their major degradation products was the object of Martins and coworkers' investigations. At the maximum tested concentration, RIM and its major degradation products were toxic neither to *Dapnia manga* nor to *Vibro fischeri* (Martins et al. 2001).

The toxicity of RIM and other sulfonylurea herbicides towards *Lemna gibba* was investigated at three pH levels, at two temperatures and two light regimes. In all cases RIM exhibited toxicity to *Lemna gibba*, however the EC50 values were different for different experimental conditions (Rosenkrantz et al. 2013).

Risk assessment of RIM and its primary degradation products indicates adverse effects on earthworms (long-term), non-target terrestrial plants and aquatic plants (EFSA 2005; Rosenbom et al. 2010).

Dinelli and coworkers (1998) ascertained, that a high concentration of rimsulfuron in soil (5 mg/kg) resulted in a transient increase of both respiration and dehydrogenase activity.

7 Metalaxyl

Metalaxyl (MET), *N*-(2,6-dimethylphenyl)-*N*-(methoxyacetyl)alanine methyl ester, is a pesticide, which belongs to phenylamide fungicides. It was first manufactured in 1977 by the Ciba-Geigy Corporation (Zuno-Floriano et al. 2012).

(−)-R-metalaxyl is the biological active enantiomer, which acts by inhibiting the synthesis of ribosomal RNA in the fungus. Since 1996 in most countries, rac-metalaxyl is replaced by metalaxyl-M ((−)-R-metalaxyl) (Xu et al. 2011).

MET is a systemic fungicide, which is adsorbed by roots, translocated and extensively metabolized by plants (Zuno-Floriano et al. 2012).

Application of MET allows protecting potato from *Peronosporales*, especially *Phytophthora infestans*, a pathogen, which affects leaves, stems and tubers of potatoes, but does not affect their growth. It inhibits uridine incorporation into RNA by interfering with transcription. Specifically it inhibits RNA polymerase-1 of the fungi and blocks rRNA synthesis (Sharma and Awasthi 1997; Sukul and Spiteller 2000; Zuno-Floriano et al. 2012). In soil MET is moderately resistant to decomposition. It is also a heat resistant and photostable compound. It exhibits high stability in aqueous solutions within 1–8.5 pH range. At higher pH (e.g. 10) it decomposes faster and degrades rapidly in water, like other fungicides (Sukul and Spiteller 2000; Katagi 2002). In water sediments systems, it is moderately fast degraded and its half-life is about 56 days.

MET does not affect reproduction of animals and is not a teratogenic or mutagenic compound but it provokes cell alterations in mouse liver at the dose of 2.5 mg/kg/day; in dogs at dose of 0.8 mg/kg/day. It alters alkaline phosphatase levels in blood and causes an increase in liver and brain weight (Zuno-Floriano et al. 2012).

Physical and chemical properties of MET, including its half-lives in soil and water are listed in Table 2 (Source: The PPDB).

7.1 Abiotic Degradation of Metalaxyl

7.1.1 Hydrolysis of Metalaxyl

MET possesses an amide linkage moiety, which is resistant to hydrolysis in aquatic systems. Nevertheless, it undergoes degradation under acidic and alkaline conditions with half-lives of 200 and 115 days, respectively (Sukul and Spiteller 2000).

According to Katagi (2002), in acidic conditions, as a result of H_2O attack on the O-protonated carbonyl carbon, a tetrahydral intermediate is formed. Water-assisted C-N fission is an effect of protonation at the amide nitrogen. Alkaline hydrolysis is initiated by nucleophilic attack of the hydroxide on the carbonyl carbon, which leads to the formation of monoanionic tetrahydral intermediate. Proposed pathways of these reactions are presented in Fig. 18.

7.1.2 Photodegradation of Metalaxyl

As MET does not absorb light above 290 nm (it has maximum absorption at 196 nm), it is resistant to sunlight (Sukul and Spiteller 2000).

Photolysis of MET in UV light leads to rearrangement of the N-acyl group into an aromatic ring, demethoxylation, N-deacylation and elimination of the methoxycarbonyl group from the molecule. In the presence of aerobic organisms dwelling in humic acids, 65 % of metalaxyl is degraded. Humic acids, present in soil, produce OH radicals, which could degrade MET. Therefore, in the absence of humic acids photodegradation of MET does not occur (Sukul and Spiteller 2000).

MET degrades in aqueous solution in pH range from 2.8 to 8.8 exposed to UV light. The lower pH of the solution the rate of degradation is higher (Sukul and Spiteller 2000).

Marucchini and coworker (2002) investigated the stereoselective degradation of MET (as racemate) and MET-M (as (−)-(R)-enantiomer) in soil. The obtained

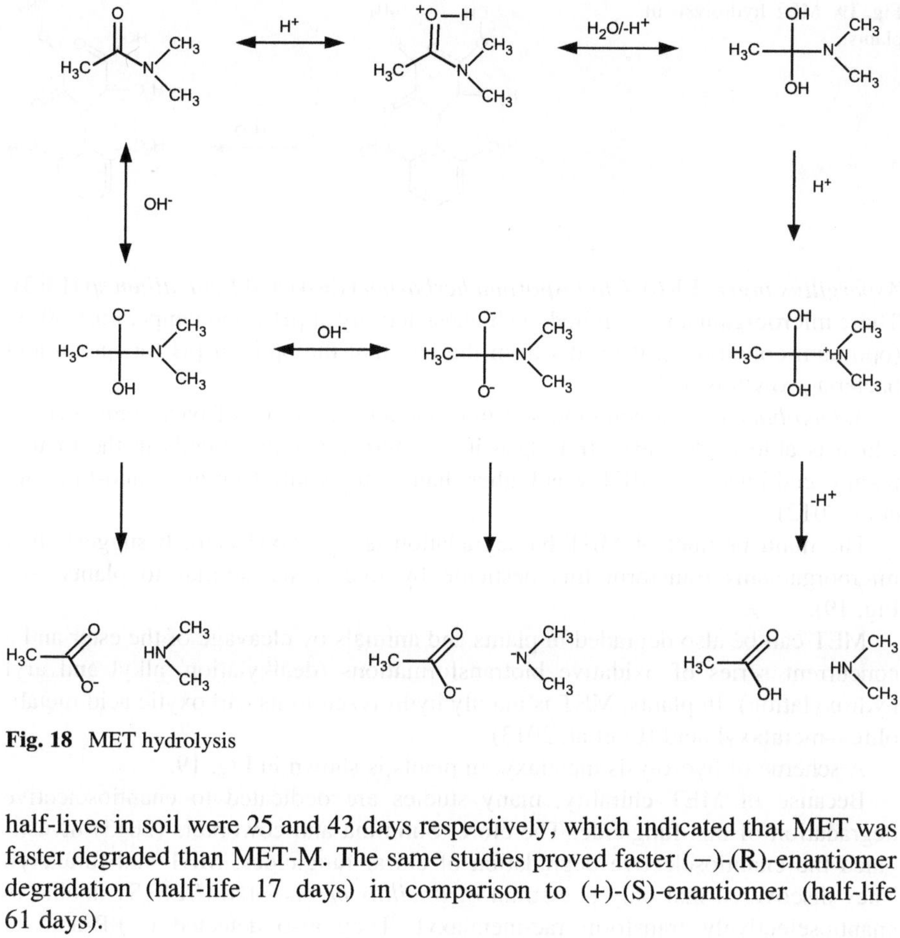

Fig. 18 MET hydrolysis

half-lives in soil were 25 and 43 days respectively, which indicated that MET was faster degraded than MET-M. The same studies proved faster (−)-(R)-enantiomer degradation (half-life 17 days) in comparison to (+)-(S)-enantiomer (half-life 61 days).

7.2 *Biodegradation of Metalaxyl*

One of the first investigations obtained by Bailey and Coffey (1985) showed that MET biodegradation occurred in soil. MET was stable in sterilized soil and its concentration was only about 10 % lower in comparison to the initial concentration 70 days earlier. In unsterilized soil it was degraded faster and the average half-life was 28 days, but for the most active soil it was 14 days. These results suggested that MET undergoes biodegradation processes.

Studies conducted by Massoud et al. (2008) showed that *Pseudomonas sp* (ER2) is the most effective bacteria for MET degradation in aquatic systems. Therefore, it is used for detoxification of water contaminated with MET. According to (Massoud et al. 2008) there were also three fungi species, that took part in MET degradation:

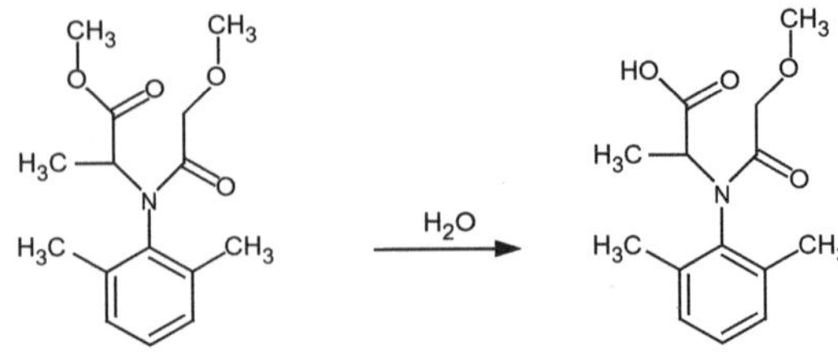

Fig. 19 MET hydrolysis in plants

Aspergillus niger (ER6), *Cladosporium herbarum* (ER4) and *Penicillium sp* (ER3). These microorganisms exhibited the highest activity at pH 7 and temperature 30 °C (optimum conditions). It results from the fact that the optimal pH for fungal and bacteria growth is 6–8.

Acinetobacter sp, a common soil microorganism, is one of bacterium species, which is able to degrade MET. Density of *Acinetobacter sp* cells in the treated potato seedlings with MET was higher than in the control group (Zuno-Floriano et al. 2012).

The main product of MET biodegradation is metalaxyl acid. It suggests that microorganisms transform this pesticide by hydrolysis, similar to plants (see Fig. 19).

MET can be also degraded in plants and animals by cleavage of the ester and a concurrent series of oxidative biotransformations (dealkylation, alkyl and aryl hydroxylation). In plants, MET is mainly hydrolyzed to its carboxylic acid metabolite—metalaxyl acid (Li et al. 2013).

A scheme of hydrolysis metalaxyl in plants is shown in Fig. 19.

Because of MET chirality, many studies are dedicated to enantioselective degradation of this fungicide. Therefore, Sulimma and coworkers (2013) investigated the enantioselective degradation of chiral fungicides: MET and furalaxyl. They discovered that only one strain of *Bacillus brevis*—INFU/LS_S1 is able to enantioselectively transform rac-metalaxyl. They also detected (−)-R-acid at higher concentration in the experimental medium (pH = 7.4) which suggests faster degradation of (−)-R-metalaxyl. It is important due to the fact that currently used agrochemicals consist of (−)-R-metalaxyl, which has higher fungicidal activity. Proposed MET metabolites are presented in Fig. 20.

Moreover, obtained results confirmed, that metabolism of MET is an enantioselective process: R-metalaxyl was degraded to pure R-metalaxyl acid and S-metalaxyl was degraded to pure S-metalaxyl acid by bacteria (Sulimma et al. 2013; Kalathoor et al. 2015). The same results were obtained by Li and coworkers (2013).

The R- and S-enantiomers of MET were also determined in cucumber, cabbage, spinach and pakchoi. The results showed that in cucumber R-enantiomer dissipated faster than S-enantiomer. Inversely, a preferential degradation of S-form was found in spinach and pakchoi. In cabbage no stereoselective degradation was observed (Wang et al. 2014).

Fig. 20 Proposed MET metabolites in soil microorganisms

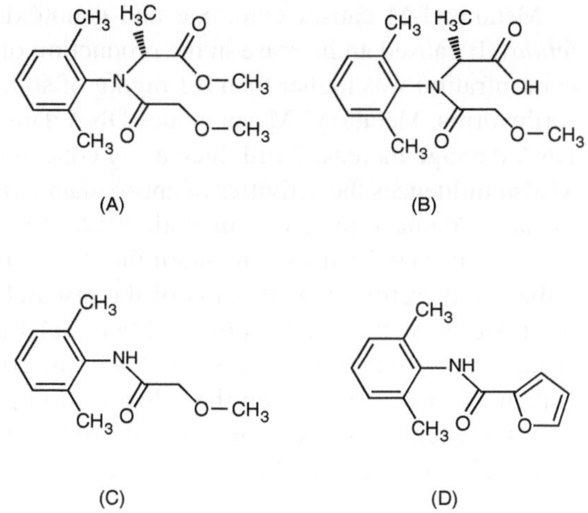

Buerge et al. (2003) observed correlations between soil pH and enantioselectivity of MET degradation. In alkaline soils MET half-life is in the range of 9–32 days, but in acidic soils it ranges between 16 and 127 days. It suggested a dependence of enantioselectivity degradation on soil pH. (−)-R-metalaxyl degraded is faster than (+)-S-metalaxyl (positive enantioselectivities) in acidic soils, but in alkaline soils an inverse situation is observed: S-enantiomer is degraded faster than R-enantiomer (negative enantioselectivities). The concentrations of both enantiomers also decrease according to first-order kinetics.

7.3 *Toxicity of Metalaxyl*

As mentioned above, MET is a chiral fungicide due to the presence of the stereogenic center in the alkyl moiety. The anti-fungicidal activity mostly originates from the R-enantiomer, which is also much more toxic than rac-metalaxyl (Chen and Liu 2008; Xu et al. 2011).

The research on the enantioselective distribution, degradation and excretion of MET after oral administration of rac-MET to mice was described in Zhang et al. (2014).

The concentration of R-MET was much higher than S-MET in heart, liver, lung, urine and feces. R-MET was also more persistent in the organism (degradation half-life: 3.0 h) then the S-MET (2.2 h). The main biotransformation reactions of MET were hydroxylation, demethylation and didemethylation. The concentration of products of these reaction was higher in urine and feces than those in tissues.

Above mentioned examples of different behavior of MET isomers indicate that the enantioselectivity of MET lies in distribution, degradation and excretion processes, which depend on the organism.

Metalaxyl-M causes cytotoxic and genotoxic effects on earthworms (*Eisenia fetida*). It caused an increase in the production of reactive oxygen species when the concentration was higher than 0.1 mg/kg of soil, which led to lipid peroxidation in earthworms. Metalaxyl-M can induce DNA damage in earthworms, and the level of DNA damage increased with increasing concentration of the fungicide. Metalaxyl-M also influences the activities of antioxidant enzymes, which results in irreversible oxidative damage in cells (Liu et al. 2014a, b).

MET is classified as Xn—harmful, N—dangerous for the environment (see Table 2), therefore, the attention of the researchers is focused rather on assessing the toxicity of the MET isomers (Chen and Liu 2008; Wilson et al. 2001; Yao et al. 2008; Kungolos et al. 2009; Xu et al. 2011; Bermúdez-Couso et al. 2013; Kalathoor et al. 2015), then their degradation products. However, to obtain complete knowledge about the influence of MET on the environment, the in-depth study on toxicity of its metabolites is necessary.

8 Influence of Pesticides and Their Degradation Products on Humans

Pesticides represent a potential threat to humans. Since pesticides are present in almost all parts of environment, the human exposure to agrochemicals can occur via a number of pathways such as indirect (e.g. through food, drinking water, residential and occupational exposure) and direct (oral, inhalation and dermal). However, the major concerns are from consumption of pesticide residues in food crops and contaminated water (Bakırcı et al. 2014). Therefore, to ensure the safety of food for consumers and to protect their health, the monitoring of pesticide residues in food products must be pursued. The levels of pesticide residues allowed in foodstuffs are legislatively controlled through setting maximum residue levels (MRLs). These MRLs limit the types and amount of pesticides that can be legally present on foods, as determined by various regulatory bodies which minimize consumer exposure to harmful or unnecessary intake of pesticides worldwide. The MRLs of pesticides residues in potatoes were established: 0.01 mg/kg for RIM, 0.02 mg/kg for LC, 0.05 mg/kg for MET, 0.2 mg/kg for DM, 0.3 mg/kg for TMX, according to European Union legislation (Commission Regulation 524/2011, 441/2012, 834/2013, 617/2014). Similar legislation applies in USA, China and other countries.

In European Union countries the concentration of pesticide residues in drinking water must not exceed the Maximum Allowable Concentration (MAC), that is established as 0.1 μg/L for a single pesticide and 0.5 μg/L for the sum of pesticides concentration (Council Directive 98/83/EC).

The specific risk is associated with contamination of water supplies that has a direct impact not only on human health and aquatic organisms, but also pesticides are translocated in the environment mainly with rainwater, which elutes them from the surface of plants and soil, and then they may be transferred with groundwater on long distances. Hereby, large areas are uncontrollably contaminated with agrochemicals.

It should be also added that the environment is contaminated not only by pesticides, but also by their degradation products. Therefore, the technologies that enable pesticide removal from water and soil purification are of great importance. There are many ways of water and soil purification, however many of these solutions have drawbacks. In fact, in some cases these procedures can leave the solution more toxic than it was originally (Lopes et al. 2008; Šojić et al. 2012; Mir et al. 2013; Zabar et al. 2012; Laberge and Rollinson 2013). As a result, aside from pesticide degradation products that are a result of chemical reactions that naturally occur in environment (hydrolysis, photolysis etc.), there are many water purification by-products.

Despite the growing awareness of the threat, which brings the usage of pesticides, replacement the old generation of pesticides with less toxic but more selective agrochemicals, or attempts to use "biopesticides", the pesticides are an inseparable part of modern agriculture.

The widespread use of pesticides causes, that these compounds are present in almost every element of the ecosystems. According to the latest studies (Bonmatin et al. 2015) the concentration of neonicotinoids, including TMX in soil is parts per billion (ppb)-parts per million (ppm) range; in water, parts per trillion (ppt)-ppb range, and plants, ppb-ppm range. According to (Starner and Goh 2012; Masiá et al. 2013; Van Dijk et al. 2013; de Lafontaine et al. 2014) the pesticides concentration in surface water has been found to exceed ecotoxicological limits imposed by national and international organizations. This provides multiple routes for chronic (and acute in some cases) exposure of non-target animals. More detailed information was presented by Main et al. (2014). Sediment and water samples (total 440) were collected before seeding, during the growing season and after the harvest of crops. At least one of the following neonicotinoids: clothianidin, thiamethoxam, imidacloprid, or acetamiprid was found in 16–91 % of the samples, depending on the time of sampling. TMX was predominant in water samples at mean concentration 40.3 ng/L; its maximum concentration was 1490 ng/L.

TMX was also determined in pollen or in pollen-derived matrices in the concentration range of 2.8–28.9 ng/g (Pohorecka et al. 2012; Stoner and Eitzer 2012; Sanchez-Bayo and Goka 2014). Depending on the source of material, TMX was determined in 1–37 % of collected samples.

Lafontaine et al. (2014) determined RIM in 24 % of surface water samples, collected in 2009–2011, at the concentration up to 233 ng/L. RIM exhibits low mammalian toxicity and short half-live in plants; therefore, its concentration in food of animal origin is often below MRL (Wu et al. 2012; Pang et al. 2016).

According to Allinson et al. (2015), MET was determined in 46 % of surface water samples at the concentration range 0.01–0.191 μg/L. MET was detected in 1 % of vegetables samples (total number of samples 300), in the concentration range: 0.019–0.034 mg/kg, thereby, the chronic and acute intake risk of this fungicide was assessed as low (Fang et al. 2016).

The monitoring studies conducted during the last 5 years revealed (Bakırcı et al. 2014; Sivaperumal et al. 2015; Szpyrka et al. 2015), that about 2–17 % of potatoes available on the European market contained MET.

Although the half-lives of pyrethroid pesticides in the environment are typically on the order of few days to weeks (Table 2), they are found in water, sediments, and biological samples (Chang et al. 2016; Corcellas et al. 2015). In 13 % of river water samples collected by Feo and coworkers (2010) DM was determined at the range of concentration 2–58.8 ng/L. These concentration levels were higher than median lethal concentration values found for DM when short time toxic effects were considered. At the same time 60 % of sediment samples contained LC at concentration of 0.42–16.6 mg/kg (Hunt et al. 2016). LC was determined in 21 % of peer samples at the concentration range 83–950 μg/kg (Han et al. 2015). DM was determined in 40 % of *Ginko biloba* samples (max. concentration 10.1 μg/kg) (Páleníková et al. 2015). LC and DM were determined in black, green and scented tea. The concentration of LC was in range 17.56–58.80 mg/kg. DM was determined in one black tea sample (2.31 mg/kg) (Cao et al. 2015).

According to above cited latest literature reports, pesticide residues are determined in less than 50 % of environmental samples, often below the maximum concentration set by legislation. Unfortunately, the studies rarely encompass also the monitoring of the pesticide degradation products, therefore the overall risk associated with pesticide residues in environment, and the corresponding human exposure, are difficult to assess.

Although, TMX has favorable safety profiles, due to its preferential affinity for nicotinic receptor (nAChR) subtypes in insects, poor penetration of the mammalian blood–brain barrier, and low application rates (Simon-Delso et al. 2015), there are more and more reports concerning its negative influence on mammals. All of these reports were summarized with critical comments by Sheets and coworkers (2015).

The results of the research on the enantioselective distribution and degradation of metalaxyl in different species of plants, invertebrates, and mammals, indicated that all these processes were enantioselective (Chen and Liu 2008; Xu et al. 2011; Li et al. 2013; Wang et al. 2014; Zhang et al. 2014). These results may have potential implications to predict toxicity and provide additional information associated with adverse health effects for risk assessment of MET.

Pyrethroids are extremely toxic to aquatic organisms (aquatic insects are sensitive to insecticide poisoning from less than 1 ppb concentrations) (Mugni et al. 2013; Antwi and Reddy 2015), but tend to be less toxic to mammals and birds than other types of insecticides, due to more effective and rapid biotransformation. However, pyrethroids are active in both insect, bird and mammalian neurons. Their potential to cause adverse effects in laboratory animals has been well studied in regulatory tests, and there are data which indicate that the neuronal effects do occur in humans. Exposure to DM leads to hepatotoxic, cardiovascular, immunosuppression nephrotoxic, neurotoxic and reproductive side effects for human and many species, including birds and fish (Bhaskar et al. 2012; Abdel-Daim et al. 2016; Doe et al. 2016).

9 Summary and Conclusion

Pesticides are essential in modern agriculture, however, their use is associated with toxicological risk. Therefore, for many years attention has been drawn to pesticide residue monitoring in the environment (soils, sediments, water, air, plant and animal tissues). Nevertheless, in many cases the information about pesticide concentrations in the environment is insufficient to assess the total risk connected with the presence of these compounds in environment. After spraying, agrochemicals undergo various transformations: they are degraded by microorganisms as well as they are decomposed in abiotic processes: hydrolysis, photolysis, redox-reactions etc. All of these processes are influenced by external (temperature, insolation, type of matrix etc.) and internal conditions (physico-chemical properties of the pesticide). As a result, a mixture of degradation products is formed. The ecotoxicological properties of pesticide by-products are usually different then the parent compounds. Therefore, it is essential to investigate the degradation pathways of pesticides and the influence of the degradation products on the environment.

For the review, we chose popular pesticides used in potato production. The potato is one of the most popular crops, and consequently, pesticides used in its cultivation are introduced into the environment in large quantities. While many studies focus on pesticide degradation processes, there are only few reports concerning degradation product toxicity.

Available results indicate both the significance (toxicity of degradation products to the investigated organisms) and the complexity of the problem (low levels of concentration, coexistence of many degradation products with a similar properties, difficulties in ensuring reproducibility of experimental conditions). A comprehensive research to determine overall pesticide impact on the environment is, however, yet to be undertaken.

References

Abdel-Daim M, El-Bialy BE, Rahman HGA, Radi AM, Hefny HA, Hassan AM (2016) Antagonistic effects of Spirulina platensis against sub-acute deltamethrin toxicity in mice: biochemical and histopathological studies. Biomed Pharmacother 77:79–85

Alalm MG, Tawfik A, Ookawara S (2015) Combined solar advanced oxidation and PAC adsorption for removal of pesticides from industrial wastewater. J Mater Environ 6(3):800–809

Allinson G, Zhang P, Bui AD, Allinson M, Rose G, Marshall S, Pettigrove V (2015) Pesticide and trace metal occurrence and aquatic benchmark exceedances in surface waters and sediments of urban wetlands and retention ponds in Melbourne, Australia. Environ Sci Pollut Res 22:10214–10226. doi:10.1007/s11356-015-4206-3

Ansari RW, Shukla RK, Yadav RS, Seth K, Pant AB, Singh D, Agrawal AK, Islam F, Khanna VK (2012) Cholinergic dysfunctions and enhanced oxidative stress in the neurobehavioral toxicity of lambda-cyhalothrin in developing rats. Neurotox Res 22:292–309. doi:10.1007/s12640-012-9313-z

Antwi FB, Reddy GVP (2015) Toxicological effects of pyrethroids on non-target aquatic insects. Environ Toxicol Pharmacol 40:915–923. doi:10.1016/j.etap.2015.09.023

Bailey AM, Coffey MD (1985) Biodegradation of metalaxyl in avocado soils. Phytopathology 75:135–137

Bakırcı GT, Acay DBY, Bakırcı F, Ötles S (2014) Pesticide residues in fruits and vegetables from the Aegean region, Turkey. Food Chem 160:379–392

Balfour NJ, Carreck NL, Blanchard HE, Ratnieks FLW (2015) Size matters: significant negative relationship between mature plant mass and residual neonicotinoid levels in seed-treated oilseed rape and maize crops. Agric Ecosyst Environ 215:85–88

Barceloux DG (2008) Potatoes, tomatoes, and solanine toxicity (*Solanum tuberosum* L., *Solanum lycopersicum* L.). In: Barceloux DG (ed) Medical toxicology of natural substances: foods, fungi, medicinal herbs, toxic plants, and venomous animals. Wiley, Hoboken, NJ, pp 77–83

Bass C, Denholm I, Williamson MS, Nauen R (2015) The global status of insect resistance to neonicotinoid insecticides. Pest Biochem Phys 121:78–87

Beketov MA, Liess M (2008) Acute and delayed effects of the neonicotinoid insecticide thiacloprid on seven freshwater arthropods. Environ Toxicol Chem 27:461–470

Bermúdez-Couso A, Nóvoa-Muñoz JC, Arias-Estévez M, Fernández-Calviño D (2013) Influence of different abiotic and biotic factors on the metalaxyl and carbofuran dissipation. Chemosphere 90:2526–2533. doi:10.1016/j.chemosphere.2012.10.090

Bhaskar N, Shahani L, Taparia N, Bhatnagar P (2012) Effect of deltamethrin containing formulation on developing chick embryo: morphological and skeletal changes. Int J Toxicol Pharmacol Res 4:81–87

Bhaskar N, Shahani L, Bhatnagar P (2015) Toxicological implications of a commercial formulation of deltamethrin (Decis®) in developing chick embryo. Hum Ecol Risk Assess. doi:10.1080/10807039.2015.1071647

Biziuk M (2001) Pestycydy—występowanie, oznaczanie i unieszkodliwianie. Wydawnictwo Naukowo-Techniczne, Warszawa

Blacquiere T, Smagghe G, Van Gestel CA, Mommaerts V (2012) Neonicotinoids in bees: a review on concentrations, side-effects and risk assessment. Ecotoxicology 21:973–992

Bonmatin J-M, Giorio C, Girolami V, Goulson D, Kreutzweiser DP, Krupke C, Liess M, Long E, Marzaro M, Mitchell AED, Noome DA, Simon-Delso N, Tapparo A (2015) Environmental fate and exposure; neonicotinoids and fipronil. Environ Sci Pollut Res 22:35–67. doi:10.1007/s11356-014-3332-7

Buerge IJ, Poiger T, Muller MD, Buser HR (2003) Enantioselective degradation of metalaxyl in soils: chiral preference changes with soil pH. Eviron Sci Technol 37:2668–2674

Burrows HD, Canle M, Santaballa JA, Steenken S (2002) Reaction pathways and mechanisms of photodegradation of pesticides. J Photochem Photobiol B Biol 67:71–108. doi:10.1016/S1011-1344(02)00277-4

Cao Y, Tang H, Chen D, Li L (2015) A novel method based on MSPD for simultaneous determination of 16 pesticide residues in tea by LC–MS/MS. J Chromatogr B 998–999:72–79. doi:10.1016/j.jchromb.2015.06.013

Carreck NL, Ratnieks FLW (2014) The dose makes the poison: have 'field realistic' rates of exposure of bees to neonicotinoid insecticide been overestimated in laboratory studies. J Apicult Res 53:607–614

Chang DC, Park CS, Kim SY, Lee YB (2012) Growth and tuberization of hydroponically grown potatoes. Potato Res 55:69–81

Chang J, Wang Y, Wang H, Li J, Xu P (2016) Bioaccumulation and enantioselectivity of type I and type II pyrethroid pesticides in earthworm. Chemosphere 144:1351–1357. doi:10.1016/j.chemosphere.2015.10.011

Chen S, Liu W (2008) Toxicity of chiral pesticide rac-metalaxyl and R-metalaxyl to *Daphnia magna*. Bull Environ Contam Toxicol 81:531–534. doi:10.1007/s00128-008-9567-6

Chen S, Lai K, Li Y, Hu M, Zhang Y, Zen Y (2011) Biodegradation of deltamethrin and its hydrolysis product 3-phenoxybenzaldehyde by a newly isolated *Streptomyces aureus* strain HP-S-0. Appl Microbiol Biotechnol 90:1471–1483

Chen S, Dong YH, Chang C, Deng Y, Zhang XF, Zhong G, Song H, Hu M, Zhang LH (2013) Characterization of a novel cyfluthrin-degrading bacterial strain *Brevibacterium aureum* and its biochemical degradation pathway. Bioresour Technol 132:16–23

Chen S, Deng Y, Chang C, Lee J, Cheng Y, Cui Z, Zhou J, He F, Hu M, Zhang LH (2015) Pathway and kinetics of cyhalothrin biodegradation by *Bacillus thuringiensis* strain ZS-19. Sci Rep 5:8784. doi:10.1038/srep08784

Chiipanthenga M, Maliro M, Demo P, Njoloma J (2012) Potential of aeroponics system in the production of quality potato (Solanum tuberosum l.) seed in developing countries. J Biotechnol 11:3993–3999

Colombo R, Ferreira TCR, Suellen AA, Carneiro RL, Lanza MRV (2013) Application of the response surface and desirability design to the Lambda-cyhalothrin degradation using photo-Fenton reaction. J Environ Manage 118:32–39

Commission Regulation 441/2012, 24 May 2012, amending Annexes II and III to Regulation (EC) No 396/2005 of the European Parliament and of the Council as regards maximum residue levels for bifenazate, bifenthrin, boscalid, cadusafos, chlorantraniliprole, chlorothalonil, clothianidin, cyproconazole, deltamethrin, dicamba, difenoconazole, dinocap, etoxazole, fenpyroximate, flubendiamide, fludioxonil, glyphosate, metalaxyl-M, meptyldinocap, novaluron, thiamethoxam, and triazophos in or on certain products

Commission Regulation 524/2011, 26 May 2011, amending Annexes II and III to Regulation (EC) No 396/2005 of the European Parliament and of the Council as regards maximum residue levels for biphenyl, deltamethrin, ethofumesate, isopyrazam, propiconazole, pymetrozine, pyrimethanil and tebuconazole in or on certain products

Commission Regulation 617/2014, 3 June 2014, amending Annexes II and III to Regulation (EC) No 396/2005 of the European Parliament and of the Council as regards maximum residue levels for ethoxysulfuron, metsulfuron-methyl, nicosulfuron, prosulfuron, rimsulfuron, sulfosulfuron and thifensulfuron-methyl in or on certain products

Commission Regulation 834/2013, 30 August 2013, amending Annexes II and III to Regulation (EC) No 396/2005 of the European Parliament and of the Council as regards maximum residue levels for acequinocyl, bixafen, diazinon, difenoconazole, etoxazole, fenhexamid, fludioxonil, isopyrazam, lambda-cyhalothrin, profenofos and prothioconazole in or on certain products

Corcellas C, Eljarrat E, Barcelo D (2015) First report of pyrethroid bioaccumulation in wild river fish: a case study in Iberian river basins (Spain). Environ Int 75:110–116

Council Directive 98/83/EC, 3 November 1998, on the quality of water intended for human consumption

Cycoń M, Żmijowska A, Piotrowska-Seget Z (2014) Enhancement of deltamethrin degradation by soil bioaugmentation with two different strains of *Serratia marcescens*. Int J Environ Sci Technol 11:1305–1316. doi:10.1007/s13762-013-0322-0

Dąbrowska D, Kot-Wasik A, Namieśnik J (2002) Degradacja związków organicznychw środowisku. Chemia i Inżynieria Ekologiczna 10:1077–1083

de Lafontaine Y, Beauvais C, Cessna AJ, Gagnon P, Hudon C, Poissant L (2014) Sulfonylurea herbicides in an agricultural catchment basin and its adjacent wetland in the St. Lawrence River basin. Sci Total Environ 479–480:1–10. doi:10.1016/j.scitotenv.2014.01.094

de Urzedo AP et al (2007) Photolytic degradation of the insecticide thiamethoxam in aqueous medium monitored by direct infusion electrospray ionization mass spectrometry. J Mass Spectrom 42:1319–1325. doi:10.1002/jms.1204

Devisri S, Iyer PR (2013) Degradation of deltamethrin by organisms isolated from Koovam river water. Int J Curr Microbiol App Sci 2(10):106–111

Dinelli G, Vicari A, Accinelli C (1998) Degradation and side effects of three sulfonylurea herbicides in soil. J Environ Qual 27:1459–1464. doi:10.2134/jeq1998.00472425002700060023x

Doe JE, Lander DR, Doerrer NG, Heard N, Hines RN, Lowit AB, Pastoor T, Phillips RD, Sargent D, Sherman JH, Tanir JY, Embry MR (2016) Use of the RISK21 roadmap and matrix:

human health risk assessment of the use of a pyrethroid in bed netting. Crit Rev Toxicol 46:54–73. doi:10.3109/10408444.2015.1082974
EFSA (2005) Conclusion regarding the peer review of the pesticide risk assessment of the active substance rimsulfuron. EFSA Sci Rep 45:1–61
Elliott M, Janes NF (1978) Synthetic pyrethroids—a new class of insecticide. Chem Soc Rev 7:473–505. doi:10.1039/CS9780700473
Europa (2013) Bees & pesticides: commission goes ahead with plan to better protect bees. http://ec.europa.eu/food/plant/pesticides/max_residue_levels/index_en.htm. Accessed 26 Jan 2016
Fang L, Zhang S, Chen Z, Du H, Zhu Q, Dong Z, Li H (2016) Risk assessment of pesticide residues in dietary intake of celery in China. Reg Toxicol Pharmacol 73:578–586
Feo ML, Ginebreda A, Eljarrat E, Barceló D (2010) Presence of pyrethroid pesticides in water and sediments of Ebro River Delta. J Hydrol 393:156–162. doi:10.1016/j.jhydrol.2010.08.012
Fetoui H, Garoui EM, Zeghal N (2009) Lambda-cyhalothrin-induced biochemical and histopathological changes in the liver of rats: ameliorative effect of ascorbic acid. Exp Toxicol Pathol 61:189–196. doi:10.1016/j.etp.2008.08.002
Grant RJ, Daniell TJ, Betts WB (2002) Isolation and identification of synthetic pyrethroid-degrading bacteria. J Appl Microbiol 92(3):534–540. doi:10.1046/j.13652672.2002.01558.x
Guo W, Zhang F, Lin C, Wang LZ (2012) Direct growth of TiO_2 nanosheet arrays on carbon fibers for highly efficient photocatalytic degradation of methyl orange. Adv Mater 24(35):4761–4764
Guzsvany V, Csandi J, Gaal F (2006) NMR study of the influence of pH on the persistence of some neonicotinoids in water. Acta Chim Slov 53:52–57
Han J, Fang P, Xu X, Li-Zheng X, Shen H, Ren Y (2015) Study of the pesticides distribution in peel, pulp and paper bag and the safety of pear bagging. Food Control 54:338–346. doi:10.1016/j.foodcont.2015.02.021
He LM, Troiano J, Wang A, Goh K (2008) Environmental chemistry, ecotoxicity, and fate of lambda-cyhalothrin. Rev Environ Contam Toxicol 195:71–91
Hem L, Park JH, Shim JH (2010) Residual analysis of insecticides (Lambda-cyhalothrin, Lufenuron, Thiamethoxam and Clothianidin) in Pomegranate using GC-ECD or HPLC-UVD. Korean J Environ Agric 29:257–265. doi:10.5338/KJEA.2010.29.3.257
Hooker WJ (1981) Compendium of potato diseases. American Phytopathological Society, St Paul
Hunt L, Bonetto C, Resh VH, Buss DF, Fanelli S, Marrochi N, Lydy MJ (2016) Insecticide concentrations in stream sediments of soy production regions of South America. Sci Total Environ 547:114–124. doi:10.1016/j.scitotenv.2015.12.140
Jalali AM, Van Leeuwen T, Tirry L, De Clercq P (2009) Toxicity of selected insecticides to the two-spot ladybird Adalia bipunctata. Phytoparasitica 37:323–326. doi:10.1007/s12600-009-0051-6
Kalathoor R, Botterweck J, Schäffer A, Schmidt B, Schwarzbauer J (2015) Quantitative and enantioselective analyses of non-extractable residues of the fungicide metalaxyl in soil. J Soil Sediment 15:659–670. doi:10.1007/s11368-014-1027-9
Katagi T (2002) Abiotic hydrolysis of pesticides in the aquatic environment. Rev Environ Contam Toxicol 175:79–261
Kearney PC, Kaufman DD (1969) Degradation of herbicides. Dekker Inc, New York
Kolo RJ, Lamai S, Ojutiku RO (2010) Subacute toxicity of Karate to Sarotherodon galilieus (Linne,1758). J Water Chem Technol 32:107–112. doi:10.3103/S1063455X10020074
Kondrakow AO, Igantev AN, Frimmel FH, Brase S, Hom H, Revelsky AI (2014) Formation of genotoxic quinones during bisphenol A degradation by TiO_2 photocatalysis and UV photolysis: a comparative study. Appl Catal B Environ 160–161:106–114
Koprucu K, Aydm R (2004) The toxic effects of pyrethroid deltamethrin on the common carp (Cyprinus carpio L.) embryos and larvae. Pestic Biochem Physiol 80:47–53. doi:10.1007/s10653-007-9108-y
Kryczyński S (2010) Wirusologia roślinna. Wydawnictwo Naukowe PWN, Warszawa

Kungolos A, Emmanouil C, Tsiridis V, Tsiropoulos N (2009) Evaluation of toxic and interactive toxic effects of three agrochemicals and copper using a battery of microbiotests. Sci Total Environ 1:4610–4615. doi:10.1016/j.scitotenv.2009.04.038

Laabs V, Amelung W, Pinto A, Altstaedt A, Zech W (2000) Leaching and degradation of corn and soybean pesticides in an Oxisol of the Brazilian Cerrados. Chemosphere 41:1441–1449

Laberge M, Rollinson R (2013) Degradation of thiamethoxam via photocatalysis: kinetics, mineralization, and toxicity. Worcester Polytechnic Institute, Worcester

Lazzarini M, Salum C, Del Bel EA (2005) Combined treatment of ascorbic acid or alpha-tocopherol with dopamine receptor antagonist or nitric oxide synthase inhibitor potentiates cataleptic effect in mice. Psychopharmacology (Berl) 181:71–79

Lee JH, Shan G, Watanabe T, Stoutamire DW, Gee SJ, Hammock BD (2002) Enzyme-linked immunosorbent assay for the pyrethroid deltamethrin. J Agric Food Chem 50:5526–5532. doi:10.1021/jf030519p

Li Y, Dong F, Liu X, Xu J, Chen X, Han Y, Cheng Y, Jian Q, Zheng Y (2013) Enantioselective separation and transformation of metalaxyl and its major metabolite metalaxyl acid in tomato and cucumber. Food Chem 141:10–17

Liquing Z, Guoguang L, Dezhi S, Kun Y (2006) Hydrolysis of thiamethoxam. Environ Contam Toxicol 6:942–949. doi:10.5902/2179460X17302

Liu P, Liu Y, Liu Q, Liu J (2009) Photodegradation mechanism of deltamethrin and fenvalerate. J Environ Sci 22:1123–1128

Liu L, Liu Z, Bai H, Sun DD (2012) Concurrent filtration and solar photocatalytic disinfection/degradation using high-performance Ag/TiO_2 nanofiber membrane. Water Res 46 (4):1101–1112

Liu PY, Li B, Liu HD, Tian L (2014a) Photochemical behavior of fenpropathrin and lambda cyhalothrin in solution. Environ Sci Pollut Res 21:1993–2001

Liu T, Zhu L, Han Y, Wang J, Wang J, Zhao Y (2014b) The cytotoxic and genotoxic effects of metalaxy-M on earthworms (Eisenia fetida). Environ Toxicol Chem 33(10):2344–2350

Lopes RP, de Urzedo APFM, Nascentes CC, Augusti R (2008) Degradation of the insecticides thiamethoxam and imidacloprid by zero-valent metals exposed to ultrasonic irradiation in water medium: electrospray ionization mass spectrometry monitoring. Rapid Commun Mass Spectrom 22:3472–3480

Maienfisch P (2001) Chemistry and biology of thiamethoxam: a second generation neonicotinoid. Pest Manag Sci 57:906–913

Maienfisch P (2006) Synthesis and properties of thiamethoxam and related compounds. Z Naturforsch 61b:353–359

Main AR, Headley JV, Peru KM, Michel NL, Cessna AJ, Morrissey CA (2014) Widespread use and frequent detection of neonicotinoid insecticides in wetlands of Canada's prairie Pothole region. PLoS One 9:e92821. doi:10.1371/journal.pone.0092821

Maloney SE, Maule A, Smith ARW (1988) Microbial transformation of the pyrethroid insecticides: permethrin, deltamethrin, fastac, fenvalerate, and fluvalinate. Appl Environ Microbiol 54:2874–2876

Manigandan G, Nelson R, Jeevan P (2013) Biodegradation of lambda cyhalothrin by Pseudomonas fluorescens and Trichoderma viridae. J Microbiol Biotechnol Res 3:42–44

Maradani A, Sabahi Q, Resekh A, Almasi A (2016) Lethal and sublethal effects of three insecticides on the aphid parasitoid, Lysiphlebus fabarum Marshall (Hymenoptera: Aphidiidae). Phytoparasitica 8:1–8

Martins JM, Marmoud A (1999) Transport of rimsulfuron and its metabolites in soil columns. Chemosphere 38(3):601–616. doi:10.1016/S0045-6535(98)00197-0

Martins JMF, Chevre N, Spack L, Tarradellas J, Mermoud A (2001) Degradation in soil and water and ecotoxicity of rimsulfuron and its metabolites. Chemosphere 45:515–522. doi:10.1016/S0045-6535(01)00040-6

Marucchini C, Zadra C (2002) Stereoselective degradation of metalaxyl and metalaxyl-M in soil and sunflower plants. Chirality 14:32–38

Masiá A, Campo J, Vázquez-Roig P, Blasco C, Picó Y (2013) Screening of currently used pesticides in water, sediments and biota of the Guadalquivir river basin (Spain). J Hazard Mater 263:95–104

Massoud AH, Derbalah AS, Brelal El-Sayed B (2008) Microbial detoxification of metalaxyl in aquatic system. J Environ Sci 20:262–267

Mickaël H et al (2012) A common pesticide decreases foraging success and survival in honey bees. Science 336:348–350. doi:10.1126/science.1215039

Mir NA, Khan A, Muneer M, Vijayalakhsm S (2013) Photocatalytic degradation of a widely used insecticide Thiamethoxam in aqueous suspension of TiO_2: adsorption, kinetics, product analysis and toxicity assessment. Sci Total Environ 458–460:388–398

Mugni H, Paracampo A, Marrochi N, Bonetto C (2013) Acute toxicity of cypermethrin to the non-target organism Hyalella curvispina. Environ Toxicol Pharmacol 35:88–92

Myresiotis KC, Vryzas Z, Papadopoulou-Mourkidou E (2012) Biodegradation of soil-applied pesticides by selected strains of plant growth-promoting rhizobacteria (PGPR) and their effects on bacterial growth. Biodegradation 23:297–310

Nageswara Rao T, Venkata-Ramasubbih A, Parvathi T (2012) Development and validation of a HPLC-UV method for simultaneous determination of five sulfonylurea herbicide residues in soybean oil followed by Matrix Solid-Phase Dispersion. Int J Chem Environ Pharm Res 3:117–121. doi:10.5402/2012/908795

Nahri-Niknafs B, Ahmadi A (2013) Photodegradation of deltamethrin and fenvalerate under simulated solar light irradiation and identification of photoproducts. Rev Chim 64(8):828–831

Nauen R, Ebbinghaus-Kintscher U, Salgado VL, Kaussmann M (2003) Thiamethoxam is a neonicotinoid precursor converted to clothianidin in insects and plants. Pestic Biochem Phys 73(2):55–69. doi:10.1016/S0048-3575(03)00065-8

Palanisamy B, Babu CM, Sundaravel S, Anandan S, Murugesan V (2013) Sol–gel synthesis of mesoporous mixed Fe_2O_3/TiO_2 photocatalyst: application for degradation of 4-chlorophenol. J Hazard Mater 252–253:233–242

Páleníková A, Martínez-Domínguez G, Arrebola FJ, Romero-González R, Hrouzková S, Frenich AG (2015) Determination of pesticides and transformation products in Ginkgo biloba nutraceutical products by chromatographic techniques coupled to mass spectrometry. Food Anal Methods 8:20194–22201

Pandey G, Dorrian SJ, Russell J, Oakeshott JG (2009) Biotransformation of the neonicotinoid insecticides imidacloprid and thiamethoxam by Pseudomonas sp. 1G. Biochem Biophys Res Commun 380:710–714. doi:10.1016/j.bbrc.2009.01.156

Pang N, Wang T, Hu J (2016) Method validation and dissipation kinetics of four herbicides in maize and soil using QuEChERS sample preparation and liquid chromatography tandem mass spectrometry. Food Chem 190:793–800

Patel N, Jaiswall R, Warang T, Scarduelli G, Dashora A, Ahuja BL, Kothan DC, Miotello A (2014) Efficient photocatalytic degradation of organic water pollutants using V–N-codoped TiO_2 thin films. Appl Catal B Environ 150–151:74–81

Pavan FA, Dallago RM, Zanella R, Martins AF (1999) Determination of deltamethrin in cattle dipping baths by high-performance liquid chromatography. J Agric Food Chem 47:174–176

Peackock TJ, Mikell AT Jr, Moore MT, Smith S Jr (2014) Application of a redox gradostat reactor for assessing rhizosphere microorganism activity on lambda-cyhalothrin. Bull Environ Contam Toxicol 92:347–351. doi:10.1007/s00128-014-1202-0

Pohorecka K, Skubida P, Miszczak A, Semkiw P, Sikorski P, Zagibajło K, Teper D, Kołtowski Z, Zdańska D, Skubida M, Bober A (2012) Residues of neonicotinoid insecticides in bee collected plant materials from oilseed rape crops and their effect on bee colonies. J Apic Sci 56:115–134

Rahman IQ, Ahmad M, Misra KS, Lohani M (2013) Effective photocatalytic degradation of rhodamine B dye by ZnO nanoparticles. Mater Lett 91:170–174

Rana S, Jindal V, Mandal K, Kaur G, Gupta VK (2015) Thiamethoxam degradation by Pseudomonas and Bacillus strains isolated from agricultural soils. Environ Monit Assess 187:300. doi:10.1007/s10661-015-4532-4

Rosenbom AE, Kjær J, Olsen P (2010) Long-term leaching of rimsulfuron degradation products through sandy agricultural soils. Chemosphere 79:830–838

Rosenkrantz RT, Cedergreen N, Baun A, Kusk KO (2013) Influence of pH, light cycle, and temperature on ecotoxicity of four sulfonylurea herbicides towards Lemna gibba. Ecotoxicology 22:33–41

Rouchaud J, Neus O, Callens D, Bulcke R (1997) Soil metabolism of the herbicide rimsulfuron under laboratory and field conditions. J Agric Food Chem 45:3283–3291

Ruan Z, Zhai Y, Song J, Shi Y, Li K et al (2013) Molecular cloning and characterization of a newly isolated pyrethroid-degrading esterase gene from a genomic library of ochrobactrum anthropi YZ-1. PLoS One 8(10):e77329. doi:10.1371/journal.pone.0077329

Sanchez-Bayo F, Goka K (2014) Pesticide residues and bees—a risk assessment. PLoS One 9: e94482

Saravanan R, Karthikeyan S, Gupta VK, Sekaran G, Narayanan V, Stephen A (2013) Enhanced photocatalytic activity of ZnO/CuO nanocomposite for the degradation of textile dye on visible light illumination. Mater Sci Eng 33:91–98

Schneider GE, Koeppe MK, Naidu MV, Horne P, Brown HM, Mucha CF (1993) Fate of rimsulfuron in the environment. J Agric Food Chem 41:2404–2410

Schwartz BJ, Sparrow FK, Heard EN, Thede BM (2000) Simultaneous derivatization and trapping of volatile products from aqueous photolysis of thiamethoxam insecticide. J Agric Food Chem 48:4671–4675. doi:10.1021/jf990966y

Scrano L, Bufo SA, Peruccci P, Meallier P, Mansour M (1999) Photolysis and hydrolysis of rimsulfuron. Pestic Sci 55:955–961. doi:10.1007/3-540-26531-7_46

Seńczuk W (2005) Toksykologia współczesna. Wydawnictwo Lekarskie PZWL, Warszawa

Sharma D, Awasthi MD (1997) Adsorption and movement of metalaxyl in soils under unsaturated flow conditions. Plant Soil 195:293–298

Sheets LP, Li AA, Minnema DJ, Collier RH, Creeke MR, Peffer RC (2015) A critical review of neonicotinoid insecticides for developmental neurotoxicity. Crit Rev Toxicol 46:153–190

Shen C-C, Shen D-S, Shentu J-L, Wang M-Z, Wan M-Y (2015a) Could humic acid relieve the biochemical toxicities and DNA damage caused by nickel and deltamethrin in earthworms (Eisenia foetida)? Environ Sci Proc Imp 17:2074–2081

Shen X, Zemin X, Zhang X, Yang F (2015b) Stable carbon isotope fractionation during the biodegradation of lambda-cyhalothrin. Sci Total Environ 532:415–419

Sikorska K, Wędzisz A (2009) Nowoczesne pestycydy—spinosad. Bromat Chem Toksykol 2:203–212

Simon-Delso N, Amaral-Rogers V, Belzunces LP, Bonmatin JM, Chagnon M, Downs C (2015) Systemic insecticides (neonicotinoids and fipronil): trends, uses, mode of action and metabolites. Environ Sci Pollut Res Int 22:5–34

Sivaperumal P, Anand P, Riddhi L (2015) Rapid determination of pesticide residues in fruits and vegetables, using ultra-high-performance liquid chromatography/time-of-flight mass spectrometry. Food Chem 168:356–365

Šojić D, Despotović V, Orcić D, Szabó E, Arany E, Armakoviś S, Illés E, Gajda-Schrantz K, Dombi A, Alapi T, Sajben-Nagy E, Palágyi A, Vágvölgyi SC, Manczinger L, Bjelica L, Abramović B (2012) Degradation of thiamethoxam and metoprolol by UV, O3 and UV/O3 hybrid processes: kinetics, degradation intermediates and toxicity. J Hydrol 472–473:314–327

Song J et al (2013) Biodegradation of nicosulfuron by a Talaromyces flavus LZM1. Bioresour Technol 140:243–248. doi:10.1016/j.biortech.2013.02.086

Song H, Zhou Z, Liu Y, Deng S, Xu H (2015) Kinetics and mechanism of fenpropathrin biodegradation by a newly isolated Pseudomonas aeruginosa sp. strain JQ-41. Curr Microbiol 71:326–332

Starner K, Goh KS (2012) Detections of the neonicotinoid insecticide imidacloprid in surface waters of three agricultural regions of California, USA, 2010–2011. Bull Environ Contam Toxicol 88(3):316–321. doi:10.1007/s00128-011-0515-5

Stoner KA, Eitzer BD (2012) Movement of soil-applied imidacloprid and thiamethoxam into nectar and pollen of squash (Cucurbita pepo). PLoS One 7:e39114

Sukul P, Spiteller M (2000) Metalaxyl: persistence, degradation, metabolism and analytical methods. Rev Environ Contam Toxicol 164:1–26

Sulimma L, Bullach A, Kusari S, Lamshoft M, Zuhlke S, Spiteller M (2013) Enantioselective degradation of the chiral fungicides metalaxyl and furalaxyl by Brevibacillus brevis. Chirality 25:336–340

Szpyrka E, Kurdziel A, Matyaszek A, Podbielska M, Rupar J, Słowik-Borowiec M (2015) Evaluation of pesticide residues in fruits and vegetables from the region of south-eastern Poland. Food Control 48:137–142

Tabaeran IV, Narahashi T (1998) Potent modulation of tetrodotoxin sensitive and tetrodotoxin resistant sodium channels by the type II pyrethroid deltamethrin. Pharmacol Therapeut 284:958–965

Taillebois E, Langlois P, Cunha T, Seraphin D, Thany SH (2014) Synthesis and biological activity of fluorescent neonicotinoid insecticide thiamethoxam. Bioorg Med Chem Lett 24:3552–3555. doi:10.1016/j.bmcl.2014.05.052

Tallur PN, Megadi VB, Ninnekar HZ (2008) Biodegradation of cypermethrin by Micrococcus sp. strain CPN 1. Biodegradation 19:77–82

The potato: tuber, International year of potato 2008. http://www.fao.org/potato-2008/en/potato/pests.html. Accessed 8 Sept 2015

The PPDB: Pesticides Properties DataBase, University of Hertfordshire. http://sitem.herts.ac.uk/aeru/ppdb/en/index.htm. Accessed 8 Sept 2015

Tuzmen N, Canadan N, Kaya E, Demiryas N (2008) Biochemical effects of chlorpyrifos and deltamethrin on altered antioxidative defense mechanism and lipid peroxidation in rat liver. Cell Biochem Funct 26:119–124. doi:10.1002/cbf.1411

University of Kentucky, College of Agriculture, Food and Environment. http://www2.ca.uky.edu/entomology/entfacts/ef304.asp. Accessed 8 Sept 2015

Ural MS, Saglam N (2005) A study on the acute toxicity of pyrethroid deltamethrin on the fry rainbow trout (Oncorhynchus mykiss Walbaum, 1792). Pestic Biochem Phys 83:124–131. doi:10.1016/j.pestbp.2005.04.004

Van den Brink PJ, Van Smeden JM, Bekele RS, Dierick W, De Gelder DM, Noteboom M, Roessink I (2016) Acute and chronic toxicity of neonicotinoids to nymphs of a mayfly species and some notes on seasonal differences. Environ Toxicol Chem 35:128–133

Van Dijk TC, Van Staalduinen MA, Van der Sluijs JP (2013) Macroinvertebrate decline in surface water polluted with imidacloprid. PLoS One 8:e62374

Wang S, Kimber SWL, Kennedy IR (1997) The dissipation of lambda-cyhalothrin from cotton production systems. J Environ Sci Health B 32:335–352

Wang M, Guiquin Y, Wang X, Yao Y, Min H, Lu Z (2011) Nicotine degradation by two novel bacterial isolates of Acinetobacter sp. TW and Sphingomonas sp. TY and their responses in the presence of neonicotinoid insecticides. J Microbial Biotechnol 27:1633–1640

Wang M, Zhang Q, Cong L, Yin W, Wang M (2014) Enantioselective degradation of metalaxyl in cucumber, cabbage, spinach and pakchoi. Chemosphere 95:241–246. doi:10.1016/j.chemosphere.2013.08.084

Webb ME, Smith AG (2011) Pantothenate biosynthesis in higher plants. Adv Bot Res 58:204

Wilson PC, Whitwell T, Klaine SJ (2001) Metalaxyl toxicity, uptake, and distribution in several ornamental plant species. J Environ Qual 30:411–417

Wu Y, Liu X, Dong F, Xu J, Zheng Y (2012) Dissipation and residues of rimsulfuron in potato and soil under field conditions. Bull Environ Contam Toxicol 89:1264–1267

Xiao Y, Chen S, Gao Y, Hu W, Hu M, Zhong G (2015) Isolation of a novel beta-cypermethrin degrading strain Bacillus subtilis BSF01 and its biodegradation pathway. Appl Microbiol Biotechnol 99:2849–2859. doi:10.1007/s00253-014-6164-y

Xu P, Diao J, Liu D, Zhou Z (2011) Enantioselective bioaccumulation and toxic effects of metalaxyl in earthworm Eisenia foetida. Chemosphere 83:1074–1079. doi:10.1016/j.chemosphere.2011.01.047

Yao K, Zhu L, Duan Z, Chen Z, Li Y, Zhu X (2008) Comparison of R-metalaxyl and rac-metalaxyl in acute, chronic, and sublethal effect on aquatic organisms: *Daphnia magna*, *Scenedesmus quadricanda*, and *Danio rerio*. Environ Toxicol 24:148–156. doi:10.1002/tox.20415

Yao W, Zhang B, Huang C, Ma C, Song X, Xu Q (2012) Synthesis and characterization of high efficiency and stable Ag_3PO_4/TiO_2 visible light photocatalyst for the degradation of methylene blue and rhodamine B solutions. J Mater Chem 22:4050–4055

Yousef MI (2010) Vitamin E modulates reproductive toxicity of pyrethroid lambda-cyhalothrin in male rabbits. Food Chem Toxicol 48:1152–1159. doi:10.1016/j.fct.2010.02.002

Yu C, Li G, Kumar S, Yang K, Jin R (2014) Phase transformation synthesis of novel Ag_2O/Ag_2CO_3 heterostructures with high visible light efficiency in photocatalytic degradation of pollutants. Adv Mater 26(6):892–898

Zabar R, Komel T, Fabjan J, Bavcon Kralj M, Trebše P (2012) Photocatalytic degradation with immobilised TiO_2 of three selected neonicotinoid insecticides: imidacloprid, thiamethoxam and clothianidin. Chemosphere 89:293–301. doi:10.1016/j

Zhang P, Zhu W, Qiu J, Wang D, Wang X, Wang Y, Zhou Z (2014) Evaluating the enantioselective degradation and novel metabolites following a single oral dose of metalaxyl in mice. Pest Biochem Physiol 116:32–39. doi:10.1016/j.pestbp.2014.09.008

Zhang H et al (2016) Biodegradation potential of deltamethrin by the Bacillus cereus strain Y1 in both culture and contaminated soil. Int Biodeter Biodegr 106:53–59

Zhao MR, Liu WP (2009) Enantioselectivity in the immunotoxicity of the insecticide acetofenate in an in vitro model. Environ Toxicol Chem 28:578–585

Zhao M, Chen F, Wang C, Zhang Q, Gan J, Liu W (2010) Integrative assessment of enantioselectivity in endocrine disruption and immunotoxicity of synthetic pyrethroids. Environ Pollut 158:1968–1973

Zheng W, Liu W (1999) Kinetics and mechanism of the hydrolysis of imidacloprid. Pest Sci 55:482–485

Zheng S, Jiang W, Cai Y, Dionysiou DD, O'Shea EK (2014) Adsorption and photocatalytic degradation of aromatic organoarsenic compounds in TiO_2 suspension. Catal Today 224:83–88

Zhou G et al (2012) Biodegradation of the neonictoinoid insecticide thiamethoxam by the nitrogen-fixing and plant-growth-promoting rhizbacterium Ensifer adhaerens strain TMX-23. Appl Microbiol Biotechnol 97:4065–4074. doi:10.1007/s00253-012-4638-3

Zuno-Floriano FG et al (2012) Effect of Acinetobacter sp on metalaxyl degradation and metabolite profile of potato seedlings (Solanum tuberosum L.) Alpha Variety. PLoS One 7(2):e31221. doi:10.1371/journal.pone.0031221

A Review on Polychlorinated Biphenyls (PCBs) and Polybrominated Diphenyl Ethers (PBDEs) in South Asia with a Focus on Malaysia

Han Yeong Kaw and Narayanan Kannan

Contents

H.Y. Kaw
Key Laboratory of Nature Resources of Changhai Mountain and Functional Molecular (Yanbian University) Ministry of Education, Jilin, China
e-mail: hanyeong.kaw@gmail.com

N. Kannan (✉)
Institute for Graduate Studies, Taylor's University (Lakeside Campus), No. 1, Jalan Taylor's, 47500, Subang Jaya, Selangor Darul Ehsan, Malaysia
e-mail: drnkannan@yahoo.com

P. de Voogt (ed.), *Reviews of Environmental Contamination and Toxicology, Volume 242*, Reviews of Environmental Contamination and Toxicology 242,
DOI 10.1007/398_2016_14

1 Introduction

The monarchy of Malaysia consists of peninsular Malaysia, located between Singapore, Thailand and Indonesia, and states of Sabah and Sarawak on the island of Kalimantan (Borneo). Malaysia has a total landmass of 329,758 km^2 and is almost as diverse as its culture (Fig. 1).

Since 1970s the leading sector of development has been a range of export orientated manufacturing industries such as textiles, electrical and electronic goods, rubber products, while Malaysia became the world's largest producer of oil palm in the 1970s (Drabble 2001). Due to the current policy of achieving high income status by 2020 and transform Malaysia into a multi-sector economy, Malaysia is attempting to move farther up in the value-added production chain by shifting its focus on high tech industries (Central Intelligence Agency 2015). Thus, 34.7 % of the total Gross Domestic Product (GDP) generated in Malaysia in 2014 was from industrial sector, followed by agriculture and other service industries which accounted for 9.3 % and 56 % respectively. In which, the electrical and electronic contributing 35 % of country's total exports between January to May 2015 (Department of Statistics Malaysia 2015).

Polychlorinated biphenyls (PCBs) are aromatic, synthetic chemicals with a chemical formula $C_{12}H_{10-n}Cl_n$, where n ranges from 1 to 10 (Fig. 2). PCBs consist of 209 congeners which exhibit wide differences in their chemical characteristics and biological effects (Kannan et al. 1989). PCBs were used widely in electrical equipments like capacitors and transformers and in hydraulic fluids, heat transfer fluids, lubricants, and plasticizers. PCBs have been used commercially since 1929 (De Voogt and Brinkman 1989) but its presence in the environment was not recognized until 1966, when Jensen identified PCBs in human and wildlife samples (Jensen 1966).

PCBs are well known environmental contaminants with persistence, bioaccumulation and human health problems (Safe 1994). Approximately two million tonnes of PCBs have been produced commercially and about 10 % of which still remains in the environment today (WHO 2003). Besides, potentially toxic coplanar PCBs have been determined in humans, rats, dogs, cats, rhesus monkeys (terrestrial), finless porpoise, Dall's porpoises, Baird's beaked whales, striped dolphins and killer whales (marine) (Kimbrough et al. 1975; Kannan et al. 1989, 1993; Golub et al. 1991; Kannan 2000). Mono- and non-*ortho* coplanar PCBs that persist in the environment and in animals cause reproductive dysfunction (Barsotti et al. 1976; Golub et al. 1991; Kannan et al. 1988). The Stockholm Convention entered into force in 2004 and recommended that all equipment containing PCBs are eliminated by the year 2025 and then manage the rest with environmentally sound policies until 2028. In fact based on the Stockholm Convention website, Malaysia has only signed the Stockholm Convention but that the country has not ratified the Convention to date, therefore Malaysia has yet to develop a National Implementation Plan on POPs. Recently, the International Agency for Research on Cancer (IARC) has classified PCBs as carcinogen Group 1 (carcinogenic to humans), which further underline the significance of PCBs monitoring (IARC 2015).

Fig. 1 Location of Malaysia in Southeast Asia

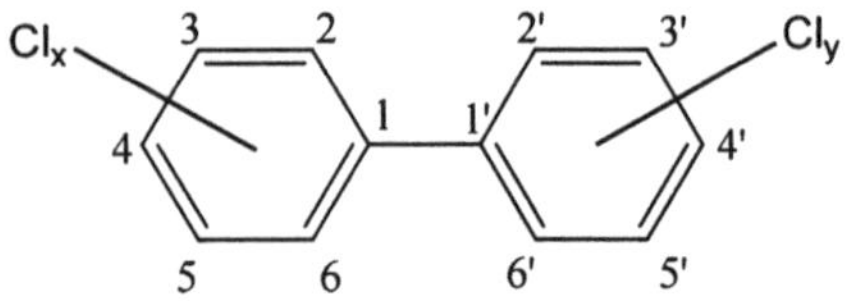

Fig. 2 The general chemical structure of chlorinated biphenyls

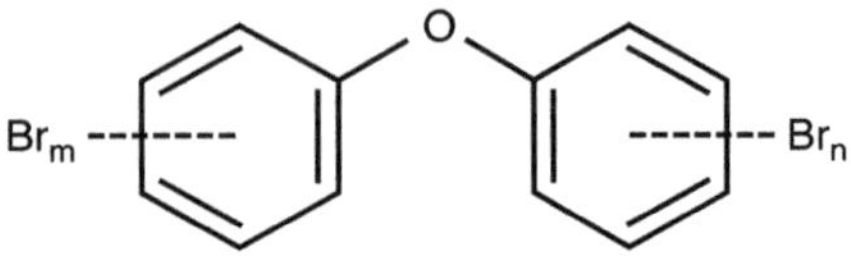

Fig. 3 Structure of polybrominated diphenyl ether (PBDEs)

Polybrominated diphenyl ethers (PBDEs) (Fig. 3) have been widely used as flame retardants (BFRs). PBDEs like PCBs are hydrophobic, lipophilic and thermally stable and are capable of bioaccumulating in food chain (Rahman et al. 2001; de Wit 2002; Kannan et al. 2005). Furthermore, the similarity between the structures of PCBs and PBDEs suggests similar toxicity to animals and humans (Ogata et al. 2009; Hong et al. 2010). The higher proportion of lower brominated PBDE congeners in the brain tissues of little brown bats compared to fat tissues indicated that lower brominated PBDE congeners could cross the blood-brain barrier more efficiently than the higher brominated PBDE congeners (Kannan et al. 2010). Studies carried out by McDonald (2002), Fowles et al. (1994), Zhou et al. (2002) and Branchi et al. (2002) show that PBDEs in humans disrupt thyroid hormones, cause neurobehavioral deficiency and possibly cause cancer. Hence in 2009, Stockholm Convention has listed commercial Penta-BDE and Octa-BDE as POPs.

Although PCBs-containing electrical equipment and other applications are not manufactured in Malaysia, they have been imported for a while until they were banned in June 1998 (Hashim 2003; Ibrahim 2007). PCBs have been used in heat exchangers and hydraulic systems while PBDEs were commonly used as flame retardants in a series of applications including textiles, television sets, plastics and wire insulation. Waste motor and transformer oils are considered as a main source of PCBs emission into the environment in Malaysia (Jahromi et al. 2014). There is still approximately 15 % of known PCBs residing in developing countries including Malaysia, mostly as a result of shipments of wastes from industrialized countries (Abel and McConnell 2009).

2 PCBs and PBDEs in E-Waste in Malaysia

Solid waste is a major environmental problem in Malaysia. In 2008, 23,000 tons of waste were produced each day in Malaysia with less than 5 % being recycled and approximately 19 % of waste being ended up in the drains, polluting the environment including PCBs and PBDEs (Wu et al. 2008; Tue et al. 2010; Desa et al. 2012).

Department of Environment (DOE) Malaysia has defined e-wastes as wastes from the used electronic and electrical (EEE) equipment such as accumulators, mercury switches, transformers, cathode-ray tubes (CRT) or PCBs-containing capacitors (DOE 2008), which are regulated in Malaysia since 2005 and listed as scheduled wastes. According to the E-waste Inventory of Malaysia, total volume of e-waste generated in 2008 was 688,000 metric tonnes and it will reach 11.2 million metric tonnes by 2020. Currently, there are 18 full recovery facilities and 128 partial recovery facilities permitted by the DOE of Malaysia for the segregation, dismantling and treatment of e-waste. Despite the law that e-waste could only be treated and recovered by the licensed facilities in Malaysia, Environmentally Sound Management (ESM) is lacking in the collection, treatment and recovery of e-waste (Suja et al. 2014).

3 PCBs in Malaysia

3.1 PCBs in Sewage Sludge

A research conducted by Ahmad et al. (2005) on palm oil and sewage sludge revealed PCB congeners such as PCB-28, 52, 101, 118, 138, 153 and 180 in oil and PCBs congeners 28, 52 and 101 in sludge, while PCBs 118, 138, 153 and 180 were not detected (3.37–70.6 ng/g). On the other hand, Halim (2007) investigated the concentrations of PCBs in sewage sludge using two types of extraction methods. The samples were extracted using supercritical fluid carbon dioxide ($SFCO_2$) and subcritical water extraction (SWE) and PCB-28, 52, 101, 118,138, 153 and 180 were determined using GC-ECD. The concentration of PCBs in the sludge samples extracted using $SFCO_2$ were comparatively higher (31.2–82.0 ng/g) than samples extracted using SWE (20.2–60.1 ng/g).

3.2 PCBs in Pellet

Plastic resin pellets (0.1–0.5 cm) are found on beaches around the world. A Japanese study (Ogata et al. 2009) analyzed samples of polyethylene pellets from 30 beaches in 17 countries including Malaysia to determine the concentration of organochlorine compounds. Sampling locations in Malaysia were Langkawi, Penang and Borneo and the median concentration of the sum of 13 PCB congeners (CB-66, 101, 110, 149, 118, 105, 153, 138, 128, 187, 180, 170, 206) recorded at these places were 8 ng/g-pellet, 12 ng/g-pellet and 10 ng/g-pellet, respectively (Fig. 4).

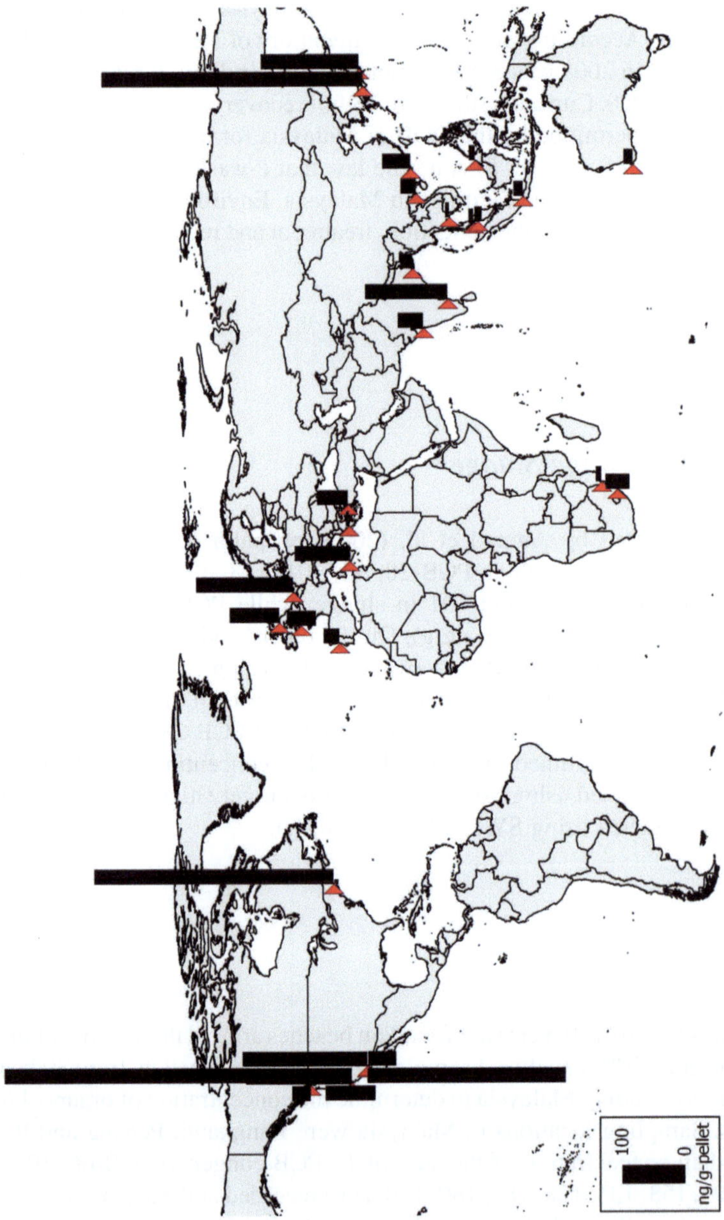

Fig. 4 Median concentration of ∑13 PCBs (ng/g-pellet) in beached plastic pellets. ∑13 PCBs is the sum of concentrations of CB# 66, 101, 110, 149, 118, 105, 153, 138, 128, 187, 180, 170, 206. (Source: Ogata et al. 2009)

3.3 PCBs in Air

PCBs were measured in air using Polyurethane foam (PUF) passive air samplers (PAS) in Malaysia, Japan, Vietnam and the Philippines by Kwan (2014). The air samplers were deployed at six locations in these Asian countries including Malaysia (Port Dickson) during dry and wet seasons for a month. CB-66, 101, 110, 149, 118, 105, 153, 138, 128, 187, 180, 170 and 206 were measured. Result showed that the atmospheric concentration of PCBs in Malaysia (less than 5 pg/m^3) was not as high as in other countries. For example, Philippines recorded a high 80 pg/m^3.

3.4 PCBs in Mussels

Bivalves such as mussels are suitable bio-indicator for monitoring trace toxic contaminants in coastal waters due to their wide distribution, sessile lifestyle, easy sampling, tolerance to a considerable range of salinity, resistance to stress and high accumulation of a wide range of chemicals (Goldberg et al. 1978). A study using green mussels (*Perna viridis*) was conducted in Penang, Malaysia. Mussel samples were collected from six stations on the island of Penang (Weld Quay Pier, Marine Depot, Pulau Jerejak, Permatang Damar Laut, Batu Maung and Gertak Sanggul) and three more stations in Singapore waters (Selatar, Serangoon and Ponggol). The concentration of PCBs analysed in Malaysia was in the range of 99.9–599.9 ng/g on a dry weight basis (Sivalingam et al. 1982).

Besides, Monirith et al. (2003) also examined the contamination of persistent organochlorines (OCs) including PCBs in mussels collected from coastal waters of Asian countries like Cambodia, China, Hong Kong, India, Indonesia, Japan, Korea, Malaysia, Philippines, Far East Russia, Singapore, and Vietnam. The sampling locations were at different parts of Malaysia (Kuala Penyu and Trayong of Sabah, Sangkar Ikan and Tanjung Rhu of Langkawi, Butterworth and Penang Bridge of Penang, Bagan Bridge of Selangor, Lukut and Pasir Panjang of Negeri Sembilan, Tanjung Batu of Malacca and Pasir Puteh of Johor Bahru). Generally the concentration of PCBs in Malaysia was in low range from <4.2 to 60 ng/g lipid weight. However, high concentration of PCBs was found at Pasir Puteh, Johor Bahru which recorded 230 and 250 ng/g lipid weight respectively. The mean and range of PCBs concentrations from Asian countries in two studies are shown in Table 1.

Similarly, there was another study that used mussel samples, including green mussels (*Perna viridis*) and blue mussels (*Mytilus edulis*) to measure the levels of PCBs in the coastal waters of Asian countries (Ramu et al. 2007). Mussels were collected at Port Dickson, Pantai Lido, Pasir Puteh, Sebatu and Penang in Malaysia and the concentrations of PCBs analyzed in this study were in between 25 to 160 ng/g lipid weight. PCBs concentrations in mussels collected from coastal waters of Asian countries are shown in Fig. 5.

Table 1 Mean and range (in parenthesis) of PCBs concentration (ng/g lipid weight) in mussels from Asian countries

Country	PCBs (ng/g lipid weight)	
	Reference: Monirith et al. (2003)	Reference: Ramu et al. (2007)
Cambodia	35 (<3.8–220)	250 (200–300)
China	120 (15–540)	61 (3.8–160)
Hong Kong	310 (40–710)	401 (30–1200)
India	340 (9.8–600)	319 (56–640)
Indonesia	87 (5.6–210)	793 (510–1400)
Japan	3000 (510–12,000)	561 (83–2000)
South Korea	170 (30–340)	262 (17–1000)
Malaysia	**56 (<4.2–250)**	**89 (25–160)**
Philippines	290 (22–640)	1500 (1100–1900)
Russia	3200 (2700–3700)	NA
Singapore	90	NA
Vietnam	160 (21–450)	99 (26–290)

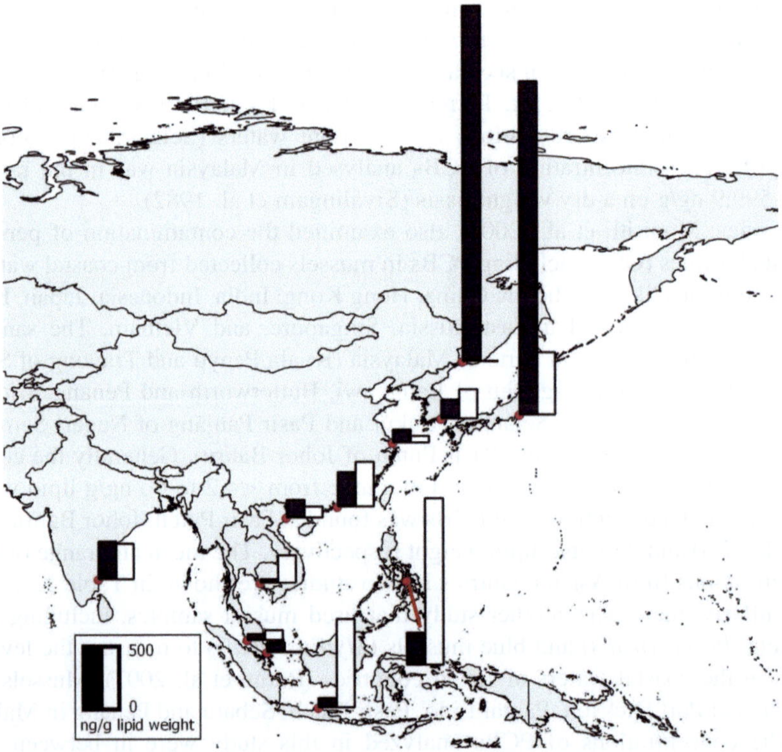

Fig. 5 Distribution of mean PCBs concentrations in mussels collected from coastal waters of Asian countries based on data from Monirith et al. (2003) (*black bar*) and Ramu et al. (2007) (*white bar*)

3.5 PCBs Concentration in Sediment, River Waters and Surface Seawaters

There was only one study each on the concentration of POPs including PCBs in river waters and surface seawaters by a Japanese group. It was found out that the total PCBs concentration was 0.45 ng/L in Selangor River water, and <5.0 ng/g of PCBs in the sediment samples from the same location (Iwata et al. 1994). Whereas the surface waters at the Straits of Malacca contained 0.02 ng/L of PCBs (Iwata et al. 1993), which was similar to the POP contamination levels at South China Sea.

3.6 PCBs Contamination in Fresh Water Fishes and Marine Organisms

Muhammad (2006) analyzed six species of fresh water fishes and seven species of marine organisms. All samples were collected from the northern region of peninsular Malaysia. The PCB concentrations in these fresh water samples ranged from 7.8 to 22.5 ng/g wet weight and the more toxic PCB-126 congener was detected only in the snakehead fish at very low concentration (0.03 ng/g wet weight). PCBs concentrations in marine organisms were generally lower than in the fresh water fishes. All the fish species in both categories and PCBs concentration are listed in Table 2.

3.7 PCBs in Human Breast Milk

PCBs were analyzed in human breast milk from non-smoking, healthy primipara mothers in the age group of 23–38 who were living in Penang and Kedah, Malaysia (Sudaryanto et al. 2005). Basically PCBs were detected in all the breast milk samples and the mono-*ortho* coplanar PCBs, were detected with an average concentration of 11 ng/g lipid weight, followed by non-*ortho* PCBs with concentration of 0.062 ng/g lipid weight. The same study also showed the result of average total TEQ levels to be 0.013 ng TEQs/g lipid weight. PCB levels in breast milk from Malaysia (Total PCBs with mean concentration of 80 ng/g lipid weight) were comparatively lower than developed countries, but slightly higher than other developing countries such as India, Vietnam, Cambodia and the Philippines (Fig. 6).

Table 2 Concentration of PCBs in fresh water fishes and marine species in Malaysia

Fresh water species	% Lipid (mean)	Concentration range of PCBs (ng/g wet weight)	Marine species	% Lipid (mean)	Concentration range of PCBs (ng/g wet weight)
Catfish (*Clarias batrachus*)	1.08	0.08–33.9	Small mackerel (*Rastrelliger sp.*)	1.20	0.03–2.5
Snakehead (*Channa striatus*)	1.03	1.69–20.4	Bigeye croaker (*Pennahia macrophthalmus*)	1.16	0.13–1.7
Gourami (*Trichogaster sp.*)	1.08	0.11–52.3	Mullet (*Mugil cephalus*)	1.35	0.09–1.9
Javanese carp (*Puntius gonionotus*)	1.04	0.04–22.2	Queenfish (*Scomberoides commersonnianus*)	1.04	0.07–5.2
Sultan fish (*Leptobarbus hoevenii*)	1.09	5.2–43.7	Catfish (*Arius sp.*)	1.20	0.18–3.3
Climbing perch (*Anabas testudineus*)	1.01	0.38–14.2	Shrimps (*Metapenaeus sp.*)	1.04	0.12–2.1
			Blood cockles (*Anadara granosa*)	1.06	0.05–3.7

Source: Muhammad (2006)

4 PBDEs in Malaysia

4.1 PBDEs in Mussels

The green mussels (*Perna viridis*) and blue mussels (*Mytilus edulis*) collected from various locations in Asia by Ramu et al. (2007) were studied for PBDEs. Based on their analysis, PBDEs were detected in all mussel samples indicating widespread contamination of these compounds in the coastal waters of Asian region. The level of contamination varied among sampling locations. Lower residue levels of PBDEs were found in samples from Cambodia, Vietnam, India and Malaysia in comparison to samples from Korea, Japan, Hong Kong, China and the Philippines. The total concentration of PBDEs in mussels collected from Malaysia ranged from 0.84 (at Sebatu) to 16 ng/g lipid weight (at Pasir Puteh) with relatively higher concentration of mono- to hepta-BDEs. On the other hand, the concentration of octa- to deca-BDEs in all mussel samples were below detection limit except at Pasir Puteh, where 0.19 ng/g lipid weight was detected. All these data are summarized in Table 3 (Fig. 7).

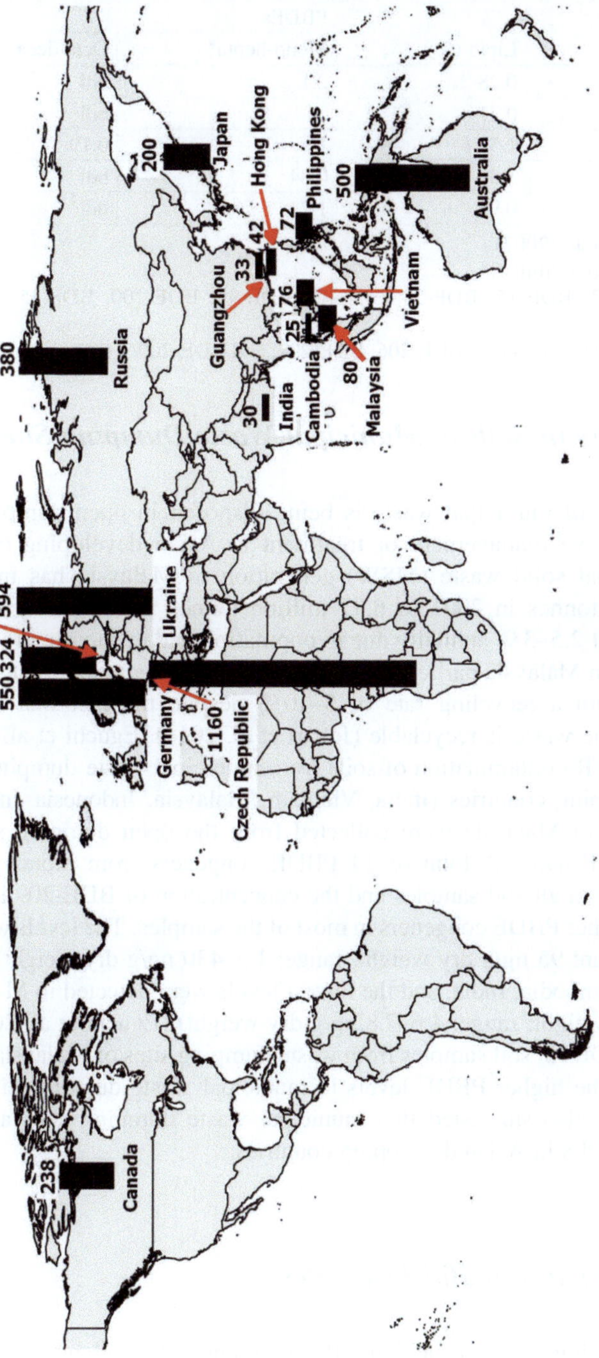

Fig. 6 Comparison of PCBs in human breast milk from developing and developed countries (ng/g lipid weight). (Source: Sudaryanto et al. 2005)

Table 3 Concentration of PBDEs (ng/g lipid wt.) in mussels from coastal waters of Malaysia

Location	Lipid (%)	PBDEs		
		Mono-hepta[a]	Octa-deca[b]	Total
Port Dickson	0.28	2.1	bdl	2.1
Pantai Lido	0.71	14	bdl	14
Pasir Puteh	1.5	16	0.19	16
Sebatu	1.8	0.84	bdl	0.84
Penang	0.95	1.2	bdl	1.2

Source: Ramu et al. (2007)
bdl below detection limit
[a]Sum of BDE-3, BDE-15, BDE-28, BDE-47, BDE-99, BDE-100, BDE-153, BDE-154, and BDE-183
[b]Sum of BDE-196, BDE-197, BDE-206, BDE-207, and BDE-209

4.2 PBDEs in Soils at Municipal Waste Dumping Sites

Huge amount of municipal waste is being disposed in open dumping sites with lack of effective management or treatment in Asian developing countries. The total municipal solid waste (MSW) generation in Malaysia has increased from 5.91 million tonnes in 2001 to 6.97 million tonnes in 2005 and is expected to increase about 2.5–3 % annually due to population and economic growth (Tarmudi et al. 2009). In Malaysia particularly, landfilling is the main option of waste disposal (90–95 %) with a recycling rate of 5–10 % despite the fact that approximately 70–80 % of the waste is recyclable (Johari et al. 2014). Eguchi et al. (2013) investigated the BFR contamination of soils from municipal waste dumping sites in five Asian developing countries (India, Vietnam, Malaysia, Indonesia and Cambodia). Soil samples in Malaysia were collected from the open dumping sites at Kuala Lumpur and Penang. A total of 14 PBDE congeners from mono- to deca-BDE were detected in all soil samples and the concentration of BDE-209 is significantly higher than other PBDE congeners in most of the samples. The levels were highest in Vietnam (mean: 95 ng/g dry weight, range: 1.2–430 ng/g dry weight), followed by Indonesia, Cambodia, India, and the lowest levels were detected in Malaysia (mean: 6.2 ng/g dry weight; range: 4.6–7.8 ng/g dry weight) (Eguchi et al. 2013). Concentration of PBDEs in soil samples from waste dumping sites of Malaysia can be found in Table 4. The higher PBDE levels in municipal waste dumping sites than those reference site also suggested that municipal waste dumping sites are a potential source of PBDEs in Asian developing countries.

4.3 PBDEs in Landfill Leachates

PBDE-containing products at landfills or municipal solid waste dumping sites (MSWDS) have been identified (Rahman et al. 2001). The concentration levels of

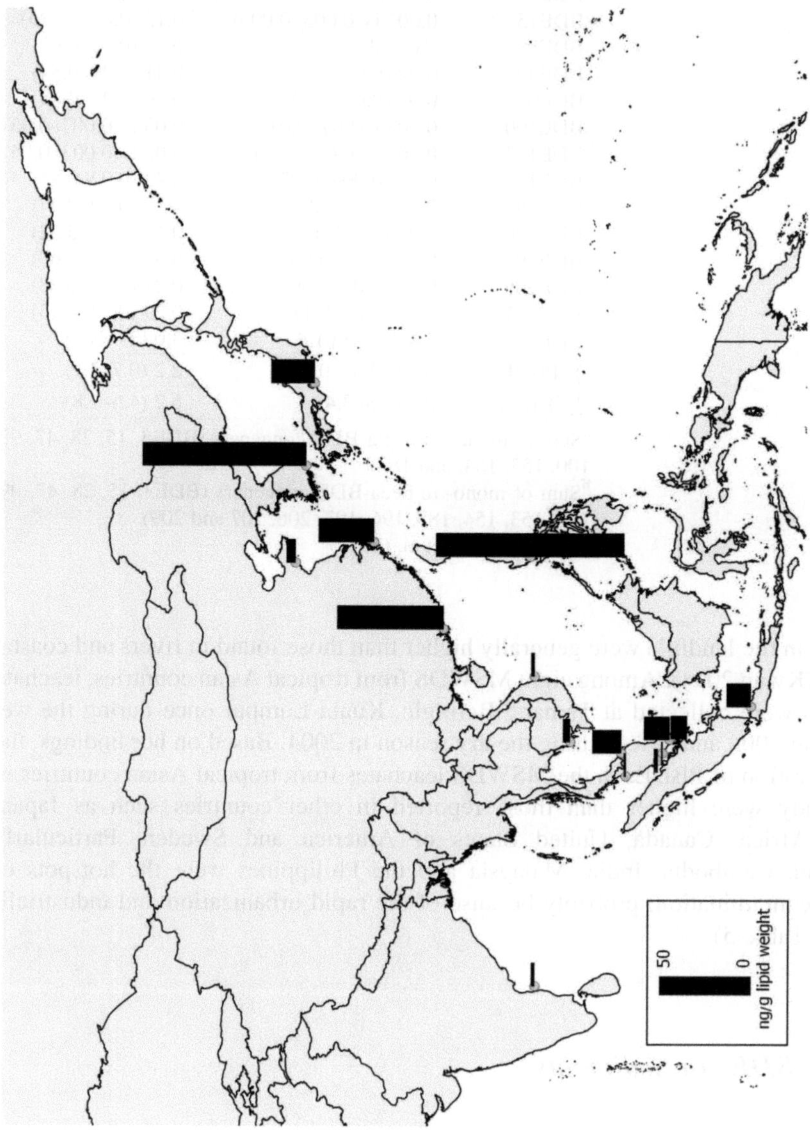

Fig. 7 Geographical distribution of PBDEs concentrations in mussels from coastal waters of Asia. (Source: Ramu et al. 2007)

Table 4 Concentration of PBDEs (ng/g dry weight) in soils of waste dumping sites in Malaysia

	Mean (min–max)	
	Reference sites	Dumping sites
TOC (%)	1.9 (0.36–3.4)	4.5 (3.3–5.7)
PBDEs		
BDE3	<0.003	<0.003
BDE15	0.003 (<0.003–0.006)	0.12 (0.007–0.24)
BDE28	<0.003	0.32 (0.02–0.62)
BDE47	0.02 (0.02)	0.41 (0.29–0.54)
BDE99	0.01 (0.01–0.02)	0.36 (0.25–0.48)
BDE100	0.008 (0.007–0.01)	0.03 (<0.003–0.05)
BDE153	0.005 (<0.003–0.01)	0.02 (<0.003–0.05)
BDE154	0.01 (0.008–0.02)	0.44 (0.02–0.85)
BDE183	0.02 (0.02–0.03)	0.53 (0.09–0.97)
BDE196	0.01 (0.006–0.02)	0.18 (0.04–0.32)
BDE197	0.02 (0.01–0.03)	0.33 (0.01–0.65)
BDE206	0.07 (0.05–0.09)	0.21 (0.14–0.28)
BDE207	0.08 (0.05–0.1)	0.29 (0.14–0.43)
BDE209	2.1 (1.1–3.0)	3.0 (2.4–3.5)
$\sum_9$PBDEs[a]	0.09 (0.07–0.11)	2.2 (0.7–3.7)
$\sum$PBDEs[b]	2.4 (1.3–3.4)	6.2 (4.6–7.8)

[a]Sum of mono- to hepta-BDE congeners (BDE3, 15, 28, 47, 99, 100, 153, 154, and 183)
[b]Sum of mono- to deca-BDE congeners (BDE3, 15, 28, 47, 99, 100, 153, 154, 183, 196, 197, 206, 207 and 209)
Source: Eguchi et al. (2013)

PBDEs in the landfills were generally higher than those found in rivers and coastal waters (Kwan 2014). Among eight MSWDS from tropical Asian countries, leachate samples were collected at Tamang Beringin, Kuala Lumpur once during the wet season in 2003 and twice during the dry season in 2004. Based on her findings, the concentration of PBDEs in the MSWDS leachates from tropical Asian countries in this study were higher than those reported in other countries such as Japan, South Africa, Canada, United States of America and Sweden. Particularly Thailand, Cambodia, India, Malaysia and the Philippines were the hotspots of PBDE contamination, probably because of the rapid urbanization and industrialization (Table 5).

4.4 PBDEs in Sediments

PBDEs being lipophilic with low solubility in water tend to accumulate in sediments or particulate matter, aiding their distribution further to marine biota. Many previous studies have shown the predominance of BDE-209 in sediments probably due to high production and usage of technical deca-BDE products (Eljarrat et al. 2005; Moon et al. 2007a; Tokarz et al. 2008; Toms et al. 2008;

Table 5 Concentration of PBDEs in the leachate samples from MSWDS Tamang Beringin, Malaysia

PBDEs	Concentration (ng/L) (min–max)	PBDEs	Concentration (ng/L) (min–max)
BDE-33/28	0.459–2.83	BDE-153	15.6–31.9
BDE-75	[a]ND	BDE-154	5.73–9.28
BDE-47	23.5–48.5	[d]Hexa-BDEs	22.3–47.5
BDE-66	0.898–3.82	BDE-183	7.27–21.8
BDE-77	ND-0.072	[e]Hepta-BDEs	11.5–31.7
[b]Tetra-BDEs	26.4–60.0	[f]Octa-BDEs	26.5–36.5
BDE-85	0.219–2.97	BDE-206	6.73–11.6
BDE-99	33.9–72.8	BDE-207	6.34–9.08
BDE-100	4.55–11.0	[g]Nona-BDEs	22.3–34.8
BDE-119	ND-0.27	BDE-209	22.3–31.2
[c]Penta-BDEs	40.3–81.8		

Source: Kwan (2014)
[a]ND = Not detected by the instrumental analysis
[b]Tetra-BDEs = ∑BDEs 47, 49, 66, 71, 75, 77
[c]Penta-BDEs = ∑BDEs 85, 99, 100, 116, 118, 119, 126
[d]Hexa-BDEs = ∑BDEs 138, 153, 154, 155, 166
[e]Hepta-BDEs = ∑BDEs 179, 181, 183, 188, 190
[f]Octa-BDEs = ∑BDEs 196, 197, 202, 203
[g]Nona-BDEs = ∑BDEs 206, 207, 208

Chen et al. 2009). This statement was further substantiated by Kwan (2014), who investigated the occurrence of PBDEs in surface sediment samples from urban areas of tropical Asian countries, including Lao PDR, Cambodia, Vietnam, India, Indonesia, Thailand, the Philippines and Malaysia. According to her analysis, nearly all the sediment samples from Lao PDR, Cambodia, Vietnam, India (Mumbai), Thailand, the Philippines and Malaysia have exhibited the predominance of BDE-209. Relative to the sediment samples from Japan (with a total PBDE geometric mean concentration (GMC) of 100 ng/g dw (dry weight)), studies from Kwan (2014) indicated the lowest total PBDE GMC in Lao PDR (9.38 ng/g dw). Compared to the GMCs of total PBDEs found in Vietnam (34.2 ng/g dw), Indonesia (42.2 ng/g dw), Cambodia (58.4 ng/g dw), India (67.3 ng/g dw), Thailand (160 ng/g dw), and the Philippines (178 ng/g dw), Malaysia (Klang river) recorded a marginally higher readings than these countries with 275 ng/g dw. In addition, Kwan (2014) has also correlated concentrations of PBDEs in the aquatic environment such as sediments to the GDP per capita and the number of MSWDS; higher population and more MSWDS will result in higher concentration of PBDEs in sediments.

5 PCBs and PBDEs in Sea Turtle

Several studies have reported that persistent organic pollutants (POPs) such as PCBs and PBDEs are found in sea turtle populations. Due to the lipophilic properties of these compounds and the mobilization of lipids by sea turtle for egg production, it is likely that POPs are transferred from nesting females to eggs and hatchlings during reproduction (Kwan 1994). van de Merwe et al. (2010) have analyzed POP contamination in a nesting population of *Chelonia mydas* in Malaysia, and investigated the maternal transfer of these chemicals to eggs and embryos as well as their impacts on hatchling development. Eggs and blood were collected from 11 adult female *Chelonia mydas* nesting at the Ma'Daerah Turtle Sanctuary, Terengganu, Malaysia in 2004 and analyzed for POP concentrations. The result showed significant correlation between maternal blood and eggs, and between eggs and hatchling, in other words, these chemicals are being transferred from nesting female *Chelonia mydas* to eggs and hatchlings. Additionally, this study also proved that the less lipophilic congeners tend to preferentially transfer from eggs to hatchlings, while the highly lipophilic congeners stay back with the mother. In general, hatchlings are at high risk from exposure to PCBs and PBDEs (Table 6).

6 Pan Asian Comparisons

6.1 Comparison of PCBs and PBDEs Concentrations in Human Breast Milk

By comparing the concentration of total PCBs (obtained by summing up the concentration of individually resolved peaks of PCB isomers and congeners) in human breast milk samples collected from mothers residing in Malaysia (Sudaryanto et al. 2005) to other Asian countries (Kunisue et al. 2002, 2004a, b, 2006; Sudaryanto et al. 2006; Wong et al. 2013). Malaysia recorded one of the highest level of PCBs contamination with 80 ng/g lipid weight. the highest concentration of PCBs around the same sampling year was detected at Payatas, Philippines with concentration of 92.5 ng/g lipid weight in 2004. The highest PCBs contaminated human breast milk sample since the year of 1989 until 2009 was collected at Hong Kong in 1989, followed by Dong Nai in 1991 and Chennai in 1988, with concentration of 640, 353 and 110 ng/g lipid weight respectively (Table 7).

Generally, PCBs contamination gradually decreased over the years as shown by readings at Chennai and Phnom Penh, which dropped from 110 to 34 ng/g lipid weight and from 39 to 20 ng/g lipid weight. The only exception is PCBs concentration in Bangkok which increased from 15 ng/g lipid weight in 1989 to 52 ng/g lipid weight in 1991.

Table 6 The concentration of PCBs and PBDEs in eggs and blood from nesting females and hatchlings of *Chelonia mydas*

	Concentration (pg/g wet mass)					
	In egg		In maternal blood		In hatchling blood	
POP compound	Mean ± SE	Range	Mean ± SE	Range	Mean ± SE	Range
PCB 99	17.4 ± 1.4	13.0–27.7	39.4 ± 4.8	22.0–64.3	81.4 ± 8.2	50.0–124.0
PCB 118	16.4 ± 1.3	11.6–23.6	33.9 ± 5.8	18.4–83.6	58.0 ± 7.1	36.0–105.8
PCB 128	14.1 ± 1.2	9.9–20.1	28.7 ± 3.8	16.3–54.4	35.4 ± 2.5	23.1–47.1
PCB 138 (+158)[a]	45.6 ± 8.9	17.0–94.1	62.9 ± 13.9	25.8–174.6	101.3 ± 20.9	46.7–226.7
PCB 153 (+132)[a]	65.8 ± 13.1	25.8–136.4	93.9 ± 24.4	30.6–288.6	171.9 ± 43.1	64.7–462.0
PCB 180 + 193	53.5 ± 7.5	33.9–100.4	114.0 ± 13.5	72.0–202.0	91.1 ± 11.9	55.0–165.2
PCB 183	14.8 ± 1.8	9.7–25.9	52.0 ± 12.1	19.8–134.2	63.8 ± 12.0	28.0–142.4
∑PCB	553.6 ± 54.6	392.8–839.4	578.9 ± 85.6	316.4–1206.5	850.8 ± 105.2	559.4–1456.6
PBDE 47	21.5 ± 1.7	11.1–28.1	13.3 ± 1.2	7.2–20.1	27.7 ± 2.7	14.9–41.4
PBDE 99	32.0 ± 3.6	12.0–54.8	21.3 ± 3.7	5.0–52.8	55.3 ± 12.1	8.9–131.8
PBDE 153	27.4 ± 1.2	20.8–35.2	86.2 ± 10.3	37.5–151.4	<LOD	–
∑PBDE	129.3 ± 8.1	61.7–163.8	120.8 ± 14.1	57.5–224.3	83.0 ± 14.4	23.8–173.2

Source: van de Merwe et al. (2010)

[a]Indicates congeners in parenthesis co-elute. However, due to their rarity in environmental samples, they are unlikely to be significantly contributing to the reported concentrations

Table 7 Comparison of PCBs and PBDEs concentrations (mean) in human breast milk between Malaysia and other Asian Countries (ng/g lipid weight)

Country	Sampling location	PCBs	PBDEs	Survey year	Reference
Vietnam	Ho Chi Minh	28	–		Schecter et al. (1989)
	Dong Nai	353	–		Schecter et al. (1991)
	Hanoi	46	0.57	2007	Tue et al. (2010)
Cambodia	Phnom Penh	39	–		Schecter et al. (1991)
	Phnom Penh	20	–	1999–2000	Kunisue et al. (2004a)
Indonesia	Jakarta	33	2.2	2001	Sudaryanto et al. (2006)
Philippines	Quezon	72	–	2000	Kunisue et al. (2002)
	Payatas	92.5	7.8	2004	Malarvannan et al. (2009)
Hong Kong	Hong Kong	640	–	1985	Ip and Phillips (1989)
		42	–	1999	Wong et al. (2002)
		74	–	1999–2000	Poon et al. (2005)
		6.4	–	2009	Wong et al. (2013)
Thailand	Bangkok	15	–		Schecter et al. (1989)
	Bangkok	52	–		Schecter et al. (1991)
India	Chennai (Madras)	110	–	1988	Tanabe et al. (1990)
	Chennai (Madras)	34	–	2000–2003	Subramanian et al. (2007)
	Kolkata	40	–	2004–2005	Devanathan et al. (2009)
	Mumbai	30	–	2005–2006	Devanathan et al. (2009)
	New Delhi	23	–	2005–2006	Devanathan et al. (2009)
China	Dalian	42	–	2002	Kunisue et al. (2004b)
	Shenyang	28	–	2002	Kunisue et al. (2004b)
	Nanjing	–	7.7	2004	Sudaryanto et al. (2008)
	Shanghai	–	8.6	2007	Ma et al. (2012)
Japan	Kanagawa	1.58	1.79	1999	Akutsu et al. (2003)
	Fukuoka	120	–	2001–2004	Kunisue et al. (2006)
	Japan (4 regions)	73.18	1.74[a]	2005	Inoue et al. (2006)
Malaysia	Penang and Kedah	80	–	2003	Sudaryanto et al. (2005)

[a]Total PBDEs concentration excluding BDE209

However, not many studies have been carried out to examine the PBDEs contamination in breast milk samples collected from Asian countries, and none has been done in Malaysia. Among all studies on PBDEs in breast milk, China recorded the highest concentration of 8.6 ng/g lipid weight and the second highest was Philippines with 7.8 ng/g lipid weight, followed by Indonesia, Japan and Vietnam.

Studies on PCBs and PBDEs concentration in breast milk samples are important because several researches proved that persistent organochlorines (OCs) compounds such as PCBs in human breast milk will be transferred to infants through breast-feeding, especially older primiparae mothers who transfer higher amounts of OCs (Kunisue et al. 2006). Besides, breast-feeding is the primary exposure route for infants, and by examining the concentration of PCBs and PBDEs in daily milk intake, the exposure risks of infants can be estimated (Ma et al. 2012) (Table 8).

Table 8 Comparison of PCBs and PBDEs concentration (mean) in bivalves collected from Asian countries (ng/g lipid weight)

Country	Sampling location	PCBs	PBDEs	Survey year	Reference
Vietnam	Various locations	160	–	1997	Monirith et al. (2003)
	Various locations	99.2	1.98	2003–2005	Ramu et al. (2007)
Cambodia	Various locations	35	–	1998	Monirith et al. (2003)
	Preab Island	250	3.8	2003–2005	Ramu et al. (2007)
Indonesia	Various locations	87	–	1998	Monirith et al. (2003)
	Jakarta Bay	792.5	13.25	2003–2005	Ramu et al. (2007)
Philippines	Philippines	290	–	1998	Monirith et al. (2003)
	Bacoor, Naic	1500	104.5	2003–2005	Ramu et al. (2007)
Singapore	Singapore	90	–	1999	Monirith et al. (2003)
	Singapore	27.52	8.85	2002	Bayen et al. (2003)
Hong Kong	Hong Kong	310	–	1998–1999	Monirith et al. (2003)
	Hong Kong	–	58.16	2004	Liu et al. (2005)
	Hong Kong	401	58.7	2003–2005	Ramu et al. (2007)
India	Various locations	340		1998	Monirith et al. (2003)
	Various locations	318.67	1.87	2003–2005	Ramu et al. (2007)
Korea	Various locations	170	–	1998	Monirith et al. (2003)
	Various locations	262.64	90.59	2005	Ramu et al. (2007)
China	Various locations	120	–	1999–2001	Monirith et al. (2003)
	Pearl River Delta	357.07	–	2001–2002	Fang (2004)
	Qingdao	75	4.9	2002	Pan et al. (2007)
	Various locations	61.08	30.51	2003–2005	Ramu et al. (2007)
Japan	Tokyo Bay	3000	–	1994	Monirith et al. (2003)
	Various locations	560.6	23.68	2003–2005	Ramu et al. (2007)
Malaysia	Malaysia	56	–	1998–1999	Monirith et al. (2003)
	Port Dickson	33	2.1	2003–2005	Ramu et al. (2007)
	Pantai Lido	76	14	2003–2005	Ramu et al. (2007)
	Pasir Puteh	160	16	2003–2005	Ramu et al. (2007)
	Sebatu	25	0.84	2003–2005	Ramu et al. (2007)
	Penang	150	1.2	2003–2005	Ramu et al. (2007)

6.2 Comparison of PCBs and PBDEs Concentrations in Bivalves

Basically two major studies that analyzed the concentration of PCBs and PBDEs in mussels as part of International Mussel Watch Programme between 1994 to 2001 and 2003 to 2005 from Asian coastal waters; one was reported by Monirith et al. (2003) and another by Ramu et al. (2007). The concentration of PCBs in mussels collected from Malaysia during 1998–1999 recorded the second lowest at 56 ng/g lipid weight after Cambodia with 35 ng/g lipid weight indicating low PCBs pollution. While the highest PCBs contamination was from Japan which recorded 3000 ng/g lipid weight, followed by India with 340 ng/g lipid weight. Similarly from mussels collected during 2003–2005, the highest PCBs concentration was found in Japanese samples (83–2000 ng/g lipid weight). Again, PCBs concentration in Malaysian coastal waters was low with an average concentration of <100 ng/g lipid weight in comparison to other heavily industrialised Asian countries such as China, Hong Kong, Korea, India, Philippines and Indonesia which showed high PCBs levels in the range of 61–1500 ng/g lipid weight.

The highest concentration of PBDEs was identified in mussel samples collected at Cavite province of Philippines which is an agriculture province. Other countries that showed high level of PBDEs contamination in mussels are Korea, Hong Kong and China. Previous studies pointed out that higher level of PCBs and PBDEs were recorded at urbanised or industrialized locations; areas with shipping activities (Moon et al. 2007a). Particularly untreated effluent that directly discharged into the environment from e-waste manufacturing or treatment plant and municipal waste treatment facility has been proved as the potential source that caused high levels of PCBs and PBDEs at Hong Kong and China (Zheng et al. 2004; Shi et al. 2009). A Korean study (Hong et al. 2009) highlights a growing pollution problem in Asia and in particular a tremendous uptrend in PBDEs pollution in Korea, in comparison to more controlled discharges and releases in Western Europe (Table 9).

6.3 Comparison of PCBs and PBDEs Concentrations in Sediments

Unfortunately there were limited studies on PCBs and PBDEs contamination in Malaysia. In fact, PCBs and PBDEs were widely found in both river and marine sediments due to its persistency. There were 195 ng/g dry weight of PCBs detected at Zhujiang River, China which was also the highest PCBs concentration recorded in Asian countries. This was confirmed in a research conducted by Fang (2004) at the same region (Pearl River Delta) for PCBs. The Pearl River delta region is one of the largest electronic and telecommunication manufacturing bases in China, hence lots of studies have been implemented in the area and previous results did indicate significant level of PCBs and PBDEs pollution within this region. Apart from that, sediment samples collected at canals in lower Mekong River Basin (Laos), Ho Chi Minh city canals (Vietnam) and a sewer system in Hanoi (Vietnam) showed high PCBs contamination with 111, 81 and 104 ng/g dry weight respectively.

On the other hand, exceptionally high concentration of PBDEs (285 ng/g dry weight) was observed at Busan Bay, Korea which is much higher than PBDEs levels in sediments from Hong Kong with 95 ng/g dry weight. In surface sediments, the average ΣPBDEs levels approached that of average ΣPCBs values. However, trends observed in the sediment core suggest that this pattern will alter over time and result in higher surface sediment PBDE concentrations than PCBs in future (Hong et al. 2010). Various diffuse and point sources for PBDEs and PCBs were

Table 9 Mean concentration of PCBs and PBDEs in sediment collected from Asian countries (ng/g dry weight)

Country	Sampling location	PCBs	PBDEs	Survey Year	Reference
Vietnam	Mekong River Delta	0.49	–	1998	Carvalho et al. (2008)
	Red River delta	4.32	–	1995–1996	Nhan et al. (1998)
	Ho Chi Minh	46.4	–	1996	Phuong et al. (1998)
	Ho Chi Minh city canals	81	–	2004	Minh et al. (2007)
	Hanoi (sewer)	104	–	2006	Hoai et al. (2010)
	Thi Nai Lagoon	2.4	1.27	2010	Romano et al. (2013)
Indonesia	Jakarta	–	0.15–130	2007	Sudaryanto et al. (2009)
Philippines	Manila Bay	<0.5–110	<0.1–18	2008	Isobe et al. (2010)
Singapore	Northeast	56.67	6.61	2003	Wurl and Obbard (2005)
	Southwest	101.5	5.54	2003	Wurl and Obbard (2005)
	Singapore	0.88	<1	2004	Bayen et al. (2005)
Hong Kong	Victoria Harbour	18.59	–	1993–1997	Wong and Poon (2003)
	Marshes Nature Reserve	2.9	–	1997	Liang et al. (1999)
	Hong Kong	17.2	–	1997–1998	Richardson and Zheng (1999)
	Hong Kong	19.2	3.03	2004–2005	Kueh and Lam (2008)
	Hong Kong	23.33	95	2006	Kueh and Lam (2008)
Thailand	Lower Mekong River Basin	10	–	2005	Sudaryanto et al. (2011)
Laos	Lower Mekong River Basin (Canal)	111	–	2005	Sudaryanto et al. (2011)
India	Sundarban	–	4.89		Binelli et al. (2007)
Taiwan	Southern Taiwan	1.43	–	2000	Doong et al. (2008)
	Northern Taiwan	11.35	–	2001, 2004	Hung et al. (2006)
	Northern Taiwan	15.52	15.52	2001, 2004	Hung et al. (2006)
Korea	Busan Bay	–	285	2003	Moon et al. (2007a)
	Various locations	–	27.8	2004	Moon et al. (2007b)
China	Pearl River Estuary	11.37	–	1997	Mai et al. (2002)
	Pearl River Delta	–	4.27		Zheng et al. (2004)
	Pearl River Delta	104.28	–	2001–2002	Fang (2004)
	Pearl River Estuary	–	3.13[a]	2004	Mai et al. (2005)
	Qingdao	5.11	1.38	1997, 1999	Pan et al. (2007)
	Zhujiang River	195.47	–	1997	Mai et al. (2002)
	Xijiang River	13.16	–	1997	Fu et al. (2001)
	Zhujiang River	–	12.9[a]	2002	Mai et al. (2005)
	Dongjiang River	–	27.3[a]	2002	Mai et al. (2005)
	Dongjiang River	–	3.94[a]	2006	Shi et al. (2009)

[a]Total PBDEs excluded BDE209

identified in this location. Shipping and other industrial activities were associated with PCB contamination while domestic and industrial waste discharges corresponded with PBDE contamination. Busan Bay is the largest harbor area in Korea which comprises more than four fifth of the country's container cargo and 42 % of the domestically produced marine products, in addition to the fact that 40 % of marine export cargo processed here (Hong et al. 2005).

Despite showing low contamination levels of PBDEs in sediment samples collected from China, concentration of BDE209 in some of these samples were extremely high. The result summarized by Mai et al. (2005) showed 12.9 ng/g dry weight of total PBDEs that consisted of BDE-28, -47, -66, -99, -100, -138, -153, -154 and -183 at Zhujiang River, while the mean concentration of BDE209 was 890 ng/g dry weight (ranged from 26.3 to 3580 ng/g dry weight). Similarly studies conducted by Shi et al. (2009) on the concentration of PBDEs in sediments from Dongjiang River, showed, for the same suite of PBDEs, a PBDEs concentration of 3.94 ng/g dry weight while the concentration of BDE209 was 1190 ng/g dry weight. The main reason for high concentration of BDE209 in Pearl River Delta was due to the manufacturing of electrical and electronic products where BDE209 is the main ingredient of BFR. Such high concentration of BDE209 in the environment poses risks and concern because it may degrade to less-brominated congeners which are more volatile and easily taken up by living organism and transferred to organisms in the higher food chain and bioaccumulate in humans. Recent studies also pointed out that BDE209 can be broken down under sunlight in sediments (Mai et al. 2005).

7 Conclusion and Recommendations

Many tropical Asian countries including Malaysia have experienced a rapid economic development and population growth during the past few decades. Industrialization and urbanization caused high waste generation and release of PCBs and PBDEs into the environment. In general, the concentrations of PCBs and PBDEs in pellets, mussels, air and soil from Malaysian samples were comparatively lower than most of neighboring Asian countries; on the other hand, their levels were high in sediment and leachate samples. Similarly, high levels PCBs were recorded in breast milk samples collected from northern Malaysia. Till now, there is no research on PBDEs in breast milk in Malaysia. Unfortunately, the samples collected in these studies were from selected locations and may not reflect the national status. The previous studies on PCBs and PBDEs contamination in Malaysia were conducted by Japanese researchers. It is surprising to note that local Malaysian universities and research institutes with analytical facilities have not generated substantial national data on these important contaminants. So far there is no nationwide study (geographical distribution) or their temporal trends in Malaysia. Additionally, the spectrum of samples were limited to bivalves, sediments, breast milk and human blood only. Other media need to be studied as well. That will provide a clearer picture of contamination status of Malaysia on a global map. Trend studies will help policymakers to amend regulations accordingly. Despite the fact that concentration of PBDEs in Malaysian air is negligible, it is still needed to be monitored. The reason is that uncontrolled waste dumping occurs in southern hemisphere countries from industrialized countries and global transboundary pollution needs to be monitored (Parolini et al. 2013). In summary, more studies are needed to fully understand POPs contamination status, their transport pathways, sources and their biological impacts on developing nations in Asia, with the emphasis of further researches in Malaysia.

8 Summary

Ever since PCBs and PBDEs were listed as persistent organic pollutant by Stockholm Convention, there are many studies around the world to study these compounds as well as their effects to human health and the environment. Despite the fact that PCBs-containing products being banned in many countries, old equipments that contain PCBs or PBDEs are still being used, especially in South Asian countries. Thus POPs will be released unintentionally into the environment from stockpiles or from waste treatment facilities or from E-waste treatment facility or area with high shipping activities. Owing to the limited research being conducted in Malaysia, the comparison of PCBs and PBDEs pollutions between Malaysia and other Asian countries were mainly focusing on mussels and human breast milk samples. PCBs in breast milk that was collected from Malaysia was the highest among Asian developing countries, while the mean concentration of PCBs in mussels in Malaysia recorded the second lowest. The concentrations of PBDEs in mussels taken from Malaysia was much lower if compared to the high concentration detected at Philippines and Korea. All in all, continual monitoring of these POPs by Malaysian researchers is needed in order to meet the target set by Stockholm Convention and assure the elimination of such compounds from the environment within the time frame.

References

Abel DC, McConnell RL (2009) Environmental oceanography: topics and analysis. Jones & Bartlett Publishers, Sudbury

Ahmad UK, Ujang Z, Zamani A, Halim A (2005) Determination of polychlorinated biphenyls in oils and sewage sludge. Malaysian J Anal Sci 9(2):280–283

Akutsu K, Kitagawa M, Nakazawa H, Makino T, Iwazaki K, Oda H, Hori S (2003) Time-trend (1973–2000) of polybrominated diphenyl ethers in Japanese mother's milk. Chemosphere 53 (6):645–654

Barsotti DA, Marlar RJ, Allen JR (1976) Reproductive dysfunction in rhesus monkeys exposed to low levels of polychlorinated biphenyls (Aroclor 1248). Food Cosmet Toxicol 14(2):99–103

Bayen S, Thomas GO, Lee HK, Obbard JP (2003) Occurrence of polychlorinated biphenyls and polybrominated diphenyl ethers in green mussels (Perna viridis) from Singapore, Southeast Asia. Environ Toxicol Chem 22(10):2432–2437

Bayen S, Wurl O, Karuppiah S, Sivasothi N, Lee HK, Obbard JP (2005) Persistent organic pollutants in mangrove food webs in Singapore. Chemosphere 61(3):303–313

Binelli A, Sarkar SK, Chatterjee M, Riva C, Parolini M, Bhattacharya BD, Bhattacharya AK, Satpathy KK (2007) Concentration of polybrominated diphenyl ethers (PBDEs) in sediment cores of Sundarban mangrove wetland, northeastern part of Bay of Bengal (India). Mar Pollut Bull 54(8):1220–1229

Branchi I, Alleva E, Costa LG (2002) Effects of perinatal exposure to a polybrominated diphenyl ether (PBDE 99) on mouse neurobehavioral development. Neurotoxicology 23:375–384

Carvalho FP, Villeneuve JP, Cattini C, Tolosa I, Thuan DD, Nhan DD (2008) Agrochemical and polychlorobyphenyl (PCB) residues in the Mekong River delta, Vietnam. Mar Pollut Bull 56 (8):1476–1485

Central Intelligence Agency (2015) The world factbook, East & Southeast Asia: Malaysia. https://www.cia.gov/library/publications/the-world-factbook/geos/my.html

Chen L, Huang Y, Peng X, Xu Z, Zhang S, Ren M, Ye Z, Wang X (2009) PBDEs in sediments of the Beijing River, China: levels, distribution, and influence of total organic carbon. Chemosphere 76:226–231

De Voogt P, Brinkman UAT (1989) Production, properties and usage of polychlorinated biphenyls. In: Kimbrough RA, Jensen AA (eds) Halogenated biphenyls, terphenyls, naphthalenes, dibenzodioxins and related products, 2nd edn. Elsevier, Amsterdam, pp 3–45

de Wit CA (2002) An overview of brominated flame retardants in the environment. Chemosphere 46(5):583–624

Department of Environment (DOE) (2008) Guidelines for the classification of used electrical and electronic equipment in Malaysia. http://www.meti.go.jp/policy/recycle/main/admin_info/law/10/pdf/malaysia_eng.pdf

Department of Statistics Malaysia (2015) Top 10 major export products, 2015. http://www.matrade.gov.my/en/malaysia-exporters-section/33-trade-statistics/3816-top-10-major-export-products-2015

Desa A, Kadir N, Yusooff F (2012) Waste education and awareness strategy: towards solid waste management (SWM) program at UKM. Procedia Social Behav Sci 59:47–50

Devanathan G, Subramanian A, Someya M, Sudaryanto A (2009) Persistent organochlorines in human breast milk from major metropolitan cities in India. Environ Pollut 157(1):148–154

Doong R, Lee S, Lee C, Sun Y, Wu S (2008) Characterization and composition of heavy metals and persistent organic pollutants in water and estuarine sediments from Gao-ping River, Taiwan. Mar Pollut Bull 57(6–12):846–857

Drabble JH (2001) An economic history of Malaysia, c.1800–1900: the transition to modern economic growth. Economic History Association, EH.NET

Eguchi A, Isobe T, Ramu K, Tue NM, Sudaryanto A, Devanathan G, Viet PH, Tana RS, Takahashi S, Subramanian A, Tanabe S (2013) Soil contamination by brominated flame retardants in open waste dumping sites in Asian developing countries. Chemosphere 90 (9):2365–2371

Eljarrat E, De La Cal A, Larrazabal D, Fernandez-Alba AR, Borrull F, March RM, Barcelo D (2005) Occurrence of polybrominated diphenylethers, polychlorinated dibenzo-p-dioxins, dibenzofurans, biphenyls in coastal sediments from Spain. Environ Pollut 136:493–501

Fang Z (2004) Organochlorines in sediments and mussels collected from coastal sites along the Pearl River Delta, South China. J Environ Sci 16(2):321–327

Fowles JR, Fairbrother A, Baecher-Steppan L, Kerkvliet NI (1994) Immunologic and endocrine effects of the flame-retardant pentabromodiphenyl ether (DE-71) in C57BL/6J mice. Toxicology 86:49–61

Fu J, Wang Z, Mai B, Kang Y (2001) Field monitoring of toxic organic pollution in the sediments of Pearl River estuary and its tributaries. Water Sci Technol 43(2):83–89

Goldberg ED, Bowen VT, Farrington JW, Harvey G, Martin JH, Parker PL, Risebrough RW, Robertson W, Schneider E, Gamble E (1978) The mussel watch. Environ Conserv 5:101–125

Golub MS, Donald JM, Reyes JA (1991) Reproductive toxicity of commercial PCB mixtures: LOAELs and NOAELs from animal studies. Environ Health Perspect 94:245–253

Halim AA (2007) Analysis of polychlorinated biphenyls in sewage sludge using supercritical carbon dioxide and subcritical water extraction. Thesis, Faculty of Science, Universiti Teknologi Malaysia, pp 18–23

Hashim R (2003) PCBs situation in Malaysia. PCB Symposium 2003 in Malaysia. http://tabemono.info/report/former/pcd/2003/malaysia/e_1.html

Hoai PM, Ngoc NT, Minh NH, Viet PH, Berg M, Alder AC, Giger W (2010) Recent levels of organochlorine pesticides and polychlorinated biphenyls in sediments of the sewer system in Hanoi, Vietnam. Environ Pollut 158(3):913–920

Hong SH, Yim UH, Shim WJ, Oh JR (2005) Congener-specific survey for polychlorinated biphenyl in sediments of industrialized bays in Korea: regional characteristics and pollution sources. Environ Sci Tech 39:7380–7388

Hong SH, Munschy C, Kannan N, Tixier C, Tronczynski J, Héas-Moisan K, Shim WJ (2009) PCDD/F, PBDE, and nonylphenol contamination in a semi-enclosed bay (Masan Bay, South Korea) and a Mediterranean lagoon (Thau, France). Chemosphere 77:854–862

Hong SH, Kannan N, Jin Y, Won JH, Han GM, Shim WJ (2010) Temporal trend, spatial distribution and terrestrial sources of PBDEs and PCBs in Masan Bay, Korea. Mar Pollut Bull 60(10):1836–1841

Hung C, Gong G, Jiann K, Yeager KM, Santschi PH, Wade TL, Sericano JL, Hsieh H (2006) Relationship between carbonaceous materials and polychlorinated biphenyls (PCBs) in the sediments of the Danshui River and adjacent coastal areas, Taiwan. Chemosphere 65 (9):1452–1461

IARC (2015) IARC monographs on the evaluation of carcinogen risks to humans. Polychlorinated and polybrominated biphenyls, Vol 107. http://monographs.iarc.fr/ENG/Monographs/vol107/mono107.pdf

Ibrahim SM (2007) Chapter 14: Persistent organic pollutants in Malaysia. In: Li A et al (eds) Development in environmental science, vol 7. Elsevier, Amsterdam, pp 629–655

Inoue K, Harada K, Takenaka K, Uehara S, Kono M, Shimizu T, Takasuga T, Senthilkumar K, Yamashita F, Koizumi A (2006) Levels and concentration ratios of polychlorinated biphenyls and polybrominated diphenyl ethers in serum and breast milk in Japanese mothers. Environ Health Perspect 114(8):1179–1185

Ip HMH, Phillips DJH (1989) Organochlorine chemicals in human breast milk in Hong Kong. Arch Environ Contam Toxicol 18(4):490–494

Isobe T, Amano A, Chang K, Maneja RH, Zamora PB, San Diego-Mcglone ML, Siringan FP, Prudente M, Miller TW, Tanabe S (2010) Retrospective assessment of environmental pollution by PBDEs and HBCDs using sediment core from Manila Bay, the Philippines. In: Isobe T, Nomiyama K, Subramanian A, Tanabe S (eds) Interdisciplinary studies on environmental chemistry—environmental specimen bank. TERRAPUB, Tokyo, pp 61–65

Iwata H, Tanabe S, Sakai N, Tatsukawa R (1993) Distribution of persistent organochlorines in the oceanic air and surface seawater and the role of ocean on their global transport and fate. Environ Sci Technol 27:1080

Iwata H, Tanabe S, Sakai N, Nishimura A, Tatsukawa R (1994) Geographical distribution of persistent organochlorines in air, water and sediments from Asia and Oceania, and their implications for global redistribution from lower latitudes. Environ Pollut 85:15

Jahromi FA, Kannan N, Zakaria MP, Aris AZ (2014) Persistent contaminants in waste oils: a short review on PCBs and PAHs as main contaminants. In: From sources to solution: Proceedings of the international conference on environmental forensic 2013. Springer, Singapore, pp 107–110

Jensen S (1966) Report of a new chemical hazard. N Scientist 32:612

Johari A, Alkali H, Hashim H, Ahmed SI, Mat R (2014) Municipal solid waste management and potential revenue from recycling in Malaysia. Modern Appl Sci 8(4)

Kannan N, Tanabe S, Tatsukawa R (1988) Toxic potential of non-*ortho* and mono-*ortho* coplanar PCBs in commercial PCB preparations: "2,3,7,8-T_4 CDD toxicity equivalence factors approach". Environ Contam Toxicol 41(2):267–276

Kannan N, Tanabe S, Ono M, Tatsukawa R (1989) Critical evaluation of polychlorinated biphenyl toxicity in terrestrial and marine mammals: increasing impact of non-*ortho* and mono-*ortho* coplanar polychlorinated biphenyls from land to ocean. Arch Environ Contam Toxicol 18:850–857

Kannan N (2000) Non-*ortho* and mono-*ortho* substituted PCBs, Chapter 6. In: Paasivirta J (ed) The handbook of environmental chemistry, vol 3, Anthropogenic compounds. Springer, Berlin, pp 127–156

Kannan N, Hong SH, Oh JR, Yim UH, Li DH, Shim WJ (2005) PYE [2-(1-pyrenyl) ethyldimethylsilylated silica] column HPLC and HR-GC-(micro) ECD in the accurate determination of toxic co-planar PCBs and polybrominated diphenyl ethers (PBDEs). Bull Korean Chem Soc 26(4):529

Kannan K, Tanabe S, Borrell A, Aguilar A, Focardi S, Tatsukawa R (1993) Isomer-specific analysis and toxic evaluation of polychlorinated biphenyls in striped dolphins affected by an epizootic in the Western Mediterranean Sea. Arch Environ Contam Toxicol 25(2):227–233

Kannan K, Yun SH, Rudd RJ, Behr M (2010) High concentrations of persistent organic pollutants including PCBs, DDT, PBDEs and PFOs in little brown bats with white-nose syndrome in New York, USA. Chemosphere 80(6):613–618

Kimbrough RD, Squire RA, Linder RE, Strandberg JD, Montalli RJ, Burse VW (1975) Induction of liver tumor in Sherman strain female rats by polychlorinated biphenyls aroclor 1260. J Natl Cancer Inst 55(6):1453–1459

Kueh CSW, Lam JYC (2008) Monitoring of toxic substances in the Hong Kong marine environment. Mar Pollut Bull 57(6–12):744–757

Kunisue T, Watanabe M, Someya M, Monirith I, Minh TB, Subramanian A, Tana TS, Viet PH, Prudente M, Tanabe S (2002) PCDDs, PCDFs, PCBs and organochlorine insecticides in human breast milk collected from Asian developing countries: risk assessment for infants. Organohalogen Comp 58:285–288

Kunisue T, Someya M, Monirith I, Watanabe M, Tana TS, Tanabe S (2004a) Occurrence of PCBs, organochlorine insecticides, tris(4-chlorophenyl)methane, and tris(4-chlorophenyl)methanol in human breast milk collected from Cambodia. Arch Environ Contam Toxicol 46(3):405–412

Kunisue T, Someya M, Kayama F, Jin Y, Tanabe S (2004b) Persistent organochlorines in human breast milk collected from primiparae in Dalian and Shenyang, China. Environ Pollut 131 (3):381–392

Kunisue T, Muraoka M, Ohtake M, Sudaryanto A, Minh NH, Ueno D, Higaki Y, Ochi M, Tsydenova O, Kamikawa S, Tonegi T, Nakamura Y, Shimomura H, Nagayaman J, Tanabe S (2006) Contamination status of persistent organochlorines in human breast milk from Japan: recent levels and temporal trend. Chemosphere 64(9):1601–1608

Kwan D (1994) Fat reserves and reproduction in the green turtle, Chelonia mydas. Wildl Res 21:257–266

Kwan CS (2014) Distribution and debromination of polybrominated diphenyl ethers (PBDEs) in tropical Asian countries. Unpublished doctoral dissertation, Tokyo University of Agriculture and Technology, Tokyo

Liang Y, Wong MH, Shutes RBE, Revitt DM (1999) Ecological risk assessment of polychlorinated biphenyl contamination in the Mai Po Marshes Nature Reserve, Hong Kong. Water Res 33(6):1338–1346

Liu Y, Zheng GJ, Yu H, Martin M, Richardson BJ, Lam MHW, Lam PKS (2005) Polybrominated diphenyl ethers (PBDEs) in sediments and mussel tissues from Hong Kong marine waters. Mar Pollut Bull 50(11):1173–1184

Ma S, Yu Z, Zhang X, Ren G, Peng P, Sheng G, Fu J (2012) Levels and congener profiles of polybrominated diphenyl ethers (PBDEs) in breast milk from Shanghai: implication for exposure route of higher brominated BDEs. Environ Int 42(1):72–77

Mai B, Fu J, Sheng G, Kang Y, Lin Z, Zhang G, Min Y, Zeng EY (2002) Chlorinated and polycyclic aromatic hydrocarbons in riverine and estuarine sediments from Pearl River Delta, China. Environ Pollut 117(3):457–474

Mai B, Chen S, Luo X, Chen L, Yang Q, Sheng G, Peng P, Fu J, Zeng EY (2005) Distribution of polybrominated diphenyl ethers in sediments of the Pearl River Delta and adjacent South China Sea. Environ Sci Technol 39(10):3521–3527

Malarvannan G, Kunisue T, Isobe T, Sudaryanto A, Takahashi S, Prudente M, Subramanian A, Tanabe S (2009) Organohalogen compounds in human breast milk from mothers living in Payatas and Malate, the Philippines: levels, accumulation kinetics and infant health risk. Environ Pollut 157(6):1924–1932

McDonald TA (2002) A perspective on the potential health risks of PBDEs. Chemosphere 46:745–755

Minh NH, Minh TB, Iwata H, Kajiwara N, Kunisue T, Takahashi S, Viet PH, Tuyen BC, Tanabe S (2007) Persistent organic pollutants in sediments from Sai Gon-Dong Nai River basin, Vietnam: levels and temporal trends. Arch Environ Contam Toxicol 52(4):458–465

Monirith I, Ueno D, Takahashi S, Nakata H, Sudaryanto A, Subramanian A, Karuppiah S, Ismail A, Muchtar M, Zheng J, Richardson BJ, Prudente M, Hue ND, Tana TS, Tkalin AV, Tanabe S (2003) Asia-Pacific mussel watch: monitoring contamination of persistent organochlorine compounds in coastal waters of Asian countries. Mar Pollut Bull 46(3):281–300

Moon H, Kannan K, Choi M, Choi HG (2007a) Polybrominated diphenyl ethers (PBDEs) in marine sediments from industrialized bays of Korea. Mar Pollut Bull 54(9):1402–1412

Moon H, Kannan K, Lee S, Choi M (2007b) Polybrominated diphenyl ethers (PBDEs) in sediment and bivalves from Korean coastal waters. Chemosphere 66(2):243–251

Muhammad SA (2006) Analysis of persistent organic pollutants in fish: health risk assessment through dietary intake. M.Sc. Thesis, Universiti Sains Malaysia, March 2006

Nhan DD, Am NM, Hoi NC, Dieu LV, Carvalho FP, Villeneuve JP, Cattini C (1998) Organochlorine pesticides and PCBs in the Red River Delta, North Vietnam. Mar Pollut Bull 36 (9):742–749

Ogata Y, Takada H, Mizukawa K, Hirai H, Iwasa S, Endo S, Mato Y, Saha M, Okuda K, Nakashima A, Murakami M, Zurcher N, Booyatumanondo R, Zakaria MP, Dung LQ, Gordon M, Miguez C, Suzuki S, Moore C, Karapanagioti HK, Weerts S, McClurg T, Burres E, Smith W, Velkenburg MV, Lang JS, Lang RC, Laursen D, Danner B, Stewardson N, Thompson RC (2009) International pellet watch: global monitoring of persistent organic pollutants (POPs) in coastal waters. 1. Initial phase data on PCBs, DDTs and HCHs. Mar Pollut Bull 58(10):1437–1446

Pan J, Yang Y, Xu Q, Chen D, Xi D (2007) PCBs, PCNs and PBDEs in sediments and mussels from Qingdao coastal sea in the frame of current circulations and influence of sewage sludge. Chemosphere 66(10):1971–1982

Parolini M, Guazzoni N, Comolli R, Binelli A, Tremolada P (2013) Background levels of polybrominated diphenyl ethers (PBDEs) in soils from Mount Meru area, Arusha district (Tanzania). Sci Total Environ 452–453:253–261

Phuong PK, Son CPN, Sauvain JJ, Tarradellas J (1998) Contamination by PCBs, DDTs, and heavy metals in sediments of Ho Chi Minh city's canals, Vietnam. Environ Contam Toxicol 60 (2):347–354

Poon BHT, Leung CKM, Wong CKC, Wong MH (2005) Polychlorinated biphenyls and organochlorine pesticides in human adipose tissue and breast milk collected in Hong Kong. Arch Environ Contam Toxicol 49(2):274–282

Rahman F, Langford KH, Scrimshaw MD, Lester JN (2001) Polybrominated diphenyl ether (PBDE) flame retardants. Sci Total Environ 275(1–3):1–17

Ramu K, Kajiwara N, Sudaryanto A, Isobe T, Takahashi S, Subramanian A, Ueno D, Zheng GJ, Lam PKS, Takada H, Zakaria MP, Viet PH, Prudente M, Tana TS, Tanabe S (2007) Asian Mussel watch program: contamination status of polybrominated diphenyl ethers and organochlorines in coastal waters of Asian countries. Environ Sci Tech 41(13):4580–4586

Richardson BJ, Zheng GJ (1999) Chlorinated hydrocarbon contaminants in Hong Kong surficial sediments. Chemosphere 39(6):913–923

Romano S, Piazza R, Mugnai C, Giuliani S, Bellucci LG, Huu CN, Vecchiato M, Zambon S, Hoai ND, Frignani M (2013) PBDEs and PCBs in sediments of the Thi Nai Lagoon (Central Vietnam) and soils from its mainland. Chemosphere 90(9):2396–2402

Safe S (1994) Polychlorinated-biphenyls (PCBs)—environmental impact, biochemical and toxic responses, and implications for risk assessment. Crit Rev Toxicol 24(2):87–149

Schecter A, Furst P, Kruger C, Meemken H, Groebel W, Constable JD (1989) Levels of Polychlorinated Dibenzofurans, Dibenodioxins, PCBs, DDT, and DDE, Hexachlorobenzene, Dieldrin, Hexachlorocyclohexane, and Oxychlordane in human breast milk from the United States, Thailand, Vietnam and Germany. Chemosphere 18(1–6):445–454

Schecter A, Furst P, Furst C, Papke O, Ball M, Dai LC, Quynh HT, Phoung NTN, Beim A, Vlasov B, Chongchet V, Constable JD, Charles K (1991) Dioxins, dibenzofurans and selected chlorinated organic compounds in human milk and blood from Cambodia, Germany, Thailand, The U.S.A., The U.S.S.R., and Vietnam. Chemosphere 23(11–12):1903–1912

Shi T, Chen S, Luo X, Zhang X, Tang C, Luo Y, Ma Y, Wu J, Peng X, Mai B (2009) Occurrence of brominated flame retardants other than polybrominated diphenyl ethers in environmental and biota samples from southern China. Chemosphere 74(7):910–916

Sivalingam PM, Allapitchay I, Kojima H, Yoshida T (1982) Mussel watch of PCBs and persistent pesticide residues in *Perna viridis Linnaeus* from Malaysian and Singapore waters. Appl Geography 2(3):231–237

Stockholm Convention (2009) PCBs overview. http://chm.pops.int/Implementation/PCBs/Overview/tabid/273/Default.aspx

Subramanian A, Ohtake M, Kunisue T, Tanabe S (2007) High levels of organochlorines in mother's milk from Chennai (Madras) city, India. Chemosphere 68(5):928–939

Sudaryanto A, Kunisue T, Tanabe S, Niida M, Hashim H (2005) Persistent organochlorine compounds in human breast milk from mothers living in Penang and Kedah, Malaysia. Arch Environ Contam Toxicol 49(3):429–437

Sudaryanto A, Kunisue T, Kajiwara N, Iwata H, Adibroto TA, Hartono P, Tanabe S (2006) Specific accumulation of organochlorines in human breast milk from Indonesia: levels, distribution, accumulation kinetics and infant health risk. Environ Pollut 139(1):107–117

Sudaryanto A, Kajiwara N, Tsydenova OV, Isobe T, Yu H, Takahashi S, Tanabe S (2008) Levels and congener specific profiles of PBDEs in human breast milk from China: implication on exposure sources and pathways. Chemosphere 73(10):1661–1668

Sudaryanto A, Setiawan IE, Ilyas M, Soeyanto E, Riadi AS, Isobe T, Takahashi S, Tanabe S (2009) Levels of brominated flame retardants in sediments and their bioaccumulation potential in biota from Jakarta Bay and its surroundings, Indonesia. In: Obayashi Y, Isobe T, Subramanian A, Suzuki S, Tanabe S (eds) Interdisciplinary studies on environmental chemistry—environmental research in Asia. TERRAPUB, Tokyo, pp 125–131

Sudaryanto A, Isobe T, Takahashi S, Tanabe S (2011) Assessment of persistent organic pollutants in sediments from Lower Mekong River Basin. Chemosphere 82(5):679–686

Suja F, Rahman RA, Yusof A, Shahbudin M (2014) e-Waste management scenarios in Malaysia. J Waste Manage 2014:609169

Tanabe S, Gondaira F, Subramanian A, Ramesh A, Mohan D, Kumaran P, Venugopalan VK, Tatsukawa R (1990) Specific pattern of persistent organochlorine residues in human breast milk from south India. J Agric Food Chem 38(3):899–903

Tarmudi Z, Abdullah ML, Tap AOM (2009) An overview of municipal solid wastes generation in Malaysia. J Technol 51:1–15

Tokarz JA, Ahn MY, Leng J, Filey TR, Nies L (2008) Reductive debromination of polybrominated diphenyl ethers in anaerobic sediment and a biomimetic system. Environ Sci Tech 42:1157–1164

Toms LML, Mortimer M, Symons RK, Paepke O, Mueller JF (2008) Polybrominated diphenyl ethers (PBDEs) in sediment by salinity and land-use type from Australia. Environ Int 34 (1):58–66

Tue NM, Sudaryanto A, Minh TB, Isobe T, Takahashi S, Viet PH, Tanabe S (2010) Accumulation of polychlorinated biphenyls and brominated flame retardants in breast milk from women living in Vietnamese e-waste recycling sites. Sci Total Environ 408(9):2155–2162

van de Merwe JP, Hodge M, Whittier JM, Ibrahim K, Lee SY (2010) Persistent organic pollutants in the green sea turtle *Chelonia mydas*: nesting population variation, maternal transfer, and effects on development. Mar Ecol Prog Ser 403:269–278

WHO (2003) Polychlorinated biphenyls: human health aspects. Concise international chemical assessment document 55. http://www.who.int/ipcs/publications/cicad/en/cicad55.pdf

Wong MH, Poon BHT (2003) Sources, fates and effects of persistent organic pollutants in China, with emphasis on the Pearl River Delta. Persist Org Pollut 3O:355–369

Wong CKC, Leung KM, Poon BHT, Lan CY, Wong MH (2002) Organochlorine hydrocarbons in human breast milk collected in Hong Kong and Guangzhou. Arch Environ Contam Toxicol 43 (3):364–372

Wong TW, Wang AHS, Nelson EAS, Qiu H, Ku SYK (2013) Levels of PCDDs, PCDFs, and dioxin-like PCBs in human milk among Hong Kong mothers. Sci Total Environ 463–464:1230–1238

Wu J, Luo X, Zhang Y, Luo Y, Chen S, Mai S, Yang Z (2008) Bioaccumulation of polybrominated diphenyl ethers (PBDEs) and polychlorinated biphenyls (PCBs) in wild aquatic species from an electronic waste (e-waste) recycling site in South China. Environ Int 34(8):1109–1113

Wurl O, Obbard JP (2005) Organochlorine pesticides, polychlorinated biphenyls and polybrominated diphenyl ethers in Singapore's coastal marine sediments. Chemosphere 58 (7):925–933

Zheng GJ, Martin M, Richardson BJ, Yu H, Liu Y, Zhou C, Li J, Hu G, Lam MHW, Lam PKS (2004) Concentrations of polybrominated diphenyl ethers (PBDEs) in Pearl River Delta sediments. Mar Pollut Bull 49(5–6):520–524

Zhou T, Taylor MM, DeVito MJ, Crofton KM (2002) Development exposure to brominated diphenyl ethers results in thyroid hormone disruption. Toxicology 66:105–116

Nitrate and Nitrogen Oxides: Sources, Health Effects and Their Remediation

Khalid Rehman Hakeem, Muhammad Sabir, Munir Ozturk, Mohd. Sayeed Akhtar, and Faridah Hanum Ibrahim

Contents

The original version of this chapter was revised. An erratum to the chapter can be found at DOI 10.1007/978-3-319-51243-3_15

K.R. Hakeem (✉)
Faculty of Forestry, Universiti Putra Malaysia, Serdang, Selangor UPM 43400, Malaysia

Department of Biological Sciences, Faculty of Science, King Abdulaziz University, Jeddah, Saudi Arabia
e-mail: kur.hakeem@gmail.com

M. Sabir
Institute of Soil and Environmental Sciences, University of Agriculture, Faisalabad, Pakistan
e-mail: cmsuaf@gmail.com

M. Ozturk
Botany Department & Centre for Environmental Studies, Ege University, Izmir, Turkey
e-mail: munirozturk@gmail.com

M.S. Akhtar
Department of Botany, Gandhi Faiz-E-Aam College, Shahjahanpur-242001, Uttar Pradesh, India
e-mail: sayeed.bot@gmail.com

F.H. Ibrahim
Faculty of Forestry, Universiti Putra Malaysia, Serdang, Selangor UPM 43400, Malaysia
e-mail: i.faridahhanum@gmail.com

P. de Voogt (ed.), *Reviews of Environmental Contamination and Toxicology, Volume 242*, Reviews of Environmental Contamination and Toxicology 242,
DOI 10.1007/398_2016_11

1 Introduction

Nitrogen (N) is a vital element for life and an essential constituent of proteins, DNA, energy compounds like ATP, chlorophyll, vitamins and hormones (Galloway et al. 2002; Ohyama 2010; Abrol et al. 2012; Santiago-Antonio et al. 2014). In terrestrial and aquatic ecosystems, N supply determines the nature and diversity of plants, the population dynamics of herbivores, and vital ecological processes such as plant productivity and the cycling of carbon and soil minerals (Cassman et al. 2002; Dobermann and Cassman 2004; Boyer et al. 2004; Ladha et al. 2005a; Abrol et al. 2012; Soussana and Lemairec 2014).

The rigorous use of N fertilizers is leading towards a detrimental impact on environment in the agricultural production systems (Xu et al. 2000; Smil 2002; Zu et al. 2002; Killebrew and Wolff 2010). Toxicity due to nitrates can be either acute or chronic (Hill 1999; Bjorne et al. 2004). Acute toxicity is expressed when nitrite enters the bloodstream in large quantities and oxidizes the ferrous iron of hemoglobin to the ferric form producing methemoglobin, which cannot transport oxygen and results in methemoglobinemia (Fewtrell 2004). The reports show that over 80 % of the total nitrate intake in humans originates from vegetables and some part from the drinking water (Shoji et al. 2001; Croen et al. 2001; Anjana Umar and Iqbal 2007; Anjana Umar et al. 2007).

Although the risks have not been fully evaluated, several developed countries have established maximum permissible levels for nitrate concentration in vegetable food crops, and other countries are considering the establishment of similar standards. As a result of increasing pollution of sources adjacent to agricultural activities, the World Health Organization has set a limit of 10 mg of nitrate nitrogen per litre in domestic water supplies (Pionke et al. 1990; McPharlin and Aylam 1995). Research has shown that high concentrations of nitrate in plant food can decrease the level of ascorbic acid, a known inhibitor of the formation of carcinogenic compounds that contain nitrite.

Nitrogen deficiencies lead to chlorosis and reduced photosynthesis, which ultimately result in lower yields, and at the same time a significant reduction in the levels of phosphoenol pyruvate carboxylase, pyruvate orthophosphate dikinase, and Rubisco and a concomitant decrease in the level of their respective mRNAs in crop plants (Dordas and Sioulas 2008; Hakeem et al. 2014). The extensive use of nitrogenous fertilizers has ensured food security to developing nations, as the cereal production has kept pace with its ever-increasing population. Application of N fertilizers has increased global food production three times over the past 50 years (Rosegrant et al. 2002; Tilman et al. 2011).

Since 1962, the commercial production of N fertilizer has increased many folds to meet the growing requirements in the agriculture sector. It has been reported that doubling the agricultural food production worldwide over the past four decades has been associated with a seven folds increase in the use of nitrogenous (N) fertilizers (Hirel et al. 2007; Santiago-Antonio et al. 2014). However, it is excruciating to know that utilization of applied fertilizer nitrogen in the field by most cereal crops does not exceed 50 %. Among these, approximately 70 % of the total nitrogenous fertilizer is applied to rice and wheat cultivation (Pathak and Nedwell 2001; Norse 2003; Mosier et al. 2004a, b; Palm et al. 2004; Abrol et al. 2012; Hakeem et al. 2012a).

Such a poor N-use efficiency is a matter of great environmental concern. Even if the efficiency of nitrogenous fertilizers remains constant at the current levels, the N-losses will increase enormously as their consumption is expected to be doubled within the next few decades. Recent and future prospects of intensive N fertilizer use will result in detrimental impacts on the diversity and functioning of the variety of ecosystems (Tilman et al. 2011; Abrol et al. 2012). Manufacture of these fertilizers involves high-cost technology that requires a whole range of feed stocks and naphtha, a petroleum product. The environmental problems associated with it include NO_3^- pollution of the ground and surface waters and emission of oxides of N (N_2O, NO, NO_2^-) which have positive radioactive forcing characteristics. Among these, NO_2^- is involved in stratospheric ozone depletion as well. Emission of unused nitrogen from agricultural fields in the form of NO_x contributes to the global warming. NO_2 has 298 times more global warming effect than CO_2 (Signor and Cerri 2013). Their excessive and injudicious use, as reported from a number of regions, further accentuates the environmental degradation besides affecting the quality of crops, human and animal health, and causes lodging in cereals which may affect crop yields and quality. In surface water, presence of high N results in growth of algae and plants, thus accelerating eutrophication, and consequently affects water quality and usage (EPA 2007; Shen et al. 2011). The incidence of stomach cancer in humans, particularly in infants, and of non-Hodgkin's lymphoma due to intake of water contaminated with nitrate has been reported (Mueller et al. 2001; Ikehata et al. 2010; Inoue-Choi et al. 2012). Nitrosamines produced from nitrite are reported to be carcinogenic; and NH_3 gas is a pollutant because of its corrosive nature due to the formation of ammonium salts. Applied nitrogen not taken up by crops, thus constitutes a huge financial loss and a severe environmental risk. It has been estimated that by a 10 % increase in NUE would result in, the annual saving of US $ 5 billion and also substantial considerable improvement in environmental condition and soil quality (Gupta and Khosla 2012). However, Prasad (2013) stated that 1 % increase in NUE in India will account for saving of 0.17 million tonnes of N or 0.4 million tonnes of urea, which is equal to the annual production of four urea plants producing 1 lakh tonnes of urea per annum.

2 Historical Perspectives of N Use in Agriculture

Nitrogen is the most crucial element which is essential for life (Abrol et al. 2012). Application of N fertilizers for improving crop yields and economic returns is historically recognized (Keeney and Hatfield 2008). N has been used for maximizing cereal grain yields in the twentieth century and cereal grain production has witnessed 40 percent increase per capita (Mosier et al. 2001). Nitrogen fertilizer use has increased tremendously over the last 50 years (Fig. 1). The major fraction of global N use is for increasing cereal grains production and N use will continue to increase by 15 % till the year 2015 (Mosier et al. 2004a, b). This unprecedented increase in the crop yield in the twenty-first century is possible only by using synthetic N fertilizers. The commercial production of synthetic N fertilizers has become possible due to the discovery of the Haber–Bosch process. However, increased use of N fertilizers has also led to increased N losses from agro-ecosystems and recovery of N in global crop

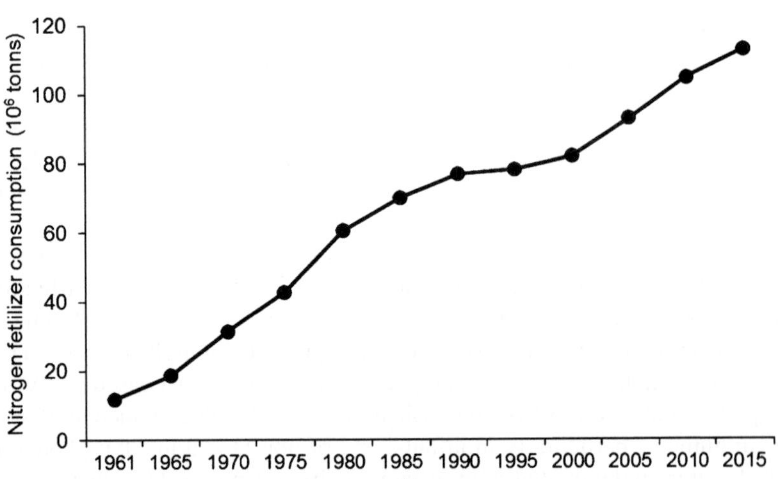

Fig. 1 Histroical and predicted nitrogen fertilizer consumption in global agriculture (fertilizer.org/ifa/ifadata/results and FAO 2011)

production system (Millar et al. 2010; Subbarao et al. 2013) and rest may be lost into the atmosphere, accumulate in the soil or enter surface or subsurface water bodies through leaching or runoff. Nitrogen is lost from the soil-plant system through denitrification in the form of oxides of N like dinitrogen, nitrous oxides, nitric oxide, in the form of ammonia through volatilization, leaching of nitrates, run off and erosion (van Cleemput and Boeckx 2005; Hijleh 2014).

3 Nitrogen Losses Through Oxides and Nitrates

3.1 *Agriculture*

The role of agricultural practices on environmental quality has been examined in the past two decades because of the links between increased agricultural productions accompanied by deterioration of environmental quality. In developing countries, intensive agricultural production systems have increased the use of N fertilizer as an effort to produce and sustain high crop yields (Fageria 2009). Consequently, N losses have also increased with passage of time (Fig. 2). Low recovery of N in annual crops is associated with its loss by volatilization, leaching, surface runoff and denitrification, which not only leads to high cost of crop production but also lead to environmental pollution (Jiao et al. 2012; Tilman et al. 2011).

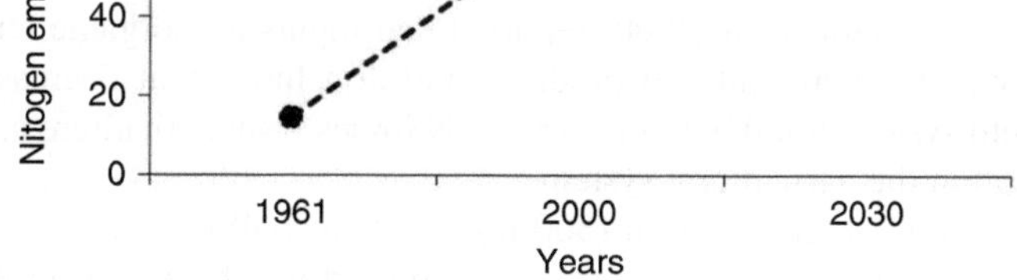

Fig. 2 Historical and predicted emission of anthropogenic reactive nitrogen in the global environment (fertilizer.org/ifa/ifadata/results and Zheng et al. 2002)

3.2 Emission of N from Fertilizers

Inorganic N applied to the soil can be absorbed by the plants, utilized by microorganism, subjected to nitrification, denitrification or volatilization (Vymazal 2007; Hakeem et al. 2012a). This represents an important source of N loss in agricultural soils under favorable conditions. Due to extensive use of N fertilizers and nitrogenous wastes, the amount of N available to plants significantly exceeds the N returned to the atmosphere by gaseous losses of N through volatilization and denitrification (Hakeem et al. 2012a). It has been reported that the concentration of N_2O in environment has increased by 16.0 percent since 1750 (Godish et al. 2015). By passing each year an average of about 0.25 percent has been added and presently, the N_2O concentrations are highest of the past 1000 years (Godish et al. 2015). A portion of this excess N has leached to ground waters as NO_3^- or has been carried in runoff waters. These are conducive conditions for N losses in agricultural soils which decrease the nitrogen use efficiency (NUE) (Hakeem et al. 2012a). Nitrate leaching to the ground water or run off into surface water bodies could lead to the levels beyond those acceptable for human consumption. Nitrate leaching into groundwater in arid regions and those containing sandy soils have been reported (Paramasivam et al. 2002; Hakeem et al. 2012a). Losses of N by leaching are affected by local variation in rainfall, water-holding capacity of soil, soil-drainage features and rates of mineralization of soil organic N (Subbarao et al. 2013). Processes such as adsorption, fixation, immobilization and microbial assimilation of added NH_4-N in soils are of great importance as they affect nitrogen use efficiency and have the corresponding environmental repercussions (Kissel et al. 2004; Hakeem et al. 2012a). The immobilization and mineralization by microorganisms are most rapid when soils are warm and moist, but not saturated with water. The quantity of inorganic nitrogen available for crop use often depends on the amount of mineralization occurring, and the balance between mineralization and immobilization (Guleryuz et al. 2010). It influences the yield and quality of the crops but has a great impact on our food security, economy, development and environment (Walley et al. 2001, 2002; Tilman et al. 2002; Wood et al. 2004). It has been reported that human activities accounted for 53 % of total annual N input into Baltic Sea (HELCOM 2015). Granstedt (2000) reported that inputs in inorganic fertilizer were related to outputs of agricultural products and concluded that decreased input of fertilizers into system could help in decrease N losses from agriculture and increased recycling within the agricultural system.

Excessive NO_3^- concentration (300 mg L^{-1}) in shallow ground waters due to application N of 500 kg N ha^{-1} has been reported in China (Zhang et al. 1996). They concluded that NO_3-N problems in the environment would increase in China because of the increasing demand for food supply.

The large increase in the size of the hypoxic zone in the Gulf of Mexico after the 1993 floods in the Midwest focused attention on the role of agriculture in nonpoint source pollution (Burkart and James 1999). Several researchers in the past concluded that mineralization of soil organic matter, commercial fertilizer and row intensity of crop production are directly related to N losses (Burkart and James 1999; Schilling and Libra 2000; Mikkelsen and Hartz 2008). However, a similar analysis of Midwestern watersheds also showed that a relationship between total N load applied and NO_3-N concentrations in Iowa and Illinois watersheds was not well-defined and there was not a simple relationship between water quality and agronomic management (Hatfield et al. 2001). Nitrate-N losses from watersheds are not isolated to the Mississippi River Basin. Throughout the world, there is increasing evidence that water quality is being degraded by the movement of nutrients from agricultural lands into water bodies. Environmental quality impacts from excess nutrients have been the major cause for the development of Mineral Accounting System (MINAS) in the Netherlands. This system provides an accounting for all nutrients entering and leaving a farm to ensure a balance of nutrients on farms (Ondersteijn et al. 2002). The MINAS approach is having a positive impact and has increased farmer awareness of the impacts that farming practices have on environmental quality and the natural resources (de Koeijer et al. 2003).

Non-agricultural sources of nitrogen contribute to less than 20 percent of the N released into the environment. About 6 percent are released from point sources (basically pipes) into water bodies, while, 14 percent is deposited from atmospheric sources. The production of reactive nitrogen worldwide has more than doubled in the last century because of human activities and population growth (Fig. 3). Advances in our understanding of the nitrogen cycle and the impacts of anthropogenic activities on regional to global scales are largely hindered by the paucity of information about nitrogen inputs from human activities in fast-developing regions of the world such as the tropics. In Brazil, the production of soybean crops over an area of less than 20 million ha, was responsible for about 3.2 Tg N or close to

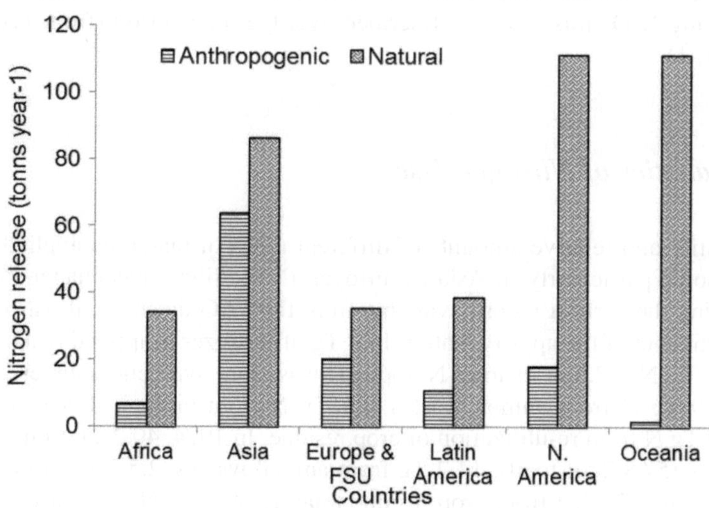

Fig. 3 Regional contribution of reactive nitrogen release into the atmosphere (fertilizer.org/ifa/ifadata/results and Boyer et al. 2004)

one-third of the N inputs from anthropogenic sources. Moreover, cattle pastures account for almost 70 percent of the estimated 280×10^6 ha of agricultural land in Brazil and potentially fix significant amounts of N when well managed, further increasing the importance of biological nitrogen fixation in the nitrogen budget. Many of these anthropogenic inputs occur in the Brazilian savannah region (*Cerrado*), while more urbanized regions such as the state of São Paulo also have high rates of nitrogenous fertilizer inputs. In the Amazon, rates of anthropogenic nitrogen inputs are relatively low, but continuing conversion of natural forests into cattle pastures or secondary forests potentially add a significant amount of new nitrogen to Brazil. Better measurements of biological fixation rates in Brazil are necessary for improving the nitrogen budgets, especially at a more refined spatial scale (Filoso et al. 2006).

In most soils, formation and emissions of N_2O to the atmosphere are enhanced by an increase in available mineral nitrogen (N) through increased rates of nitrification and denitrification (Signor and Cerri 2013). Therefore, addition of N, whether in the form of organic or inorganic compounds eventually leads to enhanced N_2O emissions. Global N_2O emissions from agricultural systems have previously been related primarily to fertilizer N input from synthetic sources (Robertson and Vitousek 2009). Little attention has been paid to N input from other N sources or to the N_2O produced from N that has moved through agricultural systems. In a new methodology used to estimate N_2O emissions on the country or regional scale, the anthropogenic N input data used include synthetic fertilizer and animal waste (feces and urine) used as fertilizer. N derived from enhanced biological N-fixation through N_2 fixing crops and crop residue is returned to the field. Using FAO (2011) database information which includes data on consumption of synthetic fertilizer, animal and crop production and estimates of N input from recycling of animal and crop, estimates of total N into Asian agricultural systems and resulting N_2O emissions are described over the time period (from 1961 to till date) (Fig. 1).

3.3 *Statistics of Nitrogen Use*

The quantity and relative amounts of different types of materials applied to agricultural soils, particularly in Asia as nitrogen (N) fertilizer has changed dramatically. Using the earliest record available from the FAO database, it was observed that in 1961, out of the approximately 15.7 Tg of fertilizer N applied to agricultural fields 2.1 Tg N (13.5 % of total N applied) was from synthetic sources, approximately 6.9 Tg N from animal wastes, 1.7 Tg N from biological N-fixation, and another 5 Tg N from re-utilization of crop residue. In 1994, 40.2 Tg from synthetic fertilizer N (57.8 % of total), 14.2 Tg from animal wastes, 2.5 Tg from biological N-fixation and 12.6 Tg from crop residue totaling 69.5 Tg N were utilized within agricultural soils in all Asian countries. The increases in N utilization have increased the emission of nitrous oxide from agricultural systems. Estimated N_2O

from agricultural systems in Asia increased from about 0.8 Tg N_2O-N in 1961 to about 2.1 in 1994. The period of time when increases in N input and resulting N_2O emissions were greatest was during 1970–1990. This evaluation of N input into Asian agricultural systems and the resulting N_2O emissions demonstrates the large change in global agriculture that has occurred in recent decades. Increase in N input is likely because of the increased need for food production. Although the rate of increase of N input and N_2O emissions during the 1990s appears to have declined, the question arises that if this had slowed the rate of increase in a general long term trend or if global food production pressures will tend to accelerate N input demand and resulting N_2O emissions as we move into the twenty-first century (Mosier and Zhaoliang 2000). However, Zheng et al. (2002) used a model (IAPN 1.0) to analyzed the anthropogenic reactive nitrogen of Asia and stated that it has been increased from ~14.4 Tg N per year in 1961 to ~67.7 Tg N per year in 2000 and most is likely to be 105.3 Tg N per year by 2030. They concluded that the lack of effective measures to improve the efficiency of N fertilizer use and prevention of NOx emissions from fossil-fuel combustion, are the key carters behind the environmental N enrichment problem. Denman et al. (2007) reported that global N_2O emissions reach about 17.7 Tg N per year, share 6.7 Tg (37.8 %) from anthropogenic sources and agricultural soils 2.8 Tg N per year (15.3 %) of the total amount of emissions. Similarly, Cerri et al. (2009) reported that the agriculture sector was responsible for the about 87.2 % of N_2O emissions, mainly from animal waste management and agricultural soils of Brazil. However, Brazil (2010) reported that total anthropogenic sources contribute about 0.350 Tg N per year in Brazil. Out of total, fossil fuel combustion and industrial processes share 0.020 Tg N per year, agricultural sectors 0.310 Tg N per year, land use change and forests share 0.010 and other sources contribute about 0.009 Tg N per year.

4 Implications of Nitrates in Plants

Agricultural fields are intensively fertilized with N fertilizers to realized yield potential of different field crop varieties. Although, N can be absorbed by the plants in NH_4^+ form, but, major fraction N is available to the plants particularly vegetables in NO_3^- forms. Moreover, absorption and distribution of NO_3^- in the crops is important with respect to environmental consequences and quality of crop produce (Chen et al. 2004). However, when the input of N exceeds the demand, plants are no longer able to absorb it, and then the N builds up in the soil (Anjana Umar and Iqbal 2007). Unutilized NO_3^- by the plants may be subjected to ground water pollution by leaching or surface water pollution through soil erosion or run off, which influences the nitrate content of plants particularly the leafy vegetables (NAAS 2005).

5 Sources of N Compounds in the Environment

5.1 Nitrates

Agriculture is considered to be the major sources of NO_3^- to the ground water as considerable amount is leached from agricultural production systems around the world. Agricultural sources of NO_3^- to the ground or surface water includes arable grassland, grazed pastures, arable cropping, mixed cropping with pasture leys, organic farming, horticultural systems, and forest ecosystems (Ribbe et al. 2008). Accumulation of NO_3^- in the soil profile coupled, with or followed by periods of high drainage are conducive conditions for leaching of NO_3^- to the ground water. Therefore, application of N in excessive amount or waste effluent application rates or at the wrong time (e.g. late autumn) of the year, ploughing pasture leys early in the autumn, or long periods of fallow ground, can all potentially lead to high NO_3^- leaching losses. Different sources follow order with respect of potential of leaching: forest < cut grassland < grazed pastures, arable cropping < ploughing of pasture < market gardens (Di and Cameron 2002). Nitrogen applied through fertilizers or manure is converted to plant-available-nitrate by bacteria living in the soil. Nitrates are subjected different fates in the soil. These may be denitrified by bacteria and release to the environment as oxides of N, may be leached to ground water and absorb by the plants (Butterbach-Bahl et al. 2013). Another source is the use of chemical fertilizers (nitrogenous fertilizers) in intensive crop producing areas and, especially when they are used in an uncontrolled manner.

Nitrates are added into the natural water bodies through numerous sources like atmosphere, geological features, anthropogenic sources, atmospheric nitrogen fixation and soil nitrogen (Gormly and Spalding 2006; Gupta and Khosla 2012). However, detailed hydrogeological investigations have indicated a heterogeneous pattern of nitrate distribution. In sandy soil with low water holding capacity and high permeability, movement of pollutants like chloride and nitrate is much quicker than in clayey soil. This is probably the main cause for high nitrates in areas with sandy soil (Khandare 2013). However, nitrates are added into ground water by different sources which include septic tanks, animal and human waste and commercial fertilizers (Gormly and Spalding 2006; Khandare 2013). Nitrogen undergoes biochemical, chemical, and physical changes in soils and waters which make it difficult or impossible to segregate different sources with any degree of certainty (Keeney and Olson 1986). Agricultural sources represent the non-point sources of NO_3^- pollution of ground water which include fertilizers, manure application and legume crops (Shabalala et al. 2013). Non-point sources of nitrogen from agricultural activities include fertilizers, manure application, and leguminous crops while the point sources include septic tanks and dairy lagoons (Almasri and Kaluarachchi 2004; Shabalala et al. 2013). Elevated level of nitrate in ground water are common around dairy and poultry operations, barnyards, and feedlots (Odiyo et al. 2014). However, increase in the nitrate concentration of groundwater has been of special concern because of the suspected harm to children due to nitrate if its concentration in drinking water exceeds permissible limits of 50 mg L^{-1} (Kapoor and Viraraghavan 1997; Thorburn et al. 2003). Elevated NO_3^- concentration in drinking water causes methemoglobinemia in infants and stomach cancer in adults (Almasri and Kaluarachchi 2004; Anayah and Almasri 2009). Some studies on contamination of groundwater with nitrate originating from fertilizers have shown little evidence for such an occurrence (Yang et al. 1998; Mueller et al. 2001; Ikehata et al. 2010) while others have indicated an increase in the NO_3 pollution of groundwater due to fertilizer application (Ju et al. 2005; Shen et al. 2011).

5.2 *Nitrous Oxide*

Nitrous oxide is primarily produced in soil by the activities of microorganisms during nitrification, and denitrification processes. The ratio of N_2O to N_2 production depends on oxygen supply or water-filled pore space, decomposable organic carbon, N substrate supply, temperature, and pH and salinity. N_2O production from soil is sporadic, both in time and space, and therefore, it is a challenge to scale up the measurements of N_2O emission from a given location and time to regional and national levels (Ussiri and Lal 2013). Estimates of N_2O emissions from various agricultural systems vary widely. For example, in flooded rice in the Riverina Plains, N_2O emissions ranged from 0.02 % to 1.4 % of fertilizer N applied, whereas in irrigated sugarcane crops, 15.4 % of the fertilizer was lost over a 4-day period. Nitrous oxide emissions from fertilized dairy pasture soils in Victoria range from 6 to 11 kg N_2O-N/ha, whereas in arable cereal cropping, N_2O emissions range from <0.01 % to 9.9 % of N fertilizer applications. Nitrous oxide emissions from soil nitrite and nitrates resulting from residual fertilizer and legumes are rarely studied, but probably exceed those from fertilizers, due to frequent wetting and drying cycles over a longer period and a larger area.

In ley cropping systems, significant N_2O losses could occur, from the accumulation of mainly nitrate-N, following mineralization of organic N from legume-based pastures. Extensive grazed pastures and rangelands contribute annually about 0.2 kg N/ha as N_2O (93 kg/ha per year CO_2-equivalent). Tropical savannas probably contribute an order of magnitude more, including that from frequent fires. Unfertilized forestry systems may emit less, but the fertilized plantations emit more N_2O than the extensive grazed pastures. However, currently there are limited data to quantify N_2O losses in systems under ley cropping, tropical savannas, and forestry in Australia. Overall, there is a need to examine the emission factors used in estimating national N_2O emissions; for example, 1.25 % of fertilizer or animal-excreted N appearing as N_2O.

The primary consideration for mitigating N_2O emissions from agricultural lands is to match the supply of mineral N (from fertilizer applications, legume-fixed N, organic matter, or manures) to its spatial and temporal needs by crops/pastures/trees. Thus, when appropriate, mineral N supply should be regulated through slow-release (urease and/or nitrification inhibitors, physical coatings, or high C/N ratio materials) or split fertilizer application. Also, N use could be maximized by balancing other nutrient supplies to plants. Moreover, non-legume cover crops could be used to take up residual mineral N following N-fertilized main crops or mineral N accumulated following legume leys. For manure management, the most effective practice is the early application and immediate incorporation of manure into soil to reduce direct N_2O emissions as well as secondary emissions from deposition of ammonia volatilized from manure and urine.

Current models such as DNDC (DeNitrification-DeComposition) and DAYCENT (Daily Century) can be used to simulate N_2O production from soil after parameterization with the local data, and appropriate modification and verification against the measured N_2O emissions under different management practices. The DNDC model was developed to assess N_2O, NO, N_2 and CO_2emissions from agricultural soils (Li et al. 2001). Recently, DAYCENT and DNDC model simulations of N_2O fluxes were used for the assessment of the impacts of climate change on the gas flux and biomass production from a humid pasture (Abdalla et al. 2011). Similarly, DAYCENT model was used to quantify the on-farm GHG emissions and N dynamics of land use conversion to N-managed switchgrass in the Southern U.S (Jim et al. 2011).

In summary, improved estimates of N_2O emission from agricultural lands and mitigation options can be achieved by a directed national research program that is of considerable duration, covers sampling season and climate, and combines different techniques (chamber and micrometeorological) using high precision analytical instruments and simulation modelling, under a range of strategic activities in the agriculture sector (Dalal et al. 2003).

5.3 Emissions During Crop Residues/Organic Matter Decomposition

Crop residue serves as the basic source for N in the agricultural practices during the process of nitrification and denitrification (Signor and Cerri 2013). It has been evident from the results of Gomes et al. (2009) that crop residues added to soil has high denitrification activity and N_2O fluxes compared to field without crops. This may lead an increase in the soil aggregation, moisture retention capacity compared to the topsoil in ecological horizon. Thus, the incorporation of residues caused an accelerated effect on the NO emissions rate and the effects of tillage on mineralization of soil organic matter directly influence the rate of NO and N_2O emissions from soils. Some earlier investigations in the past by researchers have observed that higher N_2O losses are due to high denitrification activity for no-tillage systems compared to conventional tillage (Toma and Hatano 2007). This may be related to the higher denitrification activity. In a no-tillage crop system, soil N_2O emissions were lower (Escobar et al. 2010). However, N_2O losses from no-tillage systems may be lower than from tilled soils where fields remained uncultivated for a number of years. Some studies report an increase in NO emission caused by ploughing in a temperate climate and under tropical conditions.

5.4 Non-agricultural Sources

The nutrients present in sewage and industrial effluent can be effectively used as a supplement to the chemical fertilizer if managed properly (Emongor and Ramolemana 2004). Direct discharge of septic tanks, sewage and industrial effluent causes pollution of surface water and ground water through percolation. In groundwater recharge areas with large portions of agricultural land, the nitrate concentration of well water has shown rising trends in many countries within the last two- to three decades. Nitrate leaching from agricultural land must be considered as an important non-point source for nitrate contamination of the groundwater. In developing countries, most of the urban areas have not well developed system for collection and dumping of domestic and industrial wastes. In most areas, sewerage system may be non-functional or non-existent which led to disposal of human fecal matters and industrial wastes directly into the surface drains. In other cases, most of houses have septic-soil absorption system for wastewater disposal which could lead to the contamination of ground water with NO_3^- (Ramaraju et al. 1999).

6 Atmospheric Deposition

Nitrogen is atmospherically deposited to surface water bodies downwind of anthropogenic emissions. Atmospheric deposition accounted for 10 % to over 40 % of new N loading to estuaries in eastern U.S. coast and eastern Gulf of Mexico (Paerl et al. 2002). Historically, N deposition due to urbanization (NO_x, peroxyacetyl nitrate or PAN) and agricultural expansion (NH_4^+ and possibly organic N) has increased in coastal airsheds. Ammonium is dominant form of atmospherically deposited N in western Europe due to abundance of livestock operations while in United States, dominant form of atmospheric N is oxides of N (NO_x) emitted from fossil fuel combustion (Paerl et al. 2002). These quantitative and qualitative aspects of atmospherically deposited N may promote biotic changes now apparent in estuarine and coastal waters, including the proliferation of harmful algal blooms, with cascading impacts on water quality and fisheries. In addition to these sources, oxides of N are generated through lightening and with rain water these oxides of nitrogen are available in water (Schumann and Huntrieser 2007). The atmospheric contribution was supposed to be the smallest but recent reports on dry and wet depositions of pollutants suggested a contribution to the extent of 25 % of the total load of nitrate (Gupta et al. 2007).

7 Health Risks of Nitrate and Nitrogen Oxides

The use of nitrogenous fertilizer is reported to be the major source of NO_3^- contamination of groundwater (Nolan and Stoner 2000), and if this contaminated groundwater is pumped for drinking purpose and may have a direct effect on human health. The risk of ground-water contamination by nitrogenous sources depends on the input of nitrogen input to the land surface and the degree of leaching and accumulation. Recently, an agricultural nonpoint source is held responsible for the pollution of rivers and lakes. The impact of agricultural practices on groundwater quality is of particular concern as in many cases, ground water level is shallow and thus are vulnerable to water pollution, particularly from NO_3^-. Use of high NO_3^- irrigation water could also lead to toxic levels of NO_3^- in leafy vegetables. Although current epidemiological data provide conflicting evidence regarding the potential long-term health risks of nitrate levels encountered in the diet, it is widely accepted that the reduction of dietary nitrate is a desirable preventive measure, as it may pose serious health hazards. The toxic effects of nitrate are due to its endogenous conversion to nitrite and this ion has been implicated in the occurrence of methaemoglobinemia, gastric cancer and many other diseases (Anjana Umar et al. 2007). Initially, the risk of methaemoglobinemia was limited to the infants only (Greer and Shannon 2005), but later, Gupta et al. (2008) have reported it in all age groups with high nitrate ingestion, with the infants and higher age groups (>45 years) being the most susceptible to nitrate toxicity. Apart from this, effect of nitrate consumption on human health has been studied in detail by Gupta et al. (2008). Although attention has been paid to the level of nitrate in drinking water and the extent of nitrate pollution of groundwater bodies in India due to heavy nitrogen use in India, there is a gap in our knowledge regarding the concentration of nitrate in leafy vegetables consumed regularly.

The quality of drinking water is harmfully influenced by nitrate through the possible formations of nitrite causing methemoglobinemia in babies and cancer in older persons due to the formation of nitrosamines in the digestive tract (Ward 2009). The World Health Organization (WHO) recommends that the safe level of NO_3-N in drinking water is 11.3 mg NO_3-N dm^{-3}, the unsafe level is set at 22.6 mg NO_3-N dm^{-3}. Besides the concern for the environment, the use of high amounts of fertilizer N is also prohibitive because of its high cost. The applied NO_3-N not taken up by the plant can be leached into the groundwater. It depends on soil texture, rainfall, N fertilization and type of crop (Liu et al. 2012). Several research results indicated that the fertilizer N management has a great influence on groundwater contamination, and it is suggested that with proper N management, groundwater pollution as well as economic loss can be avoided (Liu et al. 2012; Hijleh 2014).

8 Acute Toxic Effects

Exposure to high nitrate may cause acute toxicity due to acute exposure.

8.1 *In Human*

Nitrate toxicity is mainly caused due to reduction of nitrates to nitrites after ingestion. The health hazards is directly caused by nitrite that oxidizes the hemoglobin to methemoglobin (Manassaram et al. 2007; Tables 1, 2, and 3). Drinking water is the major source of nitrates intake into human body thereby contributing about < 3–21 % of the average adult intake of nitrate besides other sources like vegetables and meat (WHO 2011). The potential health hazard related to dietary nitrate is formation of N-nitroso compounds from nitrite. Nitrates are also suspected to be involved in some other human diseases. These diseases include cancer due to production of N-nitroso compounds, hypertension, increased infant mortality, central nervous system birth defects, diabetes, spontaneous abortions, respiratory tract infections, and changes to the immune system (Fewtrell 2004; Chiu and Tsai 2007). Methemoglobin (MetHb) is produced due to the oxidation of ferrous (Fe^{2+}) iron in hemoglobin to ferric (Fe^{3+}) which affects oxygen binding capacity of MetHb. This results in expression of cyanosis, stupor, and cerebral anoxia (Fewtrell 2004).

Table 1 Summary of inorganic nitrates causes and sources for the induction of methemoglobinemia

Causes of inorganic nitrates	Sources
Agricultural fertilizers	Fertilizers used in Agricultural practices
Leafy vegetables	Leafy vegetables and packed food items
Industrials salts and preservatives	Tainted well water and nitrous oxides canisters used for anesthesia
Ammonium	Industrial salts and preservatives
Nitrates	Used in burn therapy or as oxidizing agents (ammonium nitrate)

Adapted from fertilizer.org/ifa/ifadata/results and Manassaram et al. 2007

Table 2 Summary of organic causes of nitrates and sources for the induction of methemoglobinemia

Causes of organic nitrates	Sources
Amyl/sodium nitrate	Used as inhalant in cyanide antidote kit
Isobutyl nitrite	Common room deodorizer or propellants
Nitroglycerine	Used as oral/sublingual, or transdermal pharmaceuticals for treatment of angina

Adapted from fertilizer.org/ifa/ifadata/results and Manassaram et al. 2007

Table 3 Summary of other causes of nitrates and sources for the induction of methemoglobinemia

Other causes of nitrates	Sources
Acetanilide, Chloramine, Flutamide, Metoclopramide, Nitric oxide, Nitrofurans, Nitroprusside, Paraquat, Phenacetin, Zopiclone	Use for medicinal purposes
Aniline dyes	Used in laundry ink, colored wax crayons, diaper marking ink
Aminophenols	Tinted shoes or blankets
Herbal medicines	High dose of herbal medicine of *Ginkgo biloba* showed adverse effect
Nitrobenzene	Cleaning prodcuts for ammunition and industrial solvents
Nitroethane	Used in making of nail polish, resins, rubber adhesives
Benzocaine, Lidocaine, Propitocaine, Prilocaine	Used for local anesthetics purposes
Chloroquine, Copper sulphate, Naphthalene, Primaquine, Sulfonamides	Used as antibacterial and antifungal drugs
Resorcinol	Used as antiseborrheic, antipruritic, antiseptic
Chlorates	In preparation of match boxes or explosives
Combustion products	As automobile exhaust, burning fumes of plastics/ wood and fires

Adapted from fertilizer.org/ifa/ifadata/results and Manassaram et al. 2007

Human lethal doses of 4–50 g NO_3^- (equivalent to 67–833 mgNO_3^-/kg bw) have been reported. Toxic doses with methemoglobin formation as a criterion for toxicity-ranged from 2 to 5 g of NO_3 (WHO 2011). These values are equivalent to 33–83 and 100–150 mg NO_3^-/kg body weight respectively.

8.2 *In Animals*

Acute nitrate toxicity to animals varies according to species. Generally, ruminants suffer from methemoglobinemia while monogastric animals may face severe gastritis. Ingestion of nitrate also cause impaired functioning of thyroid, decreased feed consumption, and interference with vitamin A and E metabolism (Bruning-Fann and Kaneene 1993; McDowell 2012). Hematologic changes caused due to high nitrate exposure include both compensatory increases in red blood cells and anemia, along with increased neutrophils and eosinophils. Unlike nitrate, nitrite is capable of inducing methemoglobinemia in a wide range of species, i.e. cattle, sheep, swine, dogs, guinea pigs, rats, chickens and turkeys. In rats, chronic nitrite exposure causes pathologic changes in a variety of tissues, alterations in motor activity and brain electrical activity, and alters gastric mucosal absorption (Bruning-Fann and Kaneene 1993; WHO 2011). Nitrite affects the metabolism of sulfonamide drugs in animals such as the pig, guinea pig, and rat. The N-nitroso compound di-methylnitrosamine causes toxic hepatosis in cattle, sheep, mink, and fox. Nitrosamines have been reported in cow's milk and been found to pass into the milk of goats under experimental conditions (WHO 2011).

The studies relating to acute exposure of nitrate were conducted in animals. The acute oral toxicity of nitrate to laboratory animals is low to moderate. LD_{50} values of 1600–9000 mg of sodium nitrate per kg of body weight have been reported in mice, rats, and rabbits. Ruminants are more sensitive to the effects of nitrate as a result of high nitrate reduction in the rumen; the LD_{50} for cows was 450 mg of sodium nitrate per kg of body weight. Nitrite is more toxic than nitrate: LD_{50} values of 85–220 mg of sodium nitrite per kg of body weight have been reported for mice and rats (WHO 2011).

8.3 *Mechanism of Methemoglobin Formation*

The essential action in the formation of methemoglobin is an oxidation of the ferrous to the ferric ion. This oxidation may be brought about in one of the following ways (Sharma et al. 2013): by the direct action of the oxidants, the action of hydrogen donors in the presence of oxygen, and by auto oxidation. In the presence of nitrites, the ferrous ion of hemoglobin gets directly oxidized to the ferric state. Normally, the methemoglobin that is formed is reduced by the following reaction.

$$\mathrm{Hb^{+3} + Red\ cyt\ b_5 \longrightarrow Hb^{+2} + oxy\ cyt\ b_5}$$

Oxidized cytochrome b_5 (Red Cyt b_5) is regenerated by the enzyme cytochrome b_5 reductase:

$$\mathrm{oxy\ cyt\ b_5 + NADH \xrightarrow{cyt\ b_5\ reductase} Red\ cyt\ b_5 + NAD}$$

Thus, the enzyme cytochrome b_5 reductase plays a vital role in counteracting the effects of nitrate ingestion. However, permissible concentration of nitrate (50 mg L^{-1}) normally present in water or food do not cause any risk associated with the human.

8.4 *Water Bodies*

A considerable amount of N is deposited into estuarine and coastal waters due to atmospheric deposition downwind of anthropogenic emissions. Approximately, 40 % of total N added to the eastern U.S. coast and eastern Gulf of Mexico is contributed by atmospheric deposition. In receiving estuarine and coastal waters, phytoplankton community structural and functional changes, associated water quality, and trophic and biogeochemical alterations (i.e. algal blooms, hypoxia, food web, and fisheries habitat disruption) are frequent consequences of N-driven eutrophication. Increases in and changing the proportions of various new N sources regulate phytoplankton competitive interactions, dominance, and successional patterns. These quantitative and qualitative aspects of AD-N and other atmospheric nutrient sources (e.g. iron) may promote biotic changes now apparent in estuarine and coastal waters, including the proliferation of harmful algal blooms, with cascading impacts on water quality and fisheries (Paerl et al. 2002).

8.5 Ozone Depletion

Among the various factors, N_2O is the most important factor responsible for ozone-depletion and it will continue to be on the top of all the factors in twenty-first century. Decreasing the emission of N_2O would help in amelioration of ozone layer depletion (Ravishankara et al. 2009). Ozone layer in stratosphere is depleted by different chemicals of anthropogenic origin which is a serious environmental threat of this era. Oxides of N are major factors which are responsible for the depletion of the ozone layer. Depletion of the ozone layer by oxides of N is explained by the following reactions;

$$NO + O_3 \rightarrow NO_2 + O_2$$

$$O_2 + NO_2 \rightarrow NO + O_2$$

The main cause behind the depletion of the ozone layer is the stockpile of halocarbons in the stratosphere through various heterogeneous chemical reactions with aerosol particles (Solomon 1999). Amongst the various ozone depleting gases, N_2O is the most predominant gases emitted mainly by human activities (Ravishankara et al. 2009), and the emission of N_2O causes maximum hazards to the environment when released either singly or in combination with other hydro-carbon (Daniel et al. 2010). Moreover, change in the concentration of N_2O may also be directly responsible for the chemical and radioactive changes in the stratospheres (Portmann et al. 2012).

9 Remediation Measures

In the present scenario a wide range of remedial measures are available for the treatment of water bodies. Blending low-nitrate in high-nitrate water is one of the cheapest methods, but it depends upon the availability of low-nitrate water supplies in the surrounding area. In this regards, other methods such as biological denitrification, ion exchange, reverse osmosis and electro-dialysis are the most feasible alternatives. Previously researchers concluded that biological denitrification and ion exchange as the most striking options for the treatment of water bodies in the UK followed by reservoir storage, lining of boreholes and relocation of boreholes.

9.1 Concepts of Nitrogen Use Efficiency (NUE)

The term NUE has two basic components; (1) Nitrogen uptake, recovery or acquisition efficiency and (2) Nitrogen use, physiological N use, or internal N use efficiency. The terms NUE is a ratio between an output (i.e., grain yield, dry matter yield, N accumulation in grain, or N accumulation in total plant dry matter) and input (i.e. total N supply, soil N supply or fertilizer N supply). NUE is based on different parameters of efficiency, including N uptake, utilization or acquisition efficiency, and N-use (physiological/internal) efficiency. Nitrogen use efficiency and crop response to applied N are important for evaluation of N requirements for crop to achieve maximum economic yield. Globally, recovery of N in crop plants is generally $< 50\,\%$ and remaining is lost due to volatilization, leaching, surface runoff, denitrification, and plant canopy (Fageria 2009). Low recovery of N leads to higher cost of crop production and environmental pollution. Under such conditions, enhancing NUE will improve crop yields, reducing cost of production, and maintaining environmental quality. NUE can be classified into agronomic, recovery and physiological efficiency.

Agronomic efficiency of nitrogen measures is the overall efficiency or an integrative index of total economic outputs relative to the use of all sources of N (indigenous soil N and applied fertilizer N). Nitrogen recovery efficiency considers the efficiency of the plant to take up N and physiological N use efficiency is the efficiency with which the plant uses N from acquired available N to produce grain or total plant dry matter. To calculate the efficiency of soil vis-a-vis fertilizer N, the output from a non-fertilizer or control plot where the source of N is only soil should be determined. The average agronomic N efficiency had a relatively narrow compared to applied N applied, and amongst the three cultivates crops it was recorded least in maize and the maximum in rice crops. These differences in agronomic efficiency indicate that maize and rice produce larger economic outputs than wheat in terms of the use of all N sources (soil and fertilizer). The average estimates of N recovery efficiency based on grain N were similar among the three crops and were highest (44 %) in maize and lowest (35 %) in wheat. The average physiological efficiency of N ranged from 35 to 69 kg grain increase per kg of N uptake. The main reason for large physiological efficiency of N values with rice was the crop's low grain-N concentration (9–12 g kg^{-1}), especially when compared with that of maize (13–14 g kg^{-1}) and wheat (16–18 g kg^{-1}). Rice is also known to be highly efficient in mobilizing N from senescing leaves and clumps to grain as indicated by its large N-harvest index (Ladha et al. 2005a) suggested that the mode of photosynthesis (C_3 or C_4 photosynthetic pathway) and grain-N concentrations govern PEN.

In general, grain yields increase with an increase in total N uptake and some genotypes produce different grain yields with the same amount of N uptake. Differences in the efficiency of physiological N use may arise from differences in (1) internal N requirement for plant growth, (2) ability of the plant to translocate, distribute and mobilize absorbed N to and from various organs, (3) flag leaf N import/export and leaf senescence pattern and (4) plant's efficiency in converting CO_2 to carbohydrate (Ladha et al. 2005a).

Proper N-application timing and rates are critical for meeting plant needs and improving NUE. In addition, the growth stage of plants at the time of fertilizer application also determines NUE, with significant genotypic variations (Hirel et al. 2007; Hakeem et al. 2011). It was opined that the timing of fertilizer N applications has a significant effect on the uptake of fertilizer N by the crop and the resulting partitioning of added N between soil and plant. While the amount of N available to the plant can be improved by using sustained-release fertilizers, split applications, minimizing fertilizer losses and other nutrient management and crop management strategies, the inherent efficiency of the plant to utilize available N for higher productivity needs to be tackled biologically (Abrol et al. 1999). Nitrogen applications in split doses have been found to increase yield, NUE, and N uptake efficiency (NUpE), compared with fall application in hard red winter wheat under temperate conditions (Abrol et al. 1999; Sowers et al. 1994).

Nitrogen use efficiency (NUE) in the context of photosynthesis is called as photosynthetic nitrogen use efficiency (PNUE), which is determined by the rate of carbon assimilation per unit leaf nitrogen (Kumar et al. 2001). As Ribulose bisphosphate carboxylase/oxygenase (Rubisco) is a major sink for nitrogen supplied in the form of fertilizers, manipulation of crops to improve NUE would have economic and environmental benefits, reducing the pollution of water by nitrates.

9.2 Transgenomic Approaches

9.2.1 Nitrogen Use Efficiency

Various approaches have been used for the improvement of N-use efficiency (NUE) among crop plants and also to reduce N losses from agricultural fields. Application of the same amount of N in more than two splits under field conditions increases the nitrogen availability at later stages of growth to exploit the sub-optimal activity of the upper leaf laminae (Abrol et al. 1999). Differential regulations of various nitrates and ammonium transporters in plants are well known (von Wirén et al. 2000; Bu et al. 2011), however, studies regarding the effect of over expression of genes of these transporters are very rare. Studies involving transgenic over expression of a *chl1* cDNA (representing the constitutive high-affinity nitrate transporter) driven by the cauliflower mosaic virus 35S promoter in a chl1 mutant have revealed effective recovery of the nitrate uptake defect for the constitutive phase (Liu and Tsay 2003). This was however not reflected in the case of induced phase, which was consistent with the constitutive level of *chl*1 expression in the transgenic plant. In another experiment involving transgenic tobacco plants expressing the NpNRT2.1 gene (encoding high affinity nitrate transporter), a steady-state increase in its mRNA levels accompanied by an increase in the NO_3^- influx, but the NO_3^- contents were remarkably similar in wild-type and transgenic plants has been reported (Fraisier et al. 2000). These findings indicate that increasing the uptake of nitrate by genetic manipulation may not necessarily lead to a concomitant improvement in nitrate utilization or NUE, though it remains to be seen whether different plants respond differently to the overexpression of different transporters. Ammonium transporters are very well characterized in plant systems (Glass 2003; Loqué and von Wirén 2004), but the effect of their over expression on plant growth and development is yet to be elucidated.

9.2.2 Nitrate Reductase (NR)

Nitrate reductase (NR) has long been considered to be the rate-limiting step in nitrate assimilation, but transgenic manipulation of NR expression in *Nicotiana* spp. indicated the importance of other steps (Stitt 1999). Constitutive NR expression led to a twofold increase in NR activity and a 20 % decrease in foliar nitrate content along with an increase in the total amino acid contents, but without any changes in the total N, soluble sugars, starch and productivity parameters (Masclaux et al. 2001). While the NR double mutant Nia30 does not show detectable NR activity, when transformed with the nia2-cDNA, it showed decreased NR activity with higher levels of nitrate accumulation (Hänsch et al. 2001). Transformed *Nicotiana plumbaginifolia* plants constitutively expressing nitrate reductase (NR) show temporarily delayed drought-induced losses in NR activity, thereby allowing for the more rapid recovery of N assimilation following a short-term water deficit. Deregulation of NR gene expression by constitutive expression in transgenic plants resulted in reduced nitrate levels in the tissues of tobacco (Quilleré et al. 1994). Although other factors such as NO_3^- availability regulate the flux through the pathway of N assimilation, the NR transformants were better equipped in terms of available NR protein, which rapidly restore N assimilation. While no tangible effects on biomass accumulation could be attributed in the short term, under field conditions of fluctuating water availability, constitutive NR expression was able to confer a physiological advantage by preventing slowly reversible losses in N-assimilation capacity (Ferrario-Méry et al. 1998). Since NR is post-translationally regulated by phosphorylation and binding of 14-3-3 proteins, various attempts have been made to counter this inhibitory effect on NR regulation. Deletion of 56 amino acids in the amino-terminal domain of NR was previously shown to impair this type of regulation in *Nicotiana plumbaginifolia* (Provan et al. 2000). Additional reports of over-expression of NR genes from various plants have been accumulating over the last decade (Ferrario-Méry et al. 2001; Lillo et al. 2004) but none of these has had any major implication for improving the nitrogen-use efficiency.

9.2.3 Nitrite Reductase (NiR)

Overexpression of NiR genes in *Arabidopsis* and tobacco resulted in increased NiR transcript levels but decreased enzyme activity levels, which were attributed to post translational modifications (Takahashi et al. 2001). There is no evidence yet of any benefit of NiR over-expression in terms of plant NUE.

9.2.4 Glutamine Synthetase (GS)

Transgenic tobacco plants enriched or reduced in plastidic glutamine synthetase (GS2, a key enzyme in photorespiration) were constructed (Kozaki and Takeba 1996). Those transgenic plants having twice the normal amount of *GS2* had an improved capacity for photorespiration and an increased tolerance to high-intensity light, whereas those with a reduced amount of *GS2* had a diminished capacity for photorespiration and were photo-inhibited more severely by high-intensity light compared with the control plants. Reassimilation of ammonia in transformed tobacco was also studied by Ferrario-Méry et al. (2001), while, Hoshida et al. (2000) investigated the role of constructs for the over expression of *GS2* in rice plants (Hoshida et al. 2000). Ectopic expression of *GS1* has been shown to alter plant growth (Fuentes-Ramírez et al. 2001) and the over expression of *GS1* in transgenic plants could cause the enhancement of photosynthetic rates, higher rates of photorespiration and enhanced resistance to water stress (El-Khatib et al. 2004; Fuentes-Ramírez et al. 2001). The overexpression of soybean cytosolic *GS1* in the shoots of *Lotus corniculatus* was reported to accelerate plant development, leading to early senescence and premature flowering, particularly when plants were grown under conditions of high ammonium (Vincent et al. 1997). Ectopic expression of pea *GS1* in tobacco leaves was suggested to provide an additional or alternative route for the reassimilation of photorespiratory ammonium, resulting in increased efficiency of N assimilation and enhanced plant growth (Oliveira et al. 2002). Some additional empirical evidence for enhanced nitrogen-assimilation efficiency in GS1 transgenic lines were also provided (Man et al. 2005). However, differences in the degree of ectopic *GS1* expression have been reported (Fuentes et al. 2001) and attributed to positional effects, effectiveness of chimeric constructs, or differences in growth conditions. These differences could account for the lack of correlation between the enhanced expression of *GS1* and concomitant growth (Ortega et al. 2001; Vincent et al. 1997). Transgenic overexpression and antisense technology have been employed recently to modulate the expression of NADH-GOGAT in alfalfa and rice plants (Schoenbeck et al. 2000; Yamaya et al. 2002). The studies on transgenic rice plants expressing antisense RNA for either *GS1* or NADH-GOGAT point towards the possible involvement of *GS1* in the export of N via phloem in senescing leaves. On the other hand, in case of developing leaf blades and spikelets, NADH-GOGAT was implicated in the utilization of glutamine transported from senescing organs (Yamaya et al. 2002). While these genes appear to be good candidates for improving NUE in the short run, the degree of improvement may vary with the crop and cropping conditions.

9.2.5 Transcription Factors

While, identification of global nitrate-induced transcription factors in higher plants has not been successful, a recent study suggests that targeting other transcription factors may help improve N assimilation and NUE. Yanagisawa (2004) generated transgenic *Arabidopsis* lines overexpressing *Dof1*, a maize protein that belongs to the *Dof* family of plant-specific transcription factors known to activate the expression of several C-metabolizing genes associated with organic acid metabolism. The transformants showed up to 30 % higher N content, higher levels of amino acids, better growth under low-nitrogen conditions and higher levels of mRNAs and enzyme activities for PEP carboxylase and pyruvate kinase, without any reduction of NR, GS and GOGAT RNAs. The genes up regulated by *Dof1* overexpression clearly belong to the list of known nitrate-responsive genes, though it is not clear whether *Dof1* is inducible by nitrate. If *Dof1* is not nitrate-inducible, it means that multiple transcription factors may be involved in the coordinated expression of the N and C metabolizing genes. A few other attempts to manipulate signaling/regulatory proteins have been made, but did not yield any significant advantage in terms of NUE. Castaings et al. (2011) has recently reviewed the roles of nitrate transporters, transcription factors, and kinases in their association with hormones or N-containing molecules in the regulation of N assimilation.

9.2.6 Omics Approach in Searching the Candidate Genes/Proteins to Enhance Nitrogen Use Efficiency

Ove the last decade, omics approach has been used to understand the mechanism of physiological characteristics and to search the possible candidate genes/proteins responsible to various physiological processes in plants. However, It has been found that these physiological processes are showing a multitude of complex networks and interdependent pathways involving many genes, proteins, enzymes, and metabolites, than to a single pathway (Fernie and Stitt 2012).

Recently, Simons et al. (2014) reviewed the role of omics studies in understanding NUE in Maize. There are a few proteomics studies avaliable in various crop plants viz., rice (Hakeem et al. 2012b, c), wheat (Bahrmam et al. 2004; Bahrman et al. 2005), maize (Amiour et al. 2012) to investigate into the response of a plant to N limitation. Similarly, metabolomics approach has been valuable in improving our understanding of the interactions between C and N metabolism (Fait et al. 2011). however, these studies does not show any direct relationship with transcriptome or proteome studies. Hence, it is suggested that when integrated with other 'omics' data and with information from quantitative genetics, transcriptomics and interactomics, we shall be able to understanding the complex regulatory networks underlying important phenotypic traits such as yield and nutrient perception and utilization (Verma et al. 2013).

10 Conclusion and Future Prospectives

Nitrate is one of the most extensively studied anion in water because of its chronic toxic manifestations. It is one of the common pollutants of drinking water all over the world. Despite the availability of the vast literature on nitrate toxicity, the studies of its impacts on the human populations are still limited because of ethical and other considerations, and, the most toxic manifestations are available in animal studies only. Nitrogen oxides stay more than 100 years in the atmosphere, and each molecule of nitrogen oxide absorbs 200 times more radiation than carbon dioxide. This will also play an important role in global climate change (Ozturk et al. 2013). Numerous epidemiological studies have also been conducted to establish certain cause-effect relationships for many clinical manifestations of nitrate toxicity and the results of such studies are normally inconclusive or the correlations are statistically non-significant (Cristina et al. 2014). This has led to a very insufficient knowledge generation among researchers and practicing doctors about the preventive and curative aspects of different manifestations. The farming community should be made aware of the harmful effects of N_2O emission from agriculture and should feel the need to mitigate its emission through optimum crop management practices. Inexpensive soil and plant tissue testing facilities should be made available to farmers.

The world, has been caught to the desirability of nitrogen fertilizers very quickly, and faced with its bad consequences before thinking and doing research on the effects on environmental health and the possible influence on climate. As a remedy to this dangerous trend, two focal points should be considered: first, we should encourage organic farming, and reduce the use of artificial fertilizers and pesticides.

Nitrate-containing forages may be used with proper precautions. Problems with nitrate toxicity can be avoided with proper management of forage and livestock. Drought-stricken forages may contain toxic amounts of nitrates for cattle, and prevention of problems is essentially the only valid approach to avoiding economic loss. Suspected forages should be tested and results used in the feed management. Limiting forage intake and nitrate-free feeds and preventing animals from eating nitrate-containing forages, are important in avoiding cattle losses, silages suspected of toxic levels of nitrate should be allowed to aerate overnight before feeding (Ozturk et al. 2013). Ammonia, butyric acid and non-protein nitrogen levels of the silage increase when nitrates are present in the silage crop. *Enterobacteria* are mainly responsible for the degradation of nitrate in silage, *Clostridia* can reduce nitrite to ammonia, *Lactobacilli* reduce nitrate. The type of plant material and nutritional value of forage crops is closely related with the quality. There is a positive correlation between nitrogen fertilization of plants and their crude protein content, 1000 ppm or more nitrate (upto 2000 ppm) in dry matter content has been reported to be an acceptable criterion for the quality of herbs (Ozturk et al. 2013).

There is an immediate need for considering the estimation of the amount of nitrogen fertilizers to be applied to the fields or under glasshouses as kg N/ha rather than the amount of fertilizer per ha. At the same time fertilizer application and irrigation practices reducing nitrogen losses in runoff should be used. The importance of environmental education comes to the forefront here. With an increase in the level of education of the users, the importance given to the analysis increases. Such people avoid unnecessary and excessive fertilizer usage due to environmental awareness. The soil nitrogen supply status can be deduced from the previous cropping, and manurial history of the soil, and should be categorised into an index system.

The recommendations for nitrate containing fodder use are: feed containing 1000 ppm nitrate could be used securely; feeds with nitrates between 1000 and 2000 ppm are slightly risky; forages containing nitrate between 2000 and 3400 ppm should not be given to animals, if given, the ratio between non nitrate to nitrate feeds should not exceed 50 percent; forages containing nitrate between 3400 and 4000 ppm should be kept away from the pregnant animals, If given, ratio between non nitrate to nitrate feeds should not exceed 25 percent; feeds containing more than 4000 ppm nitrate forage are poisonous and should not be given to the animals; high nitrate fodders must be given at 2–3 week intervals. The nitrate levels in the forage crops growing on dry and wet soils could be reduced by late harvesting. Animals should not consume high nitrate feed when they are hungry. Qualified researchers should guide the producers about the use of fertilizers in vegetable farming in greenhouses. Testing soils for nitrogen has been a useful practice in the drier regions for many years, and in such regions fertilizer rates can be adjusted to account for NO_3^- found in the soil prior to planting. The producers should get the scientific analysis of soil and plant samples by the experts in the field and greenhouses for a sustainable production and prevention of soil pollution through irrigation (Guo et al. 2006).

Effective means of minimizing the accumulation of nitrates in vegetables include identification of low nitrate accumulating genotypes, manipulation of fertilizer regimes, exploiting natural diurnal fluctuations in nitrate concentration, and use of nitrification inhibitors.

Rates needed to achieve different yields with different crops vary, nitrate consumption through vegetables can be kept low by harvesting them at the proper time. Some crops should be harvested in the afternoon of a sunny day when nitrate concentration in the leaves is low. It has been found that the leaf nitrate concentration is at its lowest at 1:00 p.m., when the sunlight is maximum (Anjana Umar and Iqbal 2007).

Only nitrous oxide is of concern from the perspective of agroindustry. This gas may account for up to 15 percent of the total global warming potential. Its global warming potential is estimated to be more than 250 times that of carbon dioxide. Moreover, its lifetime residency in the atmosphere is estimated to be about 130 years. Nitrous oxide also leads to ozone depletion. Globally, anthropogenic activity accounts for 64 percent of the total nitrous oxide emissions of which agriculture accounts for 92 percent.

Emissions of ammonia and methane relate primarily to animal production systems. It is this addition of nitrogen to natural ecosystems, which is of environmental concern both in the terms of its implications for biodiversity and the acidification of soil and aquatic systems. For this purpose, correct nutrient application is very important.

Improving NUE in major food crops requires collaboration among farmers, agronomists, soil scientists, agricultural economists, ecologists, and policymakers (Dobermann and Cassman 2004; Galloway et al. 2002). Agriculture can contribute significantly to reducing the global reactive N load by enhancing the indigenous N supply, which, in turn, will require innovative crop and soil management practices.

The expansion and intensification of agricultural production to meet the needs of a burgeoning population have brought vast land areas under cultivation. Continued growth in cultivated areas, however, is very unlikely since in the future there will be more competition for domestic and industrial uses. This means that crop productivity has to be increased, for which fertilizer, particularly N, plays the most crucial role. The global demand for N fertilizer is increasing day by day, driven mainly by cereal production. The world presently uses around 83 Mt of N, which is about a 100-fold increase over the last 100 years. About 60 % of global N fertilizer consumption has been used to produce the world's three major cereals, rice, maize, and wheat. The projections are that 50–70 % more cereal grain would be required by 2050 to feed 9.3 billion people. This will require an increased use of N fertilizer of a similar magnitude (50–70 %) with the current state of NUE. Moreover, there will be enormous environmental penalties for the use of such a large amount of inorganic, reactive N. Therefore, to supply sufficient N to meet increased food demand and maintain a safe environment, fertilizer N must be used efficiently and prudently, that is, N use efficiency must be improved. This chapter evaluates NUE for the major cereals grown across large agroclimatic regions and analyzes various strategies available to improve NUE.

Fertilizer N use efficiency is a complex term with many components. To quantify NUE, the term most widely used is a ratio with output (biological yield or economic yield in terms of biomass or N content) as the numerator and input (N supply from soil, fertilizer, or soil plus fertilizer) as the denominator. Based on a large number of studies conducted in research fields from across the globe in a wide diversity of agroecologies, 44–55 % of the applied fertilizer N was recovered in the first crop. When fertilizer N recovery in plant and soil during succeeding seasons was considered, an additional recovery of about 7 % was estimated. A N recovery efficiency of 40 % or more has been obtained with improved N management practices.

To achieve an N use efficiency of near 100 %, similar to that of symbiotic N_2 fixing systems such as legume rhizobium, will remain a huge challenge and currently not within reach. In symbiotic N_2 fixing systems, the N requirement and supply are highly synchronized and almost no leakage (excluding N leakage through root exudation and NH3 volatilization from leaves) of N occurs. Absolute synchrony similar to that of legumes will not be attainable in fertilizer dependent cereal production systems. Nonetheless, efficiency levels close to 70–90 % obtained in several well managed controlled experiments can be targeted, at least in well managed irrigated cereal systems.

Fertilizer NUE is governed by three major factors: N uptake by the crop, N supply from soil and fertilizer, and N losses from soil–plant systems. The crop N requirement is the most important factor influencing NUE. Much research has been conducted during the past 50 years to improve NUE by trying to develop better fertilizers or improved N management practices, based mainly on a better synchronization between the supply and the uptake of N by the crop. There is significant potential to increase NUE in cereals.

Many of the strategies needed to achieve such increases have already been developed. The use of an integrated crop management strategy comprising optimal soil, water, and crop management under good climatic conditions could attain a large NUE value. In addition, cereal genotypes with a large harvest index must be used to obtain high REN and PEN because the harvest index is tightly linked to NUE. Many approaches have been suggested for increasing NUE, for example, optimal time, rate, and methods of application for matching N supply with crop demand; the use of specially formulated forms of fertilizer, including those with urease and nitrification inhibitors; the integrated use of fertilizer, manures, and = or crop residues; and optimizing irrigation management. In addition, some modern tools such as precision farming technologies, simulation modeling, decision support systems, and resource-conserving technologies also help to improve NUE.

Importantly, some of these techniques are being adopted on a large scale by many farmers. Two success stories based on crop demand based N management warrant mention. In large scale agriculture practiced in developed countries, precision farming studies have demonstrated that variablerate N fertilizer application has the potential to significantly reduce the N rate required to achieve yields similar to those obtained with Standard uniform management practices. In agriculture with small to medium farm size in developing countries, the use of a simple and inexpensive leaf color chart assists farmers in applying N when the plant needs it. The use of this simple tool, currently practiced by about 350,000 farmers globally, has reduced fertilizer N use by about 25 %.

Because some of the main factors, such as climate controlling crop needs for N, are largely outside farmer control, it will remain difficult to predict the amount of N to apply for optimum growth. Along similar lines, the amount of N that becomes available for crop uptake through net soil N mineralization varies across the landscape and is diYcult to predict. Therefore, because of those uncertainties related to the demand for and supply of N to the crop, it will always remain a daunting task to achieve significantly higher fertilizer N use efficiencies. However, our study indicated that farmers can increase NUE substantially in well managed agricultural systems by adopting appropriate technologies and carefully crafted local policies that contribute to improved N management. In any event, improved farmer knowledge regarding N management techniques is essential. A key challenge is to ensure that such techniques are cost effective and user friendly so that they provide attractive options for adoption by farmers.

References

Abdalla M, Jones M, Yeluripati J, Smith P, Burke J, Williams M (2011) Testing DayCent and DNDC model simulations of N2O fluxes and assessing the impacts of climate change on the gas flux and biomass production from a humid pasture. Atmos Environ 44:2961–2970

Abrol YP, Chatterjee SR, Kumar PA, Jain V (1999) Improvement in nitrogen use efficiency: physiological and molecular approaches. Curr Sci 76:1357–1364

Abrol YP, Pandey R, Raghuram N, Ahmad A (2012) Nitrogen cycle sustainability and sustainable technologies for nitrogen fertilizer and energy management. J Indian Inst Sci 92:17–36

Almasri MN, Kaluarachchi JJ (2004) Assessment and management of long-term nitrate pollution of ground water in agriculture-dominated watersheds. J Hydrol 295:225–245

Amiour N, Imbaud S, Clement G et al (2012) The use of metabolomics integrated with transcriptomic and proteomic studies for identifying key steps involved in the control of nitrogen metabolism in crops such as maize. J Exp Bot 63:5017–5033

Anayah FM, Almasri MN (2009) Trends and occurrences of nitrate in the groundwater of the West Bank, Palestine. Appl Geogr 29:588–601

Anjana Umar S, Iqbal M (2007) Nitrate accumulation in plants, factors affecting the process, and human health implications. A review. Agron Sustain Dev 27:45–57

Anjana Umar S, Iqbal M, Abrol YP (2007) Are nitrate concentrations in leafy vegetables within safe limits. Curr Sci 92:355–360

Bahrmam N, Gouis J, Negroni L, Amilhat L, Leroy P, Laine AL, Jamion O (2004) Differential protein expression assessed by two-dimensional gel electrophoresis for two wheat varieties grown at four nitrogen levels. Proteomics 4:709–719

Bahrman N, Gouy A, Devienne-Barret F, Hirel B, Vedele F, Le Gouis J (2005) Differential change in root protein patterns of two wheat varieties under high and low nitrogen nutrition levels. Plant Sci 168:81–87

Bjorne H, Petersson J, Phillipson M, Weitzberg E, Holm L, Lundberg JO (2004) Nitrate in saliva increases gastric mucosal blood flow and mucus thickness. J Clin Invest 113:106–114

Boyer E, Howarth RH, Galloway JN, Dentener FJ, Cleveland C, Asner GP, Green P, Vorosmarty C (2004) Current nitrogen inputs to world regions. In: Mosier AR, Syers JK, Freney JR (eds) Agriculture and the nitrogen cycle: assessing the ımpacts of fertilizer use on food production and the environment. SCOPE 65, Paris, France, pp 221–230

Brasil (2010) Ministério da Ciência e Tecnologia. Coordenação Geral de Mudanças Globais do Clima. Segunda comunicação nacional do Brasil à convenção-quadro das Nações Unidas sobre mudança do clima. Ministério da Ciência e Tecnologia, DF, Brasília

Bruning-Fann CS, Kaneene J (1993) The effects of nitrate, nitrite, and N-nitroso compounds on animal health. Vet Hum Toxicol 35:237–253

Bu Y, Takano T, Nemoto K, Liu S (2011) Research progress of ammonium transporter in rice plants. Genom Appl Biol 2:19–23

Burkart MR, James DE (1999) Agricultural-nitrogen contributions to hypoxia in the Gulf of Mexico. J Environ Qual 28:850–859

Butterbach-Bahl K, Baggs EM, Dannenmann M, Kiese R, Zechmeister-Boltenstern S (2013) Nitrous oxide emissions from soils: how well do we understand the processes and their controls. Philos Trans R Soc Lond B Biol Sci 368:20130122

Cassman KG, Dobermann A, Walters D (2002) Agroecosystems, nitrogen-use eYciency, and nitrogen management. Ambio 31:132–140

Castaings L, Marchive C, Meyer C, Krapp A (2011) Nitrogen signalling in *Arabidopsis*: how to obtain insights into a complex signalling network. J Exp Bot 62:1391–2011

Cerri CC, Ferreira Maia SM, Galdos MV, Cerri CEP, Feigl JB, Bernoux M (2009) Brazilian greenhouse gas emissions: the importance of agriculture and livestock. Sci Agr Piracicaba, Braz 66:831–843

Chen BM, Wang ZH, Li SX, Wang GX, Song HX, Wang XN (2004) Effects of nitrate supply on plant growth, nitrate accumulation, metabolic nitrate concentration and nitrate reductase activity in three leafy vegetables. Plant Sci 167:635–643
Chiu HF, Tsai SS (2007) Nitrate in drinking water and risk of death from bladder cancer: an ecological case-control study in Taiwan. J Toxicol Environ Health A 70:1000–1004
Cristina MV, Manolis K, Sylvaine C, Michael RT, Roel V, John RN, Mark JN, Patrick L (2014) Assessing exposure and health consequences of chemicals in drinking water: current state of knowledge and research needs. Environ Health Perspect 122(3):213–221
Dalal RC, Wang W, Robertson GP, Parton WJ (2003) Nitrous oxide emission from Australian agricultural lands and mitigation options: a review. Aust J Soil Res 41:165–195
Daniel JS, Fleming EL, Portmann RW, Velders GJM, Jackman CH, Ravishankara AR (2010) Options to accelerate ozone recovery: ozone and climate benefits. Atmos Chem Phys 10:7697–7707
de Koeijer TJ, Wossink GAA, Smit AB, Janssens SRM, Renkema JA, Struik PC (2003) Assessment of the quality of farmers' environmental management and its effects on resource use efficiency: a Dutch case study. Agric Sys 78:85–103
Denman KL, Brasseur G, Chidthaisong A, Ciais P, Cox PM, Dickinson RE, Hauglustaine D, Heinze C, Holland E, Jacob D, Lohmann U, Ramachandran S, da Silva Dias PL, Wofsy SC, Zhang X (2007) Couplings between changes in the climate system and biogeochemistry. In: Solomon S, Qin D, Manning M, Chen Z, Marquis M, Averyt KB, Tignor M, Miller HL (eds) Climate change 2007: the physical science basis contribution of working group I to the fourth assessment report of the intergovernmental panel on climate change. Cambridge University Press, Cambridge, pp 499–588
Di HJ, Cameron KC (2002) Nitrate leaching in temperate agroecosystems: sources, factors and mitigating strategies. Nutr Cycl Agroecosyst 46:237–256
Dobermann A, Cassman KG (2004) Environmental dimensions of fertilizer nitrogen: what can be done to increase nitrogen use eYciency and ensure global food security? In: Mosier AR, Syers JK, Freney JR (eds) Agriculture and the nitrogen cycle: assessing the ımpacts of fertilizer use on food production and the environment. SCOPE 65, Paris, France, pp 261–278
Dordas CA, Sioulas C (2008) Safflower yield, chlorophyll content, photosynthesis, and water use efficiency response to nitrogen fertilization under rainfed conditions. Ind Crop Prod 27:75–85
El-Khatib RT, Hamerlynck EP, Gallardo F, Kirby EG (2004) Transgenic poplar characterized by ectopic expression of a pine cytosolic glutamine synthetase gene exhibits enhanced tolerance to water stress. Tree Physiol 24:729–736
Emongor VE, Ramolemana GM (2004) Treated sewage effluent (water) potential to be used for horticultural production in Botswana. Phys Chem Earth 29:1101–1108
EPA (2007) Hypoxia in the northern gulf of Mexico: an update by the environmental protection agency science advisory board. Environmental Protection Agency, Washington, USA, Report no. EPA-SAB-08-003
Escobar LF et al (2010) Postharvest nitrous oxide emissions from a subtropical oxisol as influenced by summer crop residues and their management. Revista Brasileira de Ciência do Solo 34:507–516
Fageria NK (2009) Mineral nutrition versus yield of field crops. In: Frageria NK (ed) The use of nutrients in crop plants. CRC Press, Bocca Rotan, FL, pp 1–29
FAO (2011) Current world fertilizer trends and outlook to 2015. FAO, Rome, Italy
Fait A, Nesi AN, Angelovici R, Lehmann M, Pham PA, Song L et al (2011) Targeted enhancement of glutamate-to-gamma-aminobutyrate conversion in Arabidopsis seeds affects carbon-nitrogen balance and storage reserves in a development-dependent manner. Plant Physiol 157:1026–1042
Ferrario-Méry S, Valadier MH, Foyer CH (1998) Overexpression of nitrate reductase in tobacco delays drought-induced decreases in nitrate reductase activity and mRNA. Plant Physiol 117:293–302

Ferrario-Méry S, Masclaux C, Suzuki A, Valadier MH, Hirel B, Foyer CH (2001) Glutamine and a-ketoglutarate are metabolite signals involved in nitrate reductase gene transcription in untransformed and transformed tobacco plants deficient in ferredoxin-glutamine-a-ketoglutarate. Planta 213:265–271

Fewtrell L (2004) Drinking-water nitrate, methemoglobinemia, and global burden of disease: a discussion. Environ Health Perspect 112:1371–1374

Filoso S, Martinelli LA, Howarth RW, Boyer EW, Dentener F (2006) Human activities changing the nitrogen cycle in Brazil. Biogeochemistry 79:61–89

Fraisier V, Gojon A, Tillard P, Daniel-Vedele F (2000) Constitutive expression of a putative high-affinity nitrate transporter in *Nicotiana plumbaginifolia*: evidence for post-transcriptional regulation by a reduced nitrogen source. Plant J 23:489–496

Fernie AR, Stitt M (2012) On the discordance of metabolomics with proteomics and transcriptomics: coping with increasing complexity in logic, chemistry, and network interactions scientific correspondence. Plant Physiol 158:1139–1145

Fuentes SI, Allen DJ, Ortiz-Lopez A, Hernández G (2001) Over-expression of cytosolic glutamine synthetase increases photosynthesis and growth at low nitrogen concentrations. J Exp Bot 52:1071–1081

Fuentes-Ramírez LE, Bustillos-Cristales R, Tapia-Hernández A, Jiménez-Salgado T, Wang ET, Martínez-Romero E, Caballero-Mellado J (2001) Novel nitrogen-fixing acetic acid bacteria, *Gluconacetobacter johannae* sp. nov. and *Gluconacetobacter azotocaptans* sp. nov., associated with coffee plants. Int J Syst Evol Microbiol 51:1305–1314

Galloway JN, Cowling EB, Seitzinger SP, Socolow RH (2002) Reactive nitrogen: too much of a good thing? Ambio 31:60–71

Glass ADM (2003) Nitrogen use efficiency of crop plants: physiological constraints upon nitrogen absorption. Crit Rev Plant Sci 22:453–470

Godish T, Davis WT, Fu JS (2015) Air quality, 5th edn. CRC Press, Boca Raton, FL

Gomes RL, Scrimshaw MD, Lester JN (2009) Fate of conjugated natural and synthetic steroid estrogens in crude sewage and activated sludge batch studies. Environ Sci Technol 43:3612–3618

Gormly J, Spalding RF (2006) Sources and concentrations of nitrate-nitrogen in ground water of the central platte region, Nebraskaa. Ground Water 17:291–301

Granstedt A (2000) Increasing the efficiency of plant nutrient recycling within the agricultural system as a way of reducing the load to the environment–experience from Sweden and Finland. Agric Ecosyst Environ 80:169–185

Greer FR, Shannon M (2005) Infant methemoglobinemia: the role of dietary nitrate in food and water. Pediatrics 116:784–786

Guleryuz G, Gucel S, Ozturk M (2010) Nitrogen mineralization in a high altitude ecosystem in the Mediterranean phytogeographical region of Turkey. J Environ Biol 31(4):503–514

Guo M, Li H, Zhang Y, Zhang X, Lu A (2006) Effects of water table and fertilization management on nitrogen loading to groundwater. Agric Water Manag 82:86–98, ISSN 0378–3774

Gupta ML, Khosla R (2012) Precision nitrogen management and global nitrogen use efficiency. In: Harald K, Butron GMP (eds) Proceedings of the 11th international conference on precision agriculture. Indianapolis, USA

Gupta SK, Gupta RC, Chhabra SK, Eskiocak S, Gupta AB, Gupta R (2008) Health issues related to N pollution in water and air. Indian Agric Environ Health 94:1469–1477

Gupta S, Gupta RC, Gupta AB (2007) Nitrate toxicity and human health. In: Abrol YP, Raghuram N, Sachdev MS (eds) Agricultural nitrogen uses and its environmental implications. I.K. International Publication House Pvt Ltd, New Delhi, India, pp 517–547

Hakeem KR, Ahmad A, Iqbal M, Gucel S, Ozturk M (2011) Nitrogen-efficient cultivars can reduce nitrate pollution. Environ Sci Pollut Res 18:1184–1193

Hakeem KR, Chandna R, Ahmad A, Iqbal M (2012a) Reactive nitrogen inflows use efficiency in agriculture: an environment perspective. In: Ahmad A, Prasad MNV (eds) Environment adaptation and stress tolerance of plants in the era of climate change. Springer, New York, pp 217–232

Hakeem KR, Chandna R, Ahmad A, Iqbal M (2012b) Physiological and molecular analysis of applied nitrogen in rice (*Oryza sativa* L.) genotypes. Rice Sci 19(3):213–222

Hakeem KR, Chandna R, Ahmad A, Iqbal M (2012c) Physiological studies and proteomic analysis for differentially expressed proteins and their possible role in the root of N-efficient rice (*Oryza sativa* L.). Mol Breed 32(4):785–798

Hakeem KR, Sabir M, Khan F, Rehman RU (2014) Nitrogen regulation and signalling in plants. In: Hakeem KR, Rehman RU, Tahir I (eds) Plant signaling: understanding the molecular crosstalk. Springer, India, pp 117–132

Hänsch R, Fessel DG, Witt C, Hesberg C, Hoffmann G, Walch-Liu P, Engels C, Kruse J, Rennenberg H, Kaiser WM, Mendel RR (2001) Tobacco plants that lack expression of functional nitrate reductase in roots show changes in growth rates and metabolite accumulation. J Exp Bot 52:1251–1258

Hatfield JL, Sauer TJ, Prueger JH (2001) Managing soils to achieve greater water use efficiency. Agron J 93:271–280

HELCOM (2015) Updated Fifth Baltic Sea pollution load compilation (PLC-5.5). Baltic Sea Environment proceedings no. 145, pp 143

Hijleh RJA (2014) Chemical and microbial risk management assessment of drinking water in faria catchment. M.Sc thesis, An-Najah National University, Nablus, Palestine

Hill MJ (1999) Nitrate toxicity: myth or reality. Brit J Nutr 81:343–344

Hirel B, Le Gouis J, Ney B, Gallais A (2007) The challenge of improving nitrogen use efficiency in crop plants: towards a more central role for genetic variability and quantitative genetics within integrated approaches. J Exp Bot 58:2369–2387

Hoshida H, Tanaka Y, Hibino T, Hayashi Y, Tanaka A, Takabe T, Takabe T (2000) Enhanced tolerance to salt stress in transgenic rice that overexpresses chloroplast glutamine synthetase. Plant Mol Biol 43:103–111

Ikehata K, Murphy RR, Liu Y, Sun RN, Nessl MB (2010) Health effects associated with wastewater treatment, reuse, and disposal. Water Environ Res 82:2047–2066

Inoue-Choi M, Ward MH, Cerhan JR, Weyer PJ, Anderson KE, Robin K (2012) Interaction of nitrate and folate on the risk of breast cancer among postmenopausal women. Nutr Cancer 64:685–694

Jiao P, Xu D, Wang S, Wang Y, Liu K, Tang G (2012) Nitrogen loss by surface runoff from different cropping systems. Soil Res 50:58–66

Jim FC, Shelie AM, James RF (2011) Using DAYCENT to quantify on-farm GHG emissions and N dynamics of land use conversion to N-managed switchgrass in the Southern U.S. Agri Ecosys Environ 141(3-4):332–341

Ju XT, Kou CL, Zhang FS, Christie P (2005) Nitrogen balance and groundwater nitrate contamination: comparison among three intensive cropping systems on the north China plain. Environ Pollut 143:117–125

Kapoor A, Viraraghavan T (1997) Nitrate removal from drinking water-review. J Environ Eng 123:371–380

Keeney DR, Hatfield JL (2008) The nitrogen cycle historical perspective and current and potential future concerns. In: Hatfield JL, Follett RF (eds) Nitrogen in the environment: sources problems and management. Academic, Elsevier, pp 1–18

Keeney D, Olson RA (1986) Sources of nitrate to ground water. Crit Rev Environ Sci Technol 16:257–304

Khandare HW (2013) Scenario of nitrate contamination in groundwater: its causes and prevention. Int J ChemTech Res 5:1921–1926

Killebrew K, Wolff H (2010) Environmental impacts of agricultural technologies. Agricultural Policy and Statistics Team of the Bill and Melinda Gates Foundation, EPAR brief no. 65, pp 1–18

Kissel D, Cabrera M, Vaio N, Craig J, Rema J, Morris L (2004) Rainfall timing and ammonia loss from urea in a loblolly pine plantation. Soil Sci Soc Am J 68:1744–1750

Kozaki A, Takeba G (1996) Photorespiration protects C_3 plants from photooxidation. Nature 384:557–560

Kumar S, Tamura K, Jakobsen IB, Nei M (2001) MEGA2: molecular evolutionary genetics analysis software. Bioinformatics 17:1244–1245

Ladha JK, Pathak H, Krupnik TJ, Six J, van Kessel C (2005) Efficiency of fertilizer nitrogen in cereal production: retrospects and prospects. Adv Agron 87:85–156

Li C, Zhuang Y, Cao M, Crill P, Dai Z, Frolking S, Moore B, Salas W, Song W, Wang X (2001) Comparing a process-based agro ecosystem model to the IPCC methodology for developing a national inventory of N2O emissions from arable lands in China. Nut Cycl Agroecosys 60:1–3

Lillo C, Meyer C, Lea US, Provan F, Oltedal S (2004) Mechanism and importance of post-translational regulation of nitrate reductase. J Exp Bot 55:1275–1282

Liu KH, Tsay YF (2003) Switching between the two action modes of the dual-affinity nitrate transporter CHL1 by phosphorylation. EMBO J 22:1005–1013

Liu Z, Song X, Jiang L, Lin H, Xu Y, Gao X, Zheng F, Tan D, Wang M, Jing S, Shen Y (2012) Strategies for managing soil nitrogen to prevent nitrate-N leaching in intensive agriculture system. In: Hernandez Soriano MC (ed) Soil health and land use management. InTech, pp 133–154. http://www.intechopen.com/books/soil-health-and-land-usemanagement/strategies-for-managing-soil-nitrogen-to-prevent-nitrate-n-leaching-in-intensive-agriculturesystem. Accessed 18 Apr 2015

Loqué D, von Wirén N (2004) Regulatory levels for the transport of ammonium in plant roots. J Exp Bot 55:1293–1305

Man HM, Boriel R, El-Khatib R, Kirby EG (2005) Characterization of transgenic poplar with ectopic expression of pine cytosolic glutamine synthetase under conditions of varying nitrogen availability. New Phytol 167:31–39

Manassaram DM, Backer LC, Moll DM (2007) A review of nitrates in drinking water: maternal exposure and adverse reproductive and developmental outcomes. Environ Health Prospect 114:320–327

Masclaux C, Quillere I, Gallais A, Hirel B (2001) The challenge of remobilisation in plant nitrogen economy. A survey of physio-agronomic and molecular approaches. Ann Appl Biol 138:69–81

McPharlin IR, Aylam PM (1995) Nitrogen requirements of lettuce under sprinkler irrigation and fertilization on spearwood sand. J Plant Nutr 18:219–241

McDowell LR (2012) Vitamins in animal nutrition: comparative aspects to human nutrition. Academic, London

Mikkelsen M, Hartz TK (2008) Nitrogen sources for organic crop production. Better Crop 92:16–19

Millar N, Robertson GP, Grace PR, Gehl RJ, Hoben JP (2010) Nitrogen fertilizer management for nitrous oxide (N2O) mitigation in intensive corn (Maize) production: an emissions reduction protocol for US Midwest agriculture. Mitig Adapt Strat Glob Change 15:185–204

Mosier AR, Zhaoliang Z (2000) Changes in patterns of fertilizer nitrogen use in Asia and its consequences for N_2O emissions from agricultural systems. Nutr Cycl Agroecosyst 57:107–117

Mosier AR, Bleken M, Chaiwanakupt P, Ellis EC, Freney JR, Howarth RB, Matson PA, Minami K, Naylor R, Weeks KN (2001) Policy implications of human-accelerated nitrogen cycling. Biogeochemistry 52:281–320

Mosier AR, Syers JK, Freney JR (2004a) Agriculture and the nitrogen cycle: assessing the impacts of fertilizer use on food production and the environment, 2nd edn. Island Press, Washington, DC

Mosier AR, Syers JK, Freney JR (2004b) Nitrogen fertilizer: an essential component of increased food, feed, and fiber production. In: Mosier AR, Syers JK, Freney JR (eds) Agriculture and the nitrogen cycle: assessing the impacts of fertilizer use on food production and the environment. SCOPE 65, Paris, France, pp 3–15

Mueller BA, Newton K, Holly EA, Preston-Martin S (2001) Residential water source and the risk of childhood brain tumors. Environ Health Perspect 109:551–556
NAAS (2005) Policy options for efficient nitrogen use, policy paper no. 33. National Academy of Agricultural Sciences, New Delhi, pp 1–4
Norse D (2003) Fertilizers and world food demand implications for environmental stresses. Paper presented at the IFA-FAO agriculture conference on global food security and the role of sustainable fertilization, FAO, Rome, Italy, 26–28 Mar 2003
Nolan BT, Stoner JD (2000) Nutrients in ground waters of the conterminous United States, 1992–1995. Environ Sci Technol 34:1156–1165
Odiyo JO, Makungo R, Muhlarhi TG (2014) The impacts of geochemistry and agricultural activities on groundwater quality in the Soutpansberg fractured aquifers. In: Rebbia CA (ed) Water pollution XII. WIT Press, Ashurst, Southampton, UK, pp 121–132
Ohyama T (2010) Nitrogen as a major essential element of plants. In: Ohyama T, Sueyoshi K (eds) Nitrogen assimilation in plants. Signpost, Trivandrum, Kerala, India
Oliveira ALM, Urquiaga S, Döbereiner J, Baldani JI (2002) The effect of inoculating endophytic N_2-fixing bacteria on micropropagated sugarcane plants. Plant and Soil 24:205–215
Ondersteijn C, Beldman A, Daatselaar C, Giesen G, Huirne R (2002) The Dutch mineral accounting system and the European nitrate directive: implications for N and P management and farm performance. Agric Ecosyst Environ 92:283–296
Ortega JL, Temple SJ, Sengupta-Gopalan C (2001) Constitutive overexpression of cytosolic glutamine synthetase (GS1) gene in transgenic alfalfa demonstrates that *GS1* may be regulated at the level of RNA stability and protein turnover. Plant Physiol 126:109–121
Ozturk M, Gucel S, Sakcali S, Baslar S (2013) Nitrate and edible plants in the Mediterranean Region of Turkey: an overview. In: Umar S et al (eds) Nitrate in leafy vegetables—toxicity and safety measures. I.K. Intern. Publ. House Pvt. Ltd, New Delhi-Bangalore (India), pp 17–51
Paerl HW, Dennis RL, Whitall DR (2002) Atmospheric deposition of nitrogen: implications for nutrient over-enrichment of coastal waters. Estuaries 25:677–693
Palm CA, Machado PLOA, Mahmood T, Melillo J, Murrell ST, Nyamangara J, Scholes M, Sisworo E, Olesen JE, Pender J, Stewart J, Galloway JL (2004) Societal responses for addressing nitrogen fertilizer needs: balancing food production and environmental concerns. In: Mosier AR, Syers JK, Freney JR (eds) Agriculture and the nitrogen cycle assessing the impacts of fertilizer use on food production and the environment. SCOPE 65, Paris, France, pp 71–89
Paramasivam S, Alva AK, Fares A, Sajwan KS (2002) Fate of nitrate and bromide in an unsaturated zone of a sandy soil under citrus production. J Environ Qual 31:671–681
Pathak H, Nedwell DB (2001) Strategies to reduce nitrous oxide emission from soil with fertilizer selection and nitrification inhibitor. Water Air Soil Poll 129:217–228
Pionke HB, Pionke ML, Sharma KJ (1990) Hirschberg Impact of irrigated horticulture on nitrate concentrations in groundwater. Agr Ecosyst Environ 32:119–132
Portmann RW, Daniel JS, Ravishankara AR (2012) Stratospheric ozone depletion due to nitrous oxide: influences of other gases. Philos Trans R Soc B Sci 367:1256–1264
Prasad R (2013) Fertilizer nitrogen, food security, health and the environment. Proc Indian Nat Sci Acad 79:997–1010
Provan F, Aksland LM, Meyer C, Lillo C (2000) Deletion of the nitrate reductase N-terminal domain still allows binding of 14-3-3 proteins but affects their inhibitory properties. Plant Physiol 123:757–764
Quilleré I, Dufossé C, Roux Y, Foyer CHF, Caboche M, Morot-Gaudry JF (1994) The effects of deregulation of NR gene expression on growth and nitrogen metabolism of *Nicotiana plumbaginifolia* plants. J Exp Bot 45:1205–1211
Ramaraju HK, Venkatachalappa M, Ranganna G, Sadashivaiah C, Manamohan Rao N (1999) Hazards due to migration of septic tank leakages in peri-urban settlements. In: Ellis JB (ed) Impact of urban growth on surface water and groundwater quality. International Association of Hydrogeological Science, Publication no 259, pp 219–225

Ravishankara AR, Daniel JS, Portmann RW (2009) Nitrous oxide (N_2O): the dominant ozone-depleting substance emitted in the 21st century. Science 326:123–125

Ribbe L, Delgado P, Salgado E, Flüge WA (2008) Nitrate pollution of surface water induced by agricultural non-point pollution in the Pocochay watershed, Chile. Desalination 226:13–20

Robertson GP, Vitousek PM (2009) Nitrogen in agriculture: balancing an essential resource. Annu Rev Energy Environ 34:97–125

Rosegrant MW, Cai X, Cline SA (2002) World water and food to 2025: dealing with scarcity. International Food Policy Research Institute (IFPRI), Washington, DC

Santiago-Antonio G, Lizama-Gasca MG, Carrillo-Pech M, Echevarría-Machado I (2014) Natural variation in response to nitrate starvation among varieties of habanero pepper (*Capsicum chinense* Jacq.). Aust J Crop Sci 8:523–535

Schilling KE, Libra RD (2000) The relationship of nitrate concentrations in streams to row crop land use in Iowa. J Environ Qual 29:1846–1851

Schoenbeck MA, Temple SJ, Trepp GB, Blumenthal JM, Samac DA, Gantt JS, Hernandez G, Vance CP (2000) Decreased NADH glutamate synthase activity in nodules and flowers of alfalfa (*Medicago sativa* L.) transformed with an antisense glutamate synthase transgene. J Exp Bot 51:29–39

Schumann U, Huntrieser H (2007) The global lightning-induced nitrogen oxides source. Atmos Chem Phys 7:2623–2818

Shabalala AN, Combrinck L, McCrindle R (2013) Effect of farming activities on seasonal variation of water quality of Bonsma Dam, KwaZulu-Natal. South Afr J Sci 109, Art. #0052, pp 7. doi:10.1590/sajs.2013/20120052

Sharma MK, Sharma H, Bapna N (2013) Toxic effects of high nitrate intake in oesophagus and stomach of rabbits. Int J Med Res Health Sci 2:407–411

Shen Y, Lei H, Yang D, Kanae S (2011) Effects of agricultural activities on nitrate contamination of groundwater in a Yellow river irrigated region. Int Assoc Hydrogeol Sci 348:73–80

Shoji S, Delgado J, Mosier A, Miura Y (2001) Use of controlled release fertilizers and nitrification inhibitors to increase nitrogen use efficiency and to conserve air and water quality. Commun Soil Sci Plant Anal 32:1051–1070

Signor D, Cerri CEP (2013) Nitrous oxide emissions in agricultural soils: a review. Pesq Agropecu Trop 43:322–338

Simons M, Saha R, Guillard L, Clément G, Armengaud P, Cañas R, Maranas CD, Lea PJ, Hirel B (2014) Nitrogen-use efficiency in maize (Zea mays L.): from 'omics' studies to metabolic modelling. J Exp Bot 65:5657–5671

Smil V (2002) Nitrogen and food production: proteins for human diets. Ambio 31:126–131

Solomon S (1999) Stratospheric ozone depletion: a review of concepts and history. Rev Geophys 37:275–316

Soussana JF, Lemairec G (2014) Coupling carbon and nitrogen cycles for environmentally sustainable intensification of grasslands and crop-livestock systems. Agric Ecosyst Environ 190:9–17

Sowers KE, Pan WL, Miller BC, Smith JL (1994) Nitrogen use efficiency of split nitrogen applications in soft white winter wheat. Agron J 86:942–948

Stitt M (1999) Nitrate regulation of metabolism and growth. Curr Opin Plant Biol 2:178–186

Subbarao GV, Sahrawat KL, Nakahara K, Rao IM, Ishitani M, Hash CT, Kishii M, Rasouli S, Whalen JK, Madramootoo CA (2013) Review: reducing residual soil nitrogen losses from agroecosystems for surface water protection in Quebec and Ontario, Canada: best management practices, policies and perspectives. Can J Soil Sci 94:109–127

Takahashi M, Sasaki Y, Ida S, Morikawa H (2001) Nitrite reductase gene enrichment improves assimilation of NO_2 in *Arabidopsis*. Plant Physiol 126:731–741

Thorburn PJ, Biggs JS, Weier KL, Keating BA (2003) Nitrate in groundwaters of intensive agricultural areas in coastal Northeastern Australia. Agric Ecosyst Environ 94:49–58

Tilman D, Balzer C, Hill J, Befort BL (2011) Global food demand and the sustainable intensification of agriculture. Proc Natl Acad Sci U S A 108:20260–20264

Tilman D, Cassman KG, Matson PA, Naylor RL, Polasky S (2002) Agricultural sustainability and intensive production practices. Nature 418:671–677

Toma Y, Hatano R (2007) Effect of crop residue C: N ratio on N2O emissions from Gray Lowland soil in Mikasa, Hokkaido, Japan. Soil Sci Plant Nutr 53:198–205

Ussiri D, Lal R (2013) Soil emission of nitrous oxide and its mitigation. Springer, New York

van Cleemput O, Boeckx P (2005) Alteration of nitrogen cycling by agricultural activities, and its environmental and health consequences. Gayana Bot 62:98–109

Verma S, Nizam S, Verma PK (2013) Biotic and abiotic stress signalling in plants. In: Ahmad A, Sarwat M, Abdin MZ (eds) Stress signalling in plants: genomics and proteomics perspective, vol 1. Springer Science, New York, pp 25–49

Vincent R, Fraisier V, Chaillou S, Limami MA, Deleens E, Phillipson B, Douat C, Boutin JP, Hirel B (1997) Overexpression of a soybean gene encoding cytosolic glutamine synthetase in shoots of transgenic *Lotus corniculatus* L. plants triggers changes in ammonium assimilation and plant development. Planta 201:424–433

von Wirén N, Lauter FR, Ninnemann O, Gillissen B, Walch-Liu P, Engels C, Jost W, Frommer WB (2000) Differential regulation of three functional ammonium transporter genes by nitrogen in root hairs and by light in leaves of tomato. Plant J 21:167–175

Vymazal J (2007) Removal of nutrients in various types of constructed wetlands. Sci Total Environ 380:48–65

Ward MH (2009) Too much of a good thing. Nitrate from nitrogen fertilizers and cancer. Rev Environ Health 24:357–363

WHO (2011) Nitrate and nitrite in drinking-water. World Health Organization, Avenue Appia, Geneva, Switzerland

Walley F, Pennock D, Solohub M, Hnatowich G (2001) Spring wheat (Triticum aestivum) yield and grain yield protein responses to N fertilizer in topographically defined landscape positions. Can J Soil Sci 81:505–514

Walley F, Yates T, Van Groenigen JW, Van Kessel C (2002) Relationships between soil nitrogen availability indices, yield, and nitrogen accumulation of wheat. Soil Sci Soc Am J 66:1549–1561

Wood S, Henao J, Rosegrant M (2004) The role of nitrogen in sustaining food production and estimating future nitrogen fertilizer needs to meet food demand. In: Mosier AR, Syers JK, Freney JR (eds) Agriculture and the nitrogen cycle: assessing the ımpacts of fertilizer use on food production and the environment. SCOPE 65, Paris, France, pp 245–260

Xu XK, Wang YS, Zheng XH, Wang MX, Wang ZJ, Zhou LK, Van Cleemput O (2000) Methane emission from a simulated rice field ecosystem as influenced by hydroquinone and dicyandiamide. Sci Total Environ 263:243–253

Yamaya T, Obara M, Nakajima H, Sasaki S, Hayakawa T, Sato T (2002) Genetic manipulation and quantitative-trait loci mapping for nitrogen recycling in rice. J Exp Bot 53:917–925

Yanagisawa S (2004) Dof domain proteins: plant-specific transcription factors associated with diverse phenomena unique to plants. Plant Cell Physiol 45:386–391

Yang CY, Cheng MF, Tsai SS, Hsieh YL (1998) Calcium, magnesium, and nitrate in drinking water and gastric cancer mortality. Japan J Cancer Res 89:124–130

Zhang W, Tian Z, Zhang N, Li X (1996) Nitrate pollution of groundwater in northern China. Agric Ecosyst Environ 59:223–231

Zheng X, Fu C, Xu X, Huang Y, Han S, Hu F, Chen G (2002) The Asian nitrogen cycle case study. Ambio 31:79–87

Zu X, Boeckx P, Wang Y, Huang Y, Zheng X, Hu F, van Cleemput O (2002) Nitrous oxide and methane emissions during rice growth and through rice plants: effect of dicyandiamide and hydroquinone. Biol Fertil Soils 36:53–58

Erratum to: Nitrate and Nitrogen Oxides: Sources, Health Effects and Their Remediation

Khalid Rehman Hakeem, Muhammad Sabir, Munir Ozturk, Mohd. Sayeed Akhtar, and Faridah Hanum Ibrahim

Erratum to: Chapter 5 in: P. de Voogt (ed.), *Reviews of Environmental Contamination and Toxicology, Volume 242*, Reviews of Environmental Contamination and Toxicology 242, DOI 10.1007/398_2016_11

The original version of this chapter was inadvertently published with incorrect list of authors. The correct author list should read Khalid Rehman Hakeem, Muhammad Sabir, Munir Ozturk, Mohd. Sayeed Akhtar, and Faridah Hanum Ibrahim. The authors Muhammad Ashraf and Muhammad Sajid Aqeel Ahmad were removed from the list upon their own request and with consent from all co-authors.

The original chapter was corrected.

The updated original online version for this chapter can be found at http://dx.doi.org/10.1007/398_2016_11

P. de Voogt (ed.), *Reviews of Environmental Contamination and Toxicology, Volume 242*, Reviews of Environmental Contamination and Toxicology 242, DOI 10.1007/978-3-319-51243-3_15

Index

P. de Voogt (ed.), *Reviews of Environmental Contamination and Toxicology, Volume 242*, Reviews of Environmental Contamination and Toxicology 242,
DOI 10.1007/978-3-319-51243-3

MIX
Papier aus verantwortungsvollen Quellen
Paper from responsible sources
FSC® C105338

FSC
www.fsc.org

If you have any concerns about our products,
you can contact us on
ProductSafety@springernature.com

In case Publisher is established outside the EU,
the EU authorized representative is:
Springer Nature Customer Service Center GmbH
Europaplatz 3, 69115 Heidelberg, Germany

Printed by Libri Plureos GmbH
in Hamburg, Germany